EIGHTH EDITION

SOCIOLOGY

A Global Perspective

Joan Ferrante

Northern Kentucky University

Australia • Brazil • Japan • Korea • Mexico • Singapore • Spain • United Kingdom • United States

Sociology: A Global Perspective,
Eighth Edition
Joan Ferrante

Sponsoring Editor: Erin Mitchell
Developmental Editor: Rebecca Dashiell
Assistant Editor: Linda Stewart
Editorial Assistant: Mallory Ortberg
Media Editor: Melanie Cregger
Marketing Manager: Andrew Keay
Marketing Communications Manager: Laura Localio
Content Project Manager: Cheri Palmer
Design Director: Rob Hugel
Art Director: Caryl Gorska
Print Buyer: Karen Hunt
Rights Acquisitions Specialist: Roberta Broyer
Production Service: Jill Traut, MPS Limited, a Macmillan Company
Text Designer: Ellen Pettengell
Photo Researcher: Bill Smith Group
Text Researcher: Sue Howard
Copy Editor: Heather McElwain
Illustrator: MPS Limited, a Macmillan Company
Cover Designer: Caryl Gorska
Cover Image: © SOCCER/Balan Madhavan/ Alamy
Compositor: MPS Limited, a Macmillan Company

Library of Congress Control Number: 2011928896

Student Edition:

ISBN-13: 978-1-111-83390-9

ISBN-10: 1-111-83390-7

Loose-leaf Edition:

ISBN-13: 978-1-111-83527-9

ISBN-10: 1-111-83527-6

Wadsworth
20 Davis Drive
Belmont, CA 94002-3098
USA

Cengage Learning is a leading provider of customized learning solutions with office locations around the globe, including Singapore, the United Kingdom, Australia, Mexico, Brazil, and Japan. Locate your local office at **www.cengage.com/global.**

Cengage Learning products are represented in Canada by Nelson Education, Ltd.

To learn more about Wadsworth, visit **www.cengage.com/wadsworth.**

Purchase any of our products at your local college store or at our preferred online store **www.cengagebrain.com.**

Printed in the United States of America
1 2 3 4 5 6 7 15 14 13 12 11

To my mother, Annalee Taylor Ferrante and in memory of my father, Phillip S. Ferrante (March 1, 1926–July 8, 1984)

BRIEF CONTENTS

CONTENTS

3 CULTURE 54
With Emphasis on North and South Korea

4 SOCIALIZATION 76
With Emphasis on Israel and The Palestinian Territories

5 SOCIAL INTERACTION 104
With Emphasis on the Democratic Republic of the Congo

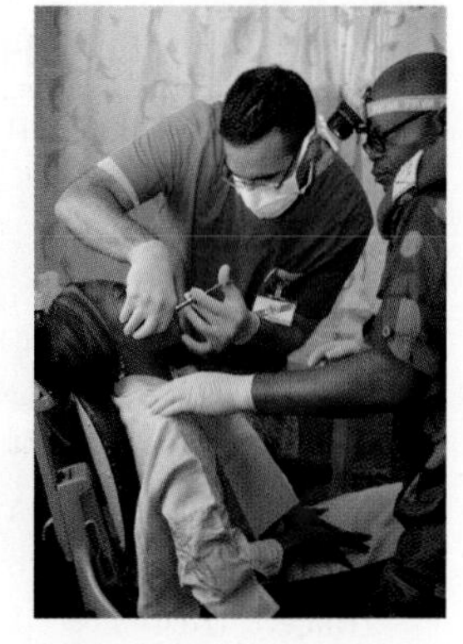

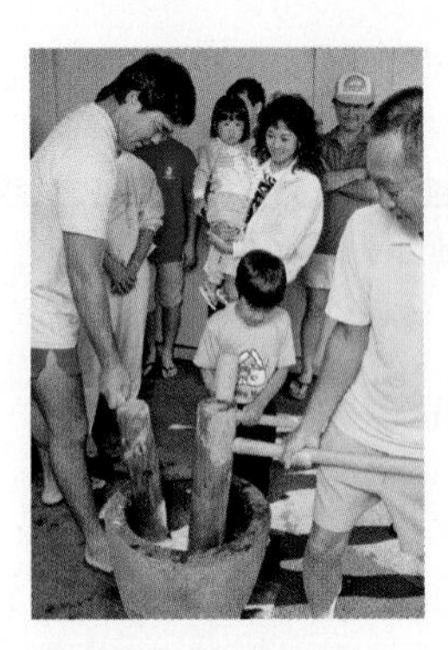

BOXES

The Sociological Imagination

The Global Comparisons

No Borders, No Boundaries

Working For Change

PREFACE

This eighth edition of *Sociology: A Global Perspective* retains its distinctive integrative approach. Each chapter pairs a sociological topic with a country or global-scale issue that becomes the focus of sociological analysis. The chosen pairings give the concepts and theories presented in each chapter a purpose. That purpose is to:

1. showcase sociology as offering the conceptual tools for understanding global-scale issues and changes and their impact on daily lives; and
2. strengthen the way sociology is presented so that readers, inundated by information explosion, will come away with a sense of satisfaction in taking the time to read, of all things, a textbook.

Major Changes to This Edition

These goals are especially evident in the descriptions of changes to the five most heavily revised chapters that follow:

- The first chapter—The Sociological Imagination—now features the mobile phone as the object of sociological analysis. The chapter opens with this question: Have you ever tried to imagine going one day without your phone? How would your life change? The sociological perspective takes readers beyond their personal dependency on mobile phones to think about how that innovation has revolutionized relationships. As one measure of the mobile phone's importance, consider that at this time there are an estimated 5.3 billion mobile phones in a world of 7 billion people. I imagine what six classic theorists—Comte, Marx, Weber, Durkheim, DuBois, and Addams—would say about the mobile phone if each were alive today. My hope is that opening the book with a focus on something deeply personal will build readers' confidence in sociology as a discipline that is worth their time to learn.
- Chapter 6, Formal Organizations with emphasis on McDonald's, now places that well-known global corporation in the context of the industrial food system. Keep in mind that McDonald's does not produce or even cook the food it sells to customers. Like other fast food corporations, it relies on suppliers from around the United States and the world to process menu items. The rise of the industrial food system has spawned a variety of slow food movements and organizations, each aiming to address some related issue, ranging from animal rights/welfare to school lunch reform.
- In the previous edition, Chapter 9—Race and Ethnicity—emphasized the global story behind the peopling of the United States. This edition shifts that emphasis to Brazil because that country's ideas of race and ethnicity offer an interesting contrast to the United States. Brazilian racial identity is built around the idea that everyone is multiracial and no racial group is distinct from another. This is in sharp contrast to the United States, where the races are still viewed as distinct categories. We use sociological concepts and theories to think about how ideas of race are constructed and the effects race has on life chances, race relations, and identity. With regard to ethnicity, the United States government recognizes only two ethnic groups: Hispanic and non-Hispanic. In contrast, the Brazilian government only recognizes the ethnic identities of the indigenous peoples.
- Chapter 11—Economics and Politics—has shifted its emphasis from Iraq to India. India is one of two rising powers (the other is China) expected to challenge the global power and influence of the United States. India is best known in the United States as an outsourcing destination, most notably of customer service and IT jobs. For all its publicity, however, only about 2.5 million of its 450 million-strong labor force works in high-tech and service industries. We apply the sociological concepts to compare and contrast the Indian and American economic and political systems and, in the process, gain insights about the major forces shaping employment opportunities in the two countries.
- Chapter 15 changes title from "Population and Urbanization with Emphasis on India:" to "Birth, Death, and Migration with Emphasis on Extreme Cases." The title change more accurately reflects the three key human experiences that determine population size, growth and other dynamics. In this chapter, we compare extreme cases or those countries that experience the highest and the lowest birth rates, including the teen birth rate; death rates including infant and maternal, and in- and out-migration rates.

We emphasize extreme cases because doing so allows us to frame the end points on the continuum of human experience. For example, if we know the teen birth rate in

Niger is 199 per 1,000 teens, we know that there are 199 births for every 1,000 teen females each year. To put it another way, each year, 20 percent or one in five teens have a baby. If we know that the teen birth rate in South Korea is 1 per 1,000, we know that there is 1 birth for every 1,000 teen females each year. What do these rates suggest about the lives of females who are teenagers in each country? If you were a teenage girl, how would you see your future if you lived in Niger versus South Korea? The point is that rates help us to think about these experiences and allow us to think more deeply about our own and others' lives.

Other Changes

Before elaborating on the changes specific to the other chapters, I will point out less obvious changes common to all chapters. First, I worked hard to further strengthen the rationale underlying the connection between chosen country/global issue and the chapter topic with which it is paired. This book's purpose, which is to showcase sociology's power to help people think about the most compelling issues of our time, is realized when that connection is crystal clear. Second, I now open and close each chapter with online poll questions asking readers to think about something going on in their lives, a decision they have made or some behavior in which they have engaged. The questions help to link seemingly personal concerns to larger sociological processes described in that chapter. Students can go online to see how their answers compare to those of other students using this textbook. Third, I reviewed and updated all statistics including text, tables, and charts, making sure that each statistic enhanced student understanding of the sociological material covered. Fourth, I critically reviewed photographs, dropping those that were outdated and adding photos that effectively represented the sociological processes described.

My revision plan also involved changes to the four types of chapter boxes. The **Sociological Imagination boxes** have been revised with the aim of further clarifying the intersection between seemingly personal experiences and remote and impersonal social forces. **Global Comparison boxes** are written with greater emphasis on how life chances are shaped in part by the country in which one happens to live. **No Borders, No Boundaries boxes** are especially important for illustrating the ever-increasing flow of goods, services, money, people, information, and culture across national borders. Finally, almost all the **Working for Change boxes** from the previous edition have been replaced. This box focuses on solutions—specifically, on people who are doing or have done something to change the system or who have persuaded others to change their behavior in ways that benefit society. Most of the Working for Change boxes were written by sociology majors at Northern Kentucky enrolled in a Senior Seminar course I taught in Fall 2010. Those students and their contributions are listed in the acknowledgement section of this preface.

There are also a number of chapter-specific changes. For these chapters, the country/global issue of emphasis has not changed but new sections have been added or existing sections have been updated:

Chapter 2 (Theoretical Perspectives and Methods of Social Research: With Emphasis on Mexico) focuses on the 700 miles of constructed fences, walls, and other barriers along the U.S.–Mexico border. The conflict and symbolic-interactionist perspective sections have been heavily revised to reflect the most recent sociological literature on the meanings and effects of these barriers. The section formerly titled "Relationships between Independent and Dependent Variables" has changed to "Proving Cause." The corresponding shift in emphasis makes this complex idea more accessible to students.

Chapter 3 (Culture: With Emphasis on North and South Korea) includes a number of important changes that help to illustrate cultural diffusion and other processes, including the debut of Korean pop artists on the global stage (cultural diffusion); the new U.S. military policy of allowing its servicemen and women to live among the Korean population instead of on military bases; and the recent reunions of North and South Korean relatives who have not seen or otherwise communicated with one another since 1953.

Chapter 4 (Socialization: With Emphasis on Israel and the Palestinian Territories) now reflects changes related to the Israeli blockade of Gaza and the long-standing efforts to reach an agreement between Palestinians and Israelis.

Chapter 5 (Social Interaction: With Emphasis on the Democratic Republic of the Congo) still considers how HIV/AIDS is an ongoing global story that cannot be understood apart from European colonization of the DRC and other African countries. Although HIV/AIDS seems like an old story now, today's students are connected to that story when they are warned of HIV/AIDS in abstinence and sex education classes. In this edition, I tried to emphasize this student connection and also to show how HIV/AIDS can be related to "conflict minerals," most notably the minerals needed to manufacture electronic products upon which the world has come to depend (for example, wireless phones and computers). I have added a section on social structure and institutions and used these concepts to illustrate how the exploitation of DRC labor and resources has been institutionalized.

As in the previous edition of Chapter 7 (Deviance, Conformity, and Social Control: With Emphasis on the People's Republic of China), a key sociological assumption about deviance—what is considered deviant varies across time and place—is illustrated by contrasting conceptions of deviance during and after the Cultural Revolution. This chapter has been updated to reflect the dramatic changes in the Chinese economy. A late 1980s study on preschools in China and the United States is replaced with a study conducted in 2008. I have also added a new section on Foucault's disciplinary society.

Chapter 8 (Social Stratification: With Emphasis on the World's Richest and Poorest) now includes a key table that contrasts selected types of consumption (auto, wireless phones, bottled water) in the wealthiest and poorest countries. A revised discussion of social class presents a creative study in which researchers ask respondents to assess a series of hypothetical wealth distributions without telling them the distribution that applies to the United States. Ninety percent of respondents indicated that they would not want to live in a country where the wealthiest 20 percent control 80 percent of the wealth. The United States, of course, has such a distribution.

The focus of Chapter 10 (Gender) remains American Samoa. This chapter now includes a section on sexuality. It also includes a new section on documenting and explaining the global subordination of women. Finally, a section on the concept of intersectionality has been added.

Chapter 12 (Family: With Emphasis on Japan) updates data and information on Japan's response to two demographic trends having a profound effect on family life in that country. Those trends are the low fertility rate and the aging of its population. The chapter also includes a new section on caregiving. This phenomenon is relevant to families, because they are assuming greater caregiving responsibilities as populations age.

Chapter 13 (Education: With Emphasis on the European Union) now includes two new sections featuring Randall Collins's credential society and Pierre Bourdieu's theory of economic and cultural capital.

Ancillary Materials

Sociology: A Global Perspective is accompanied by a wide array of supplements prepared to create the best learning environment inside and outside the classroom for both the instructor and the student.

Student Resources

Study Guide. This student learning tool, written by Joan Ferrante and Kristie Vise (Northern Kentucky University), includes 15 to 25 study questions for each chapter to guide reading, 5 concept application scenarios, practice tests containing 20 to 25 multiple-choice and 5 to 10 true–false questions, suggested Internet resources to enhance chapter material, applied research activities, and additional background information on the focus country, territory, or theme for each chapter.

Practice Tests. Written by Margaret Weinberger of Bowling Green State University, the practice tests booklet contains 50 unique questions per chapter, including multiple-choice, true–false, and fill-in-the-blank questions, giving students a greater opportunity to study for quizzes and exams.

Instructor Resources

Test Bank. Written by Joan Ferrante and Kristie Vise, this enhanced and updated test bank consists of 90 to 100 multiple-choice questions and 30 to 35 true–false questions per chapter, all with answers, page references, question type (knowledge, comprehension, or applied), and sources (indications of whether the questions are new or also appear in the study guide). The test bank also includes 5 concept application questions, 25 to 40 short-answer questions, and 3 to 5 essay questions per chapter.

PowerLecture™ with ExamView®. PowerLecture instructor resources are a collection of book-specific lecture and class tools on either CD or DVD. The fastest and easiest way to build powerful, customized, media-rich lectures, PowerLecture assets include chapter-specific PowerPoint® presentations, images, animations and video, instructor manuals, test banks, useful web links, and more. PowerLecture media-teaching tools are an effective way to enhance the educational experience.

Instructor's Resource Manual. The instructor's manual offers the instructor detailed chapter outlines, Teaching Tips that correspond with specific sections of the book, online polling questions, and background notes on each chapter's country of emphasis. Written by Joan Ferrante and Kristie Vise the fully updated and revised manual will help you teach the global perspective with confidence.

WebTutor™ with eBook. Jumpstart your course with customizable, rich, text-specific content within your Course Management System:

- *Jumpstart*—Simply load a WebTutor cartridge into your Course Management System
- *Customizable*—Easily blend, add, edit, reorganize, or delete content
- *Content*—Rich, text-specific content, media assets, ebook, quizzing, weblinks, discussion topics, interactive games and exercises, and more

CengageNOW. CengageNOW™ is an online teaching and learning resource that gives you more control in less time and delivers better outcomes—NOW.

Sociology CourseMate. *Sociology: A Global Perspective*, Eighth Edition, includes Sociology CourseMate, which helps you make the grade.Sociology CourseMate includes:

- an interactive eBook, with highlighting, note taking, and search capabilities
- interactive learning tools including:
  - Quizzes
  - Flash cards
  - Videos
  - Games
  - and more!

Login to CengageBrain.com to access these resources related to your text in Sociology CourseMate.

CourseReader. Easy-to-use and affordable access to primary and secondary sources, readings, and audio and video selections for your courses with this customized online reader. CourseReader for Sociology helps you to stay organized and facilitates convenient access to course material, no matter where you are.

Acknowledgments

The eighth edition update builds on the efforts of those who helped me with this and previous editions. Four people stand out as particularly influential: Sheryl Fullerton (the editor who signed this book in 1988), Serina Beauparlant (the editor who saw the first and second editions through to completion), and Chris Caldeira (the editor who developed the revision plan and guided me through the writing of the seventh edition). Chris is now enrolled in the PhD sociology program at the University of California–Davis. A glance at photo credits reveals that Chris took many of the photos (49 to be exact) for this edition. Chris was also a consultant on this edition. In this regard, we had many important conversations that set the goals guiding this revision. I also benefited from numerous discussions with her about how to present sociology in ways that both engage instructors and those new to the discipline. I am fortunate to have Erin Mitchell as my current editor. I most appreciate Erin's generosity, enthusiasm, and belief in my approach to sociology. I am also grateful to her for coming up with a revision plan for this edition that gave me great flexibility in selecting photographs. Erin understands my vision and this book's purpose. In short, she is a key and valued advocate.

Of course, any revision plan depends on thoughtful, constructive, and thorough reviewer critiques. In this regard, I wish to extend my deepest appreciation to those who have reviewed this edition and/or its update:

Brian Aldrich, *Winona State University*

Shaheen A. Chowdhury, *College of Dupage*

Kay Coder, *Richland College*

Janine DeWitt, *Marymount University*

Monique Diderich, *Shawnee State University*

Sara J. Fisch, *Scottsdale Community College*

Mary Grigsby, *University of Missouri*

Liza Kuecker, *Western New Mexico University*

Joyce Mumah, *Utah State University*

Cristina Stephens, *Kennesaw State University*

Greg Walker, *Lock Haven University of Pennsylvania*

When only one name—the author's—graces the cover of this textbook, it is difficult to convey just how many people were involved with its production. Their names appear in the most unassuming manner on the copyright page, belying the significant role they played in shaping the book. Perhaps the least recognized of those named on the copyright pages are production editors. For this edition, I was fortunate to work with Cheri Palmer and Jill Traut who take care of an overwhelming number of details associated with the book, including coordinating the work of the copyeditor, photo researcher, designer, proofreader, author, and others into a textbook ready to go to press. Both handled this pressure in ways that seemed effortless. But then such a style is a sign of true professionals—making something very few people can do seem effortless.

Apart from the support I received from Wadsworth/Cengage Learning on this updated and past editions, I also received ongoing support and interest from many sociology faculty at my university who have either read or used my textbook on occasion; Prince Brown, Jr. (emeritus professor), Kris Hehn, Boni Li, J. Robert Lilly (who was my undergraduate professor), Jamie McCauley, and Kristie Vise. Kristie is now the co-author on the test bank, student study guide and instructor's resource manual. I am grateful that she has accepted this role. I would also like to thank NKU professor of Visual Arts, Barbara Houghton and her husband Rick Farley for contributing photographs for the India and other chapters.

For the past three editions, I have had the privilege of working with Missy Gish, who has a bachelor's degree in sociology and a master's in liberal studies. Missy worked behind the scenes taking photos for the book, assisting with photo research, updating tables and charts, checking references, and preparing chapters for production. On the surface, Missy's job description may seem simple, but I must emphasize that these tasks require an alertness, attention to detail, and ability to handle the stress associated with meeting deadlines that very few people possess.

I would like to thank the following NKU sociology majors who contributed boxes for this edition: They are

- Keram J. Christensen, Doing Good by Being Bad
- Devon Cowherd, The Cell Phone as a Revolutionary Tool
- Jessica Ezell, Care to Learn
- Brooke Goerman, Transgender Day of Remembrance

- Victoria Michel, Measuring Happiness
- Margaret Muench, A South Korean Helps Bring a University to China and North Korea
- Ashley Novogroski, Positive Deviants
- Ashley Novogroski, The Buy Local Movement
- Dayna Schambach, Building Green, Affordable, High-Quality Design Homes
- Gennifer Toland, Becoming Athletes—Expanding Conceptions of Self
- Staci Wood, Putting Down the Cell Phone to Learn Firsthand

I wish to express my deepest appreciation to my mother, Annalee Taylor Ferrante—who keeps my files, alerts me to news and other media reports that inform my thinking, and helps me in updating the text. My mother, who is 81 years old, cooks full-course dinners for my husband and me several times a week. The care with which she prepares food and the exquisite results has no parallel. This is no easy feat in a world dominated by heavily processed and prepackaged foods and ingredients.

As always, I also express my love for and gratitude to my husband, colleague, and friend Robert K. Wallace who is without a doubt my greatest supporter. In closing, I acknowledge, as I have done in all editions of this and other books, the tremendous influence of Horatio C Wood IV, MD, on my academic career and philosophy of education. Dr. Wood died on May 28, 2009. His death only served to intensify the warmth and gratitude I continue to feel for him. In reflecting on the important mentoring role Dr. Wood has played in my life, I cannot help but wonder why there seem to be few, if any, explicit acknowledgments of the deep emotions felt for those who have the greatest influence on our work. The emotions I felt for Dr. Wood were an important component of what was, by any measure, a constructive relationship. These emotions allowed me to gauge his specialness and they gave purpose, excitement, and direction to my learning, writing, and teaching.

THE SOCIOLOGICAL IMAGINATION

1 With Emphasis on the MOBILE PHONE

Have you ever tried to imagine going one day without your phone? How would your life change? If you have a mobile phone, you own one of 5.3 billion mobile phones in the world (U.S. Central Intelligence Agency 2011). It is likely that the vast majority of owners have come to need and depend on them just as you do. Sociologists see mobile phones as a social force that shapes human interaction and activity. In studying this phone, they consider how that invention shapes the way people see the world and interact with others. They also consider how people use, adapt to, and respond to this invention. Finally, sociologists study how mobile phones affect the way people relate to groups, organizations, and institutions.

Why Focus On MOBILE PHONES?

CORE CONCEPT 1
***Sociology* is the scientific study of human activity in society. More specifically, it is the study of the social forces that affect the things people do with and to one another.**

The activities sociologists study are too many to list but, as you will learn from reading this textbook, sociologists study human activities as they relate to immigration, racial classification, religion, education, and much more. Specifically, sociologists are interested in social forces like the mobile phone that shape and change any human activity, including the ways people think about themselves and others and the things they do to and with one another. **Social forces** are anything humans create that influence or pressure people to interact, behave, respond, or think in certain ways. This textbook is about those many social forces that shape our lives.

The mobile phone is a perfect vehicle for illustrating the sociological perspective. For one, it is a human-created technology that has transformed or will transform every aspect of life. The revolutionary feature of the mobile phone is that it frees people from being in a specific physical space when they communicate with others. While the landline phone allowed us to communicate with others in faraway places, all parties had to arrange to be in a fixed location—an office, at home, in a telephone booth. In addition, people waited until they got home to some fixed location to tell someone what had happened in their day. Now people can share what is happening as it happens

With mobile phones, people are no longer confined to a fixed location. They can communicate while driving or walking to class. It doesn't matter if they are skydiving, on a mountaintop, at a party, or at church. In addition the technology is such that the mobile phone can function as a miniature laptop with Internet access and as a multimedia device allowing people to exchange photographs, text messages, music, videos, and anything that can be digitalized. Mobile phones also allow people to choose from millions of applications that facilitate social and business transactions and that meet needs for just about any kind of information or entertainment.

Online Poll

At what age did you get your first cell phone?

- ○ Before I was 10 years old
- ○ 10 to 11 years old
- ○ 12 to 13 years old
- ○ 14 to 15 years old
- ○ 16 to 17 years old
- ○ 18 or older
- ○ Don't own a mobile phone

To see how other students responded to these questions, go to **www.cengagebrain.com.**

The Sociological Imagination

CORE CONCEPT 2 The sociological imagination is a quality of mind that allows people to see how remote and impersonal social forces shape their life story or biography.

A **biography** consists of all the day-to-day activities from birth to death that make up a person's life. Social forces are considered remote and impersonal because, for the most part, people have no hand in creating them, nor do they know those who did. Think about how old you were (or if you were even alive) when the mobile phone first appeared on the market in 1984. How old were you when it first became popular in 1995? What was your age in 2005, the year the number of mobile phones exceeded the number of landlines (Hanson 2007)? It is likely that you had no direct involvement in the invention and development of the mobile phone but you become involved any time you decide to use a mobile phone. You become involved any time you check for messages upon waking up in the morning and before going to bed or anytime you text while driving (or refrain from doing so). Likewise, you become involved when you decide to turn off your mobile phone to give someone your complete attention or you choose to ask a stranger for directions rather than rely on the GPS function on your phone. The point is that when people respond to social forces in their lives, they become part of that force. People can embrace social forces, be swept along, be bypassed by them, and most importantly, challenge them.

Why is it important to develop the **sociological imagination**—a point of view that allows us to identify seemingly remote and impersonal social forces and connect them to our biographies? The payoff for those who acquire the sociological imagination is that they can better understand their own biography, recognize that choices exist, and that their choices have larger consequences for others. The concept *social facts* is useful for conceptualizing social forces that impact our lives.

Missy Gish

Missy Gish

The sociological imagination allows you to consider how larger social forces—the time in history, the place one lives, the technologies at hand—affect the individual biography. How might the mobile phone and land line phone shape a child's view of the world and sense of self in relation to others? For one thing, the child who uses a mobile may not learn to associate a person they are talking to with a specific place.

The Study of Social Facts

What happens when you leave the house without your phone and you get to your destination realizing that you left it behind? Does your mind race with thoughts like "what will my mother, partner, friend, or boss think if they call or text and I don't reply within, say 15 minutes? How will I let people know where I am and why they haven't heard from me?" French sociologist Émile Durkheim defined **social facts** as ideas, feelings, and ways of behaving "that possess the remarkable property of existing outside the consciousness of the individual" (Durkheim 1982, p. 51). That is, for the most part, social facts do not originate with the people experiencing them. From the time we are born, the people around us seek to impose upon us ways of thinking, feeling, and acting that we had no hand in creating. The words and gestures people use to express thoughts; the monetary and credit system used to pay debts; the rules governing mobile phone use and

sociology The scientific study of human activity in society.

social forces Anything humans create or take notice of that influence or pressure people to interact, behave, respond, or think in certain ways.

biography All the day-to-day activities from birth to death that make up a person's life.

sociological imagination A point of view that allows us to identify seemingly remote and impersonal social forces and connect them to our biographies.

social facts Ideas, feelings, and ways of behaving "that possess the remarkable property of existing outside the consciousness of the individual."

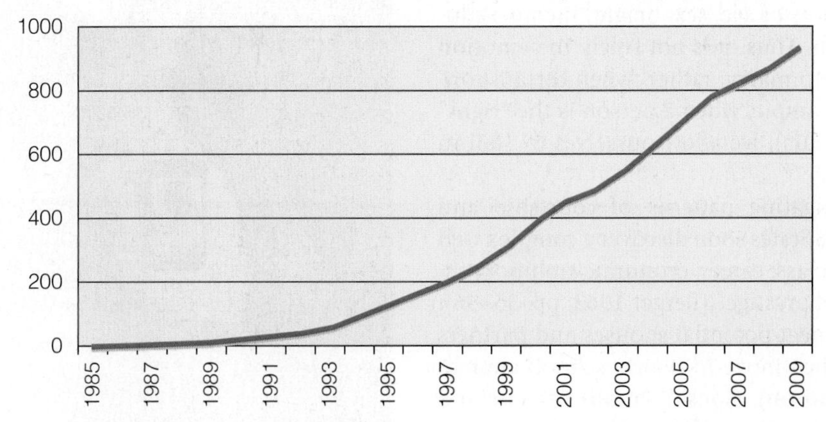

FIGURE 1.1 Durkheim argues that rates of behavior offer a window into the pressures to behave or think in particular ways. If few people around you are saving money from their paychecks each month, you will feel less pressure to save in general. If most people turn their mobile phone off in a particular situation, others will feel social pressure to turn their phones off as well. The chart shows the number of mobile phones per 1,000 people, from 1985 to 2009. We can use these rates as rough indicators of the urgency people may feel to own one. When sociologists study rates, they also think about the forces behind this increase in mobile phone use.

Source: Data from U.S. Central Intelligence Agency (2011)

etiquette; the beliefs and rituals of the religions people follow—all were created seemingly without their input. Thus social facts have a life that extends beyond the individuals who carry them out.

Not only do social facts exist outside individuals, but they also have coercive power. When people freely and unthinkingly conform to social facts, that power "is not felt or felt hardly at all" (Durkheim 1982, p. 55). Only when people resist do they come to know and experience the power of social facts. Durkheim wrote that he was not forced to speak French or to use the legal currency, but it was impossible for him to do otherwise. "If I tried to escape the necessity, my attempt would fail miserably. . . . Even when in fact I can struggle free from these rules and successfully break them, it is never without being forced to fight against them" (Durkheim 1982, p. 51). In other words, even when people challenge them, social facts make their power known by the difficulty people experience trying to do things and think in different ways. Still, it is impressive that most of us can probably think of at least one example in which we resisted following social facts. People who decide to not use a mobile phone face the inconvenience, the questioning, and even anger from others that comes with that decision (for example, "How am I supposed to get a hold of you if I need you?" "Are you serious?" "You'll lose your connection to the world!").

For Durkheim, social facts also included what he called **currents of opinion**, the state of affairs with regard to some way of being. The intensity of these currents is broadly reflected in rates summarizing various behaviors—for example, marriage, suicide, or birth rates. Durkheim believed the rates at which people around us marry, take their own life, or give birth to children both influence and reflect others' thinking and behavior on these matters. The intensity of that current of opinion shapes the behavior of people who live in the society.

The Sociological Consciousness

Sociologist Peter L. Berger offers one of the best descriptions of the sociological imagination or what he calls sociological consciousness. Berger (1963) equates sociologists with curious observers walking the neighborhood streets of a large city, fascinated with what they cannot see taking place behind the building walls. The buildings themselves offer few clues beyond hinting at the architectural tastes of the people who built the structures and who may no longer live there. According to Berger, the wish to look inside and learn more is analogous to the sociological perspective.

Berger offers the following example to illustrate his point. In the United States, we assume that people marry because they are in love. Popular belief states that love is a violent, irresistible emotion that strikes at random. Upon investigating the characteristics of people who marry,

currents of opinion The state of affairs with regard to some way of being.

however, we find that Cupid's arrow actually seems to be guided by considerations of age, sex, height, income, education, race, and so on. Thus, it is not solely the emotion of love that causes us to marry; rather, when certain conditions are met (for example, when a person is the "right" height in relation to ours), we allow ourselves to "fall in love."

Sociologists investigating patterns of courtship and marriage in the United States soon discover a complex web of motives related to "class, career, economic ambition, aspirations of power and prestige" (Berger 1963, pp. 35–36). For example, people meet potential spouses and partners in the social circles they move in (work, school, church, neighborhoods, and so on). Considerations of class and education are obviously present if a couple meet on a college campus where the tuition is $1,600 per year versus $40,000. For those who date via the Internet, social considerations come into play as well when dating service subscribers narrow the search to include only men or women of a particular age, race, ethnicity, height, sexual orientation, geographic location, profession, income, and even eye color.

Once people decide to marry, the next steps to follow have already been laid out for them, and "though there is some leeway for improvisations, too much ad-libbing is likely to risk the success of the whole operation" (Berger 1963, p. 87). Note that neither the future bride nor the future groom "invented" these steps. Friends, family, clergy, jewelers, florists, caterers, musicians, wedding planners, and others serve as the guardians of tradition and help ensure that the steps are followed. These guardians do not have to exert much pressure on the couple, because these expectations "have long been built into their own projections of the future—they want precisely what society expects of them" (p. 87). As Berger notes, the "miracle of love now begins to look somewhat synthetic" (p. 36). This statement does not mean that sociologists dismiss the emotion of love as wholly irrelevant. Rather, they look beyond the popular meanings and publicly approved interpretations of why people say they marry.

The discipline of sociology offers us theories, concepts, and methods needed to look beyond popular meanings and explanations of what is going on around us. Berger (1963) points out that sociologists, by the very logic of their discipline, are driven to debunk the social systems they study. One should not mistake this drive as being located in a sociologist's temperament or personal inclination. Apart from their field of study, sociologists may be "disinclined to disturb the comfortable assumptions" about what is going on around us (p. 29). Nevertheless, the sociological perspective compels sociologists to explore levels of reality that dig below the surface. The logic of the discipline presupposes a "measure of suspicion about the way in which human events are officially interpreted by the authorities, be they political, juridical or religious in character" (p. 29). In their effort to explore levels of reality that dig below the surface, sociologists also make distinctions between troubles and issues.

Photo courtesy of Conceptual Photo Illustration/Released

Chris Caldeira

Although sociologists do not dismiss the role of love in shaping a couple's decision to form a marriage or partnership, they do focus on the social considerations (such as age, sex, height, income, sexual orientation, education, and race) that must be present before people allow themselves to fall in love. In the case of the two couples pictured, what social considerations appear to be at work?

Troubles and Issues

> CORE CONCEPT 3 **Sociologists distinguish between troubles, which can be resolved by changing the individual, and issues, which can be resolved only by addressing the social forces that created them.**

C. Wright Mills (1959) defines **troubles** as personal needs, problems, or difficulties that can be explained as individual shortcomings related to motivation, attitude, ability, character, or judgment. The resolution of a trouble, if it can indeed be resolved, lies in changing the individual in some way. Mills states that when only one man or woman is unemployed in a city of 100,000, that situation is his or her personal trouble. For its relief, we properly look to that person's character, skills, and immediate opportunities (that is, we think, "She is lazy," "He has a bad attitude," "He didn't try very hard in school," or "She had the opportunity but didn't take it").

By comparison, an **issue** is a matter that can be explained only by factors outside an individual's control and immediate environment. When 24 million men and women are unemployed or underemployed in a nation with a workforce of 156 million, that situation is an issue. Clearly, we cannot hope to solve this kind of employment crisis by focusing on the character flaws of 24 million individuals. According to Mills, an accurate description of the problem and of the possible solutions to it requires us to think beyond individual shortcomings and to consider the underlying social forces that create them.

Mills argues that many people cannot see the intricate connection between their personal situations or troubles and the larger social forces. Mills also argues that most people cannot (or do not want to) see how their gains and successes connect to others' so-called losses and failures in life (see Sociological Imagination). For example, it is hard to recognize that a person's success at connecting 24/7 with peers is often achieved at the loss of meaningful contact with parents and others. Mills believes that most people lack this awareness because they do not possess a quality of mind that enables them to grasp the interplay between self and world and between biography and the large social forces pressuring them to think and behave as they do.

In *The Sociological Imagination*, Mills (1959) asks, "Is it any wonder that ordinary people feel they cannot cope with the larger worlds with which they are so suddenly confronted?" (pp. 4–5). Is it any wonder that people often feel trapped by the social forces that affect them? As Mills pointed out, we live in a world in which information dominates our attention and overwhelms our capacity to make sense of all we hear, see, and read every day. Consequently, we may be exhausted by the struggle to learn from that information about the forces that shape our daily lives. According to Mills, people need "a quality of mind that will help them to use information" to think about "what is going on in the world and what may be happening within themselves" (p. 5). Mills calls this quality of mind the *sociological imagination*. The payoff for those who possess a sociological imagination is that they can better understand their own experiences and fate by locating them in a larger historical, cultural, and social context; they can recognize the responses available to them by becoming aware of the many individuals who share their situations.

The sociological imagination is evident in the work of the earliest and most influential sociologists. In fact, one can make the case that the discipline of sociology emerged as part of an effort to understand how a social force known as the Industrial Revolution changed people's lives in countless ways.

The Industrial Revolution

> **CORE CONCEPT 4 Sociology emerged in part as a reaction to the Industrial Revolution, an ongoing and evolving social force that transformed society, human behavior, and interaction in incalculable ways.**

The Industrial Revolution is the name given to the changes in the way goods were produced, food was grown, people got from one place to another, extracted resources from the earth, and communicated and interacted with one another. In short, the Industrial Revolution transformed virtually every aspect of society. The defining feature of the Industrial Revolution was **mechanization**, the process of replacing human and animal muscle as a source of power with external sources derived from burning wood, coal, oil, and natural gas. Before mechanization, goods were produced and distributed at a human pace. New sources of power eventually replaced hand tools with power tools, sailboats with freighters, and horse-drawn carriages with trains. Mechanization changed how goods were produced and how people worked. It turned workshops into factories, skilled workers into machine operators, and handmade goods into machine-made goods. With industrialization, products previously crafted by a few skilled people were now standardized and assembled by many relatively unskilled workers, each involved in part of the overall production process. Now no one person could say, "I made this; this is a unique product of my labor." The factory owners gained power over the artisans as machines rendered them obsolete. Now people with little or no skill could do the artisan's work—and at a faster pace (Thrall 2007).

Industrialization did more than change the nature of work; it changed notions of time and space. A social order that had existed for centuries vanished, and a new order—familiar in its outline to us today—appeared (Lengermann 1974). A series of developments—the

troubles Personal needs, problems, or difficulties that can be explained as individual shortcomings related to motivation, attitude, ability, character, or judgment.

issue A matter that can be explained only by factors outside an individual's control and immediate environment.

mechanization The process of replacing human and animal muscle as a source of power with external sources derived from burning wood, coal, oil, and natural gas.

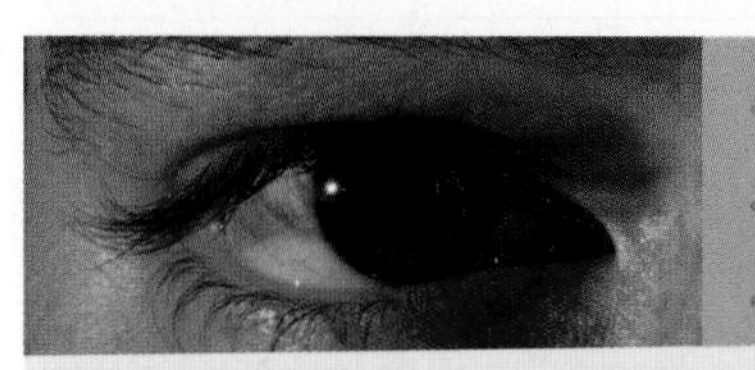

The Sociological Imagination Making Connections between Troubles and Issues

We know that there is a connection between troubles and issues when we can demonstrate that a seemingly personal problem would not exist if the person lived at another time in history or another place in the world. You should not interpret this fact to mean that people have no control over what happens to them or that they have no choices. Rather it is important to see that better choices can be made if people understand the larger context. We show four examples of seemingly personal troubles that would likely not exist if the people affected lived elsewhere in the world or at a different time.

Missy Gish

Megan and Michael met while in college and married soon after they graduated. They brought student loans to the relationship –a combined total of $80,605.70. They pay $525.58 per month, which represents 8.5 percent of their income. But Megan and Michael are not alone. In the United States, 66 percent of graduates leave college with an average loan to repay valued at $23,200, a loan that may take 20 to 30 years to pay off (Bernard 2009). The size of the debt burden American students assume is not one that students in the European Union (EU) face. This is because a greater share of the cost of higher education in Europe is taxpayer subsidized, often to the level that it is free. But even in those EU countries where a college education is not free, the cost is much lower than in the United States.

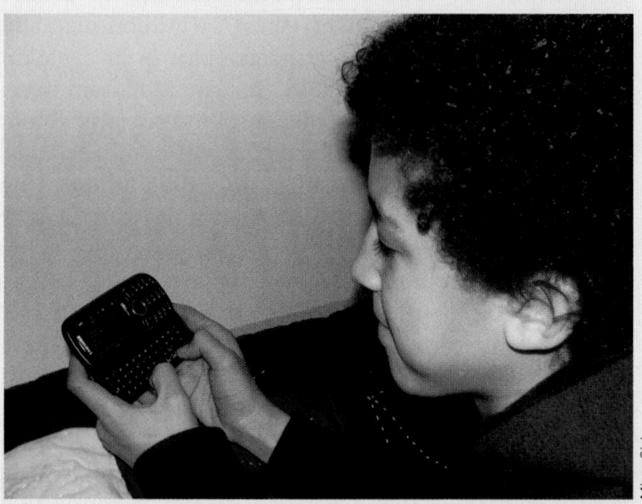

Missy Gish

A parent feels like she knows little to nothing about her son's friends or her son's life for that matter. She routinely checks her son's mobile phone to see who he calls or texts. Her son does not like this. The problem she faces can be traced in part to the fact that wireless phones are viewed as a personal item not to be shared with others, even family members. Because of this, callers do not have to speak to a third party before reaching the person of interest. There is no need to make "small talk" to a third party answering the phone. On the one hand, mobile phones expand individuality and privacy. On the other hand, parents talk to their children directly and in effect miss opportunities to connect with those, who in the past might have answered a landline phone. Is it any wonder some parents can feel left out?

railroad, the steamship, the cotton gin, the spinning jenny, running water, central heating, electricity, the telegraph, and mass-circulation newspapers—transformed how people lived their daily lives and with whom they interacted. Coal-powered trains, for example, turned a monthlong trip by stagecoach into a daylong one. These trains permitted people and goods to travel day and night; in rain, snow, or sleet; and across smooth and rough terrain. Railroads increased the human and freight traffic to and from previously remote and unconnected areas. The railroad caused people to believe they had annihilated time and space (Gordon 1989). In addition, railroads facilitated an unprecedented degree of economic interdependence, competition, and upheaval. Now people in one area could be priced out of a livelihood if people in another area could provide goods and materials at a lower price (Gordon 1989).

The Industrial Revolution drew people from even the most remote parts of the globe into a process that produced unprecedented quantities of material goods. The historical period known as the Age of Imperialism (1880–1914) involved the most rapid industrial and colonial expansion in history. During this time, rival European powers (such as Britain, France, Germany, Belgium, Portugal, the Netherlands, and Italy) competed to secure colonies and, by extension, exploit the labor and natural resources within those colonies. By 1914, for example, all of Africa had been divided into European colonies. By that year, 84 percent of the world's land area had been affected by colonization,

Missy Gish

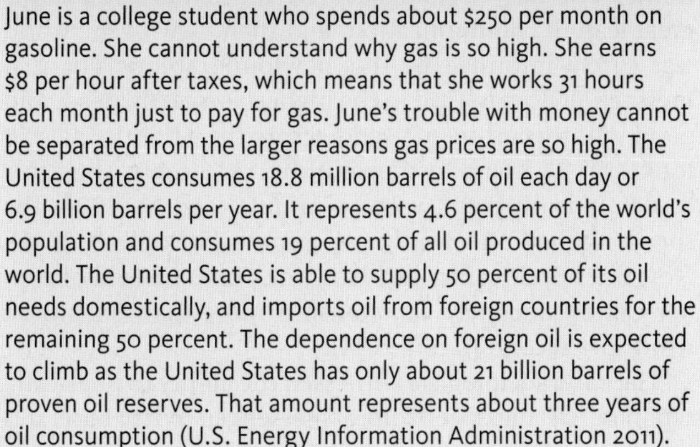

June is a college student who spends about $250 per month on gasoline. She cannot understand why gas is so high. She earns $8 per hour after taxes, which means that she works 31 hours each month just to pay for gas. June's trouble with money cannot be separated from the larger reasons gas prices are so high. The United States consumes 18.8 million barrels of oil each day or 6.9 billion barrels per year. It represents 4.6 percent of the world's population and consumes 19 percent of all oil produced in the world. The United States is able to supply 50 percent of its oil needs domestically, and imports oil from foreign countries for the remaining 50 percent. The dependence on foreign oil is expected to climb as the United States has only about 21 billion barrels of proven oil reserves. That amount represents about three years of oil consumption (U.S. Energy Information Administration 2011).

Sgt. Mark Faylioga

Although he is the son of a white mother and a black father, President Barack Obama (2008) is referred to as the first black president of the United States. During his campaign, he spoke about some of the problems he encountered: "At various stages in the campaign, some commentators have deemed me either 'too black' or 'not black enough.' . . . The press has scoured every exit poll for the latest evidence of racial polarization. . . ." Obama's experiences are not simply personal. Sociologists see race as a socially created way of categorizing people. As such, it is a social force of immense significance. As we will learn, early on in U.S. history, its lawmakers decided that a parent and his or her biological offspring can be classified as different races. Any discernible evidence of African ancestry made someone black. Obama's experience would be very different if he lived in Brazil, where he would likely be assumed multiracial and classified as brown and perhaps even white.

and an estimated 500 million people were living as members of European colonies (*Random House Encyclopedia* 1990).

The Industrial Revolution changed everything—the ways in which goods were produced, the ways in which people negotiated time and space, the relationships between what were once geographically separated peoples, the ways in which people made their livings, the density of human populations (for example, urbanization), the relative importance and influence of the home in people's lives, access to formal education (the rise of compulsory and mass education), and the emergence of a consumption-oriented economy and culture. The accumulation of wealth became a valued and necessary pursuit. In *The Wealth of Nations*, Adam Smith argued that the invisible hand of the free market (capitalism) embodied in private ownership and self-interested competition held the key to a nation's advancement and prosperity. The unprecedented changes caught the attention of the early sociologists who wrote in the nineteenth and early twentieth centuries. In fact, sociology emerged out of their efforts to document and explain the effects of the Industrial Revolution on society.

CORE CONCEPT 5 Early sociologists were witnesses to the transforming effects of the Industrial Revolution. They offered lasting conceptual frameworks for analyzing the ongoing social upheavals.

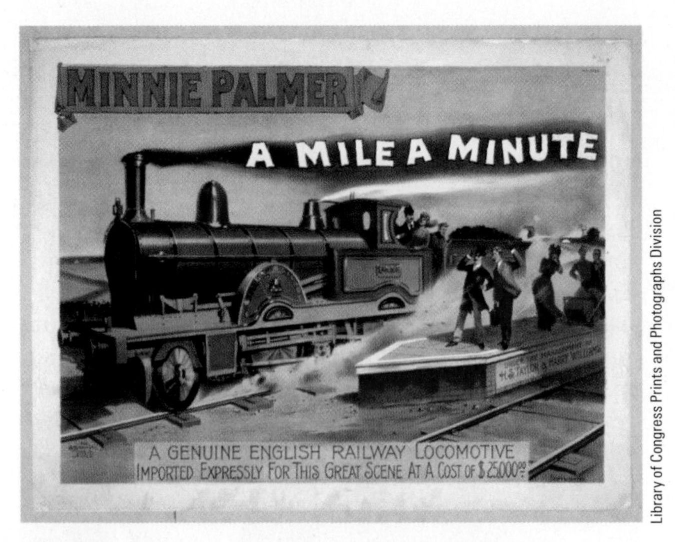

Library of Congress Prints and Photographs Division

Coal-powered locomotives celebrated in this 1891 poster permitted people to travel day and night; in rain, snow, or sleet; across smooth and rough terrain—turning monthlong trips into daylong ones. Railroads increased opportunities for personal mobility and boosted the freight and passenger traffic to and from previously remote areas.

Sociology emerged as an effort to understand the dramatic and almost immeasurable effects of the Industrial Revolution on human life across the globe. Although the early sociologists wrote in the nineteenth and early twentieth centuries, their observations remain relevant. Because most of us living today know only an industrialized life, we lack the insights that came from living through the transformation. To grasp the significance of these observations, consider the following anecdote. In a recent interview, a scientist maintained that we are close to understanding the mechanisms governing aging, and that people might soon live to be 150 years old. If aging mechanisms are in fact controlled, the first people to witness the change will have to make the greatest adjustment. In contrast, people born after this discovery will know only a life in which they can expect to live 150 years. If these post-discovery humans are curious, they may wish to understand how living to age 150 shapes their lives. To fully understand this subject, they will have to look to those who recorded life before the change and who made sense of their adjustments to the so-called advancement. So it is with industrialization: To understand how it shaped and continues to shape human life (and how it has shaped sociology), we can look to six of the early sociologists.

Three of the six sociologists covered are nicknamed the "big three." Those three are Karl Marx, Émile Durkheim, and Max Weber. Sociologists universally agree that these three are the giants in the field and that their writings form the heart of the discipline. It is safe to say that all sociologists who have come after Marx, Durkheim, and Weber have been deeply influenced by their ideas even as they expand, refine, and challenge them (Appellrouth and Edles 2007).

We also include three other central figures: Auguste Comte, because he gave sociology its name, and Jane Addams and W.E.B. DuBois. DuBois focused attention on the color line, and Jane Addams championed sympathetic knowledge or knowledge gained from living and working among those being studied. The color line and sympathetic knowledge are certainly core ideas within sociology.

Upon discussing the work of each of these six sociologists, we consider how each would write about the transformative nature of the mobile phones on thinking, interaction, and human activity. Keep in mind that the early sociologists witnessed the introduction of many inventions that annihilated space and time including the telegraph (1838) and cablegram, which allowed messages to be sent to countries on opposite sides of the Atlantic (1866) and the Pacific Oceans (1902). These inventions freed communication from having to travel with humans or animals across space (for example, via pony express or trains). The early sociologists were the first to witness this great leap in communication, and their view of the world was surely informed by this newfound ability. The first message ("Glory to God in the highest; on earth, peace and good will toward men") send across the Atlantic in 1858 took over 17 hours to transmit; by 1866 cable, the cable could transmit eight words a minute; and by 1900, the speed reached 120 words per minute (SchoolNet 2011). Now fiber-optic cables lie below the ocean instead of copper cables, and they can transmit more than 84 billion words per second (Geere 2011).

The early sociologists witnessed the ability to physically separate the message from a messenger. What separates their time in history from today is that we live in a time when people do not have to be in a fixed space/place to receive or send a message. Still, the early sociologists give us a language and perspective that helps us to frame and think about revolutions in communication.

Auguste Comte (1798–1857)

The French philosopher Auguste Comte, known as the father of positivism, gave sociology its name in 1839. **Positivism** holds that valid knowledge about the world can be derived only from *sense experience* or knowing the world through the senses of sight, touch, taste, smell, and hearing and from making empirical associations (for example, evidence of cause and effect must be observed). Comte advanced the "law of three stages," which maintains that societies develop according to three stages: (1) the theocratic, (2) the metaphysical, and (3) the

positivism A theory stating that valid knowledge about the world can be derived only from *sense experience* or knowing the world through the senses of sight, touch, taste, smell, and hearing, and from empirical associations.

Library of Congress Prints and Photographs

Lisa Southwick

The illustration celebrates the first message sent via underwater cable on August 17, 1858, between Newfoundland (Canada) and Valentia Bay (Ireland). The two men grasp hands as the distance between them is shortened. One might imagine that those who sent cablegram messages felt they had conquered time and space, much as we feel today when we bridge distance using e-mail, texting and social networking platforms. (Chiles 1987).

positive. In the theocratic stage, people explain the events going on in the world as the work of personified deities—those deities may be objects such as the sun or trees, a variety of gods, or a supreme deity. Deities possess supernatural qualities that allow them to exert their will over humans and nature. In the metaphysical stage, people draw upon abstract and broad concepts to define features of reality that cannot be observed through the senses or direct experience. Metaphysics deals with big philosophical questions such as the nature of the human mind, the meaning of life, and good versus evil. In the positive stage (stage 3)—the conceptually superior stage according to Comte—people use scientific explanations grounded in observation and experimental designs to understand the world. Comte placed sociology in this third stage of thinking; he maintained that sociologists were scientists who studied the results of the human intellect (DeGrange 1939). What did he mean by this?

First, sociology is a science and only those sociologists who follow the scientific method can presume to have a voice in describing and guiding human affairs. The scientific method rejects personal opinions and political agendas in favor of disciplined and objective strategies in thinking about and addressing social issues and making effective policies. Second, sociologists study the things humans have created: an idea, an invention, or a way of behaving and the effects those creations have on society.

Comte recommended that sociologists study **social statics**, the forces that hold societies together such that they endure over time, and **social dynamics**, the forces that cause societies to change. Comte's preoccupation with forces of order and change is not surprising given that he was writing at a time when the Industrial Revolution was transforming society in unprecedented ways.

Comte on the Mobile Phone If Auguste Comte were alive today, he would emphasize the dramatic and far-reaching changes associated with the mobile phone. At the same time, he would also consider that, in spite of thousands of changes to the ways in which people relate to one another, specific relationships do not break down into something beyond recognition. Some news headlines illustrate this point:

- "Local Drivers Pulled Over during Cell Phone Sweep: Cell Phone Crackdown Last November Netted 1,000" (*San Diego News* 2011).
- "Man Used Roommate's Cell Phone to Lure Robbery Victim with a Text Message Asking Him to Come Over" (WAFB.com 2011).
- "Starbucks Releases New Payment System for Mobile Phones" (Ferri 2011).
- "4th Annual Cell Phone Film Festival" (India PRWire 2011).

All four headlines describe a way the mobile phone has changed the way people relate—a change in reasons police pull over drivers, in the way a robber might lure a potential victim, in the way customers pay for a product, and in how people watch a film. Although these four changes are quite dramatic, there is still a timeless element to each event—police have always engaged in traffic sweeps, criminals

social statics The forces that hold societies together such that they endure over time.

social dynamics The forces that cause societies to change.

always look for ways to lure unsuspecting victims, people have always had to pay for products and services, and films have been around for some time. The mobile phone did not break down, beyond recognition, relationships between police and drivers, robbers and victims, consumers and coffee venders, and moviegoers and filmmakers.

Karl Marx (1818–1883)

The political philosopher Karl Marx was born in Germany but spent much of his professional life in London, working and writing in collaboration with Friedrich Engels. One of Marx and Engels's most influential writings is *The Communist Manifesto*, a 23-page pamphlet issued in 1848, and translated into more than 30 languages (Marcus 1998). Upon reading it today, more than 150 years later, one is "struck by the eerie way in which its 1848 description of capitalism resembles the restless, anxious and competitive world of today's global economy" (Lewis 1998, p. A17).

The *Manifesto* includes these famous lines: "The workers have nothing to lose but their chains; they have a whole world to gain. Workers of all countries, unite." In an essay marking the 150th anniversary of *The Communist Manifesto*, John Cassidy (1997) wrote that "in many ways, Marx's legacy has been obscured by the failure of Communism, which wasn't his primary interest. In fact, . . . Marx was a student of capitalism, and that is how he should be judged" (p. 248).

Marx sought to analyze and explain how **conflict** drives social change. The character of conflict is shaped directly and profoundly by the means of production, the resources (land, tools, equipment, factories, transportation, and labor) essential to the production and distribution of goods and services. Marx viewed every historical period as characterized by a system of production that gave rise to specific types of confrontation between an exploiting class and an exploited class. For Marx, class conflict was the vehicle that propelled people from one historical epoch to another.

From Marx's perspective, the system of production accompanying the Industrial Revolution gave rise to two distinct classes: the **bourgeoisie**, the profit-driven owners of the means of production, and the **proletariat**, those individuals who must sell their labor to the bourgeoisie. Marx devoted his life to documenting and understanding the causes and consequences of this fundamental and unequal divide. Marx expressed profound moral outrage over the plight of the proletariat, who, at the time of his writings, were unable to afford the products of their labor and suffered from deplorable living conditions.

Karl Marx believed that the bourgeoisie's pursuit of profit was behind the explosion of technological innovation and the never-before-seen increase in the amount of goods and services produced during the Industrial Revolution. In a capitalist system, profit is the most important measure of success. Marx described class conflict as an antagonism that grows out of the opposing interests held by these two parties. The bourgeoisie's interest lies with making a profit and the proletariat's with increasing wages. To maximize profit, the bourgeoisie work to cut labor costs by investing in labor-saving technologies, employing the lowest-cost workers, and finding the cheapest materials to make products.

The capitalist system is a vehicle of change in that it requires technology and products to be revolutionized constantly. Marx believed that capitalism was the first economic system capable of maximizing the immense productive potential of human labor and ingenuity. He also felt, however, that capitalism ignored too many human needs and that too many people could not afford to buy the products of their labor. Marx believed that if this economic system were in the right hands—the hands of the workers or the proletariat—public wealth would be more than abundant and would be distributed according to need. Instead, according to Marx (1887), capitalism survived and flourished by sucking the blood of living labor. The drive is a "boundless thirst—a werewolf-like hunger—that takes no account of the health and the length of life of the worker unless society forces it to do so" (p. 142). That thirst for profit "chases the bourgeoisie over the whole surface of the globe" in search of the lowest-cost labor and resources to make products (Marx 1881, p. 531).

Marx also named a third class, the finance aristocracy, who lived in obvious luxury among masses of starving, low-paid, and unemployed workers (Bologna 2008, Proudhon 1847). The finance aristocracy includes bankers and stockholders seemingly detached from the world of "work." Marx (1856) described this source of income as "created from nothing—without labor and without creating a product or service to sell in exchange for wealth." The finance aristocracy speculates or employs financial advisors to speculate for them. Although some speculation has a power of inventiveness, "it is at the same time also a gamble and a search for the 'easy life'; as such it is the art of getting rich without work . . . without giving anything equivalent in exchange; it is the cancer of production, the plague of society and of states" (Bologna 2008, Proudhon 1847).

Marx on Mobile Phones If Karl Marx were alive today, he would certainly focus his attention on means of production as it relates to wireless phones. He would be particularly interested in the fact that more than 5 billion

conflict The major force that drives social change.

bourgeoisie The profit-driven owners of the means of production.

proletariat Those individuals who must sell their labor to the bourgeoisie.

wireless phones are in use around the world and that billions of phones have been discarded since they became popular in 1995. Marx would ask who manufactured these billions of phones? To answer this, Marx would draw upon the research of social scientist Pun Ngai (2005), who worked in a Chinese electronics factory with a dormitory for workers.

In *Made in China: Women Factory Workers in a Global Workplace,* Ngai writes of female workers who labor 12 or more hours per day on five to six hours of sleep in a closed factory environment (windows sealed and covered with plastic) so that employees do not lose concentration by looking outside. The work was repetitive, mindless, and never ending. Watched by an electronic eye, the workers earned about 45 cents per hour. Although the workers did not know where the cameras were placed, they were nevertheless aware of their relentless gaze. As one woman described it, "I don't need to use my mind anymore. I have been doing the same thing for two years. Things come and go, repeating every second and minute. I can do it with my mind closed" (p. 83).

Ngai found that almost every task on the assembly line involved working with or exposure to highly toxic chemicals. Coupled with a stressful work environment and lack of sleep, not surprisingly, chronic pain was pervasive among the factory workers. Among the most common symptoms reported were headaches, backaches, eye strain, sore throats, nausea, and menstrual pain.

Marx, who wrote more than 110 years ago about the exploitation of labor, could not have anticipated the complex supply chain through which a product such as the mobile phone passes, beginning with extraction of raw materials to eventually reaching the consumers' hands. In spite of its small size, a mobile phone can contain from 600 to 1,000 parts, depending on brand. Marx would still argue

If Karl Marx were alive today, he would most certainly focus on those who assemble and test wireless phones, their low earnings, the long hours worked per week, and adverse working conditions.

that factory labor and the labor of those who extract and process raw materials for the wireless phones are the backbone of the industry. Handset makers such as RIM, Nokia, Samsung, Apple, and Sony, and carriers such as AT&T and Verizon could not exist without this labor.

Because firms "guard information about pricing deals they have negotiated and often compel the silence of their suppliers and contractors through nondisclosure agreements," it is virtually impossible to secure financial records documenting income, revenue, and profits of all the parties involved in the production and distribution of wireless phones (Dedrick, Kraemer, and Linden 2010, p. 7). There is some financial data on some of the major mobile phone companies as it relates to gross profits and manufacturing costs, including the cost of assembly and testing. Table 1.1 shows that the assembly and test costs are very low relative to the revenue earned per phone.

TABLE 1.1 Selected Financial Indicators for Four Major Headset Suppliers of Smart Phones

Financial Indicator	Motorola Razr	Palm Treo 650	RIM Curve	Nokia
Total Amount Received From Consumer Including Subsidy from Servicer Provider	$349.00	$418.00	$426.00	N/A
Cost of 600 to 1,000 Parts, Components	$137.92	$178.99	$100.30	$182.44
Assembly and Testing	$5.81	$11.58	$7.95	$7.51
Gross Profit*	$206.00	$228.00	$238.00	N/A
Return on Assets**	13%***	7%	No estimate	N/A

*_Gross profit_ is the difference between revenue from product and the cost of making it before things like overhead, payroll, and taxes are deducted.

**_Return on assets_ is calculated by dividing a corporation's annual earnings by its total assets to estimate a return on investment.

***For the year researchers estimated this number, the actual percentage was 23%; the company had received a substantial tax break. Without this tax break, the return would have been 13%.

N/A = Not applicable.

Source: Data from Dedrick, Kraemer, and Linden 2010.

Émile Durkheim (1858–1917)

To describe the Industrial Revolution and its effects, Frenchman Émile Durkheim focused on the division of labor and solidarity. The division of labor is the way a society divides and assigns day-to-day tasks. Durkheim was interested in how the division of labor affected **solidarity**, the system of social ties that connects people to one another and to the wider society. This system of social ties acts as cement binding people to each other and to the society. Durkheim observed that industrialization changed the division of labor from relatively simple to complex and, by extension, changed the nature of solidarity. Durkheim believed that the sociologist's task is to analyze and explain the solidarity. As you will see later in this textbook, Durkheim's preoccupation with the ties that bind is evident in his writings on education, deviance, the division of labor, and suicide. By way of introduction to Durkheim's emphasis on the ties that bind, we turn to his writings on suicide.

In *Suicide*, Durkheim argued that it is futile to study the immediate circumstances that lead people to kill themselves, because an infinite number of such circumstances exist. For example, one person may kill herself in the midst of newly acquired wealth, whereas another kills herself in the lap of poverty. One may kill himself because he is unhappy in his marriage and feels trapped, whereas another kills himself because his unhappy marriage has just ended in divorce. In one case, a person kills himself after losing a business; in another case, a lottery winner kills herself because she cannot tolerate her family and friends fighting one another to share in her newfound fortune. Because almost any personal circumstance can serve as a pretext for suicide, Durkheim concluded that there is no situation that could not serve as an occasion for someone's suicide.

Durkheim also reasoned that no central emotional quality was common to all suicides. We can point to cases in which people live on through horrible misfortune whereas others kill themselves over seemingly minor troubles. Moreover, we can cite examples in which people renounce life at times when it is most comfortable or at times of great achievement. Given these conceptual difficulties, Durkheim offered a definition of suicide that goes beyond its popular meaning (the act of intentionally killing oneself). This definition takes the spotlight off the victim and points it outward toward the ties that bind (or fail to bind) people to others in the society. In short, Durkheim viewed suicide as a severing of relationships. To make his case, he argued that every group has a greater or lesser propensity for suicide. The suicide rates for various age, sex, and race groups in the United States, for example, show that suicide is more prevalent for some categories of people—the elderly, males, 15- to 19-year-olds—than for other categories. From a sociological point of view, these differences in suicide rates cannot be explained by pointing to each victim's immediate circumstances.

Instead, Durkheim examined the social ties that bind or fail to bind social categories to others. For example, all people who suddenly find themselves in the unemployed category must adjust to life without a job. That adjustment may entail finding a way to live on a reduced budget, trying to stay cheerful while hunting for a job, or feeling uncomfortable around friends who have a job. According to Durkheim, it is inevitable that a certain number of those in the unemployed category will succumb to social pressures and choose to sever the relationships from which such pressures emanate.

Durkheim identified four types of social ties, each of which describes a different kind of relationship to the group: egoistic, altruistic, anomic, and fatalistic. **Egoistic** describes a state in which the ties attaching the individual to others in the society are weak. When individuals are detached from others, they encounter less resistance to suicide. The lives of the chronically ill, for example, are often characterized by excessive individuation if friends, family, and other acquaintances avoid interacting with the ill person out of fear of upsetting the patient or themselves.

Altruistic describes a state in which the ties attaching the individual to the group are such that the person has no life beyond the group. In these situations, a person's sense of self cannot be separated from the group. When such people commit suicide, it is on behalf of the group they love more than themselves. The classic example is members of a military unit: The first quality of soldiers is a sense of selflessness. Soldiers must be trained to place little value on the self and to sacrifice themselves for the unit and its larger purpose.

Anomic describes a state in which the ties attaching the individual to the group are disrupted due to dramatic changes in social circumstances. Durkheim gave particular emphasis to economic circumstances such as a recession, a depression, or an economic boom. In all cases, a reclassification occurs that suddenly casts individuals into a lower or higher status than before. When people are cast into a lower status, they must reduce their requirements, restrain their needs, and practice self-control. When individuals are cast into a higher status, they must adjust to increased prosperity, which unleashes aspirations and

solidarity The system of social ties that connects people to one another and to the wider society. This system of social ties acts as "cement" binding people to each other and to the society.

egoistic A state in which the ties attaching the individual to others in the society are weak.

altruistic A state in which the ties attaching the individual to the group are such that he or she has no life beyond the group.

anomic A state in which the ties attaching the individual to the group are disrupted due to dramatic changes in economic circumstances.

expands desires to an unlimited extent. A thirst to acquire goods and services arises that cannot be satisfied.

Fatalistic describes a state in which the ties attaching the individual to the group involve discipline so oppressive it offers no chance of release. Under such conditions, individuals see their futures as permanently blocked. Durkheim asked, "Do not the suicides of slaves, said to be frequent under certain conditions, belong to this type?" (1951, p. 276).

Durkheim on the Mobile Phone If Durkheim were alive today, he would focus on the ways mobile phones affect solidarity or the ties that bind people to each other and society. He would seek to understand how mobile phones affect the nature of the social bonds people form. Broadly speaking, the mobile phone (unlike the landline phone) allows people to communicate with others regardless of location—that is, people with mobile phones do not need to be in a fixed location, such as in their home at an agreed-upon time, to communicate with others. And as mobile phones increasingly allow Internet access, the potential to connect with others will increase dramatically. Callers or texters may now intrude into just about any situation (a football game, a public bathroom, a classroom, a doctor's office). In addition, people can take advantage of applications (apps) and/or browse the Internet looking for immediate answers to questions (for example, directions, places to eat) and track friends and family through social networking sites. Durkheim would ask how this technological breakthrough might strengthen, weaken, or facilitate current and new social ties. He would find that, depending on how people use them, the mobile phone has the potential to:

- *strengthen and cultivate ties.* The ability to contact someone 24/7 regardless of location can work to "strengthen the formation and maintenance of deep bonds." People communicate, not to conduct business, but for the purpose of expressing affection, offering support, giving advice, and otherwise validating the centrality of the relationship even when physically separated (Geser 2004). In this sense, mobile phones facilitate nomadic intimacy, an ability to remain connected to those who make up the core of someone's personal network even at a distance and while on the move (Fortunati 2000). Depending on the frequency of calls and intensity of the social bond, the mobile phone reduces the probability that people are receptive to information outside personal networks because they are increasingly absorbed and distracted with communicating with those in core networks (Geser 2004).
- *increase the chances that behavior can be monitored.* Unless a mobile phone is turned off, it is possible to determine geographic location of the person in possession of it. It is also possible to obtain records of texts, calls, and Internet use and to even watch what is going on in a distance location through surveillance apps. Parents, partners, police, and others can check up on children, partners, friends, or potential criminals. Likewise, the extremely dependent can "stalk" another with an onslaught of texts and calls demanding an immediate response.
- *weaken ties with those sharing a physical space.* When people use their mobile phone to text, call, or browse the Internet in public spaces, it signals to other parties in that physical space that the mobile phone user is preoccupied. This preoccupation substantially reduces the chances that a person on the phone will notice or interact with those sharing the physical space (for example, a park, a classroom, a waiting room, an airport). The activities and people are relegated to the background. When a mobile phone rings or signals a message while someone is engaged in a face-to-face conversation, it can send the message to conversation partners that they are not significant or important enough to deserve exclusive attention (Geser 2004). When the person answers the phone, conversation partners must assume the role of "hanging bystander" or engage in a "waiting strategy," left to think about whether or not to continue with the interaction (Ling 1997). Finally when everyone in a family or other primary group owns a personal phone, "knowledge about each other's communication networks declines. Specifically, we may find that each family member has many acquaintances and ongoing interactions unknown to the other family members" (Geser 2004).

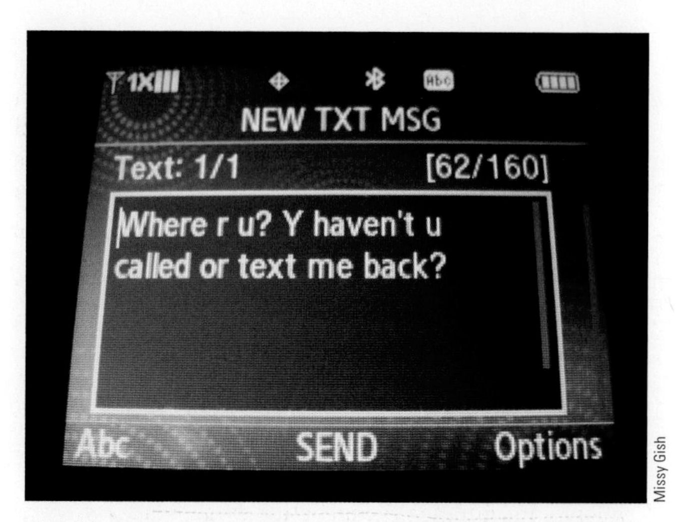

Missy Gish

If Durkheim were alive today, he would focus on the ways mobile phones shape relationships between people in new ways. What does this text message say about the relationship between sender and receiver?

fatalistic A state in which the ties attaching the individual to the group involve discipline so oppressive it offers no chance of release.

- *organize like-minded people to protest, demonstrate, celebrate, or otherwise participate in some public event.* The mobile phone can function as a community-organizing tool allowing leaders and participants to send out text messages in mass, or to post and check messages on Facebook. Participants can repost posts and forward messages to others they know. Finally, interested parties can check news coverage and reports as it happens.

Max Weber (1864–1920)

The German scholar Max Weber made it his task to analyze and explain how the Industrial Revolution affected **social actions**—actions people take in response to others—with emphasis on the forces that motivate people to act. In this regard, Weber suggested that sociologists focus on the broad reasons that people pursue goals, whatever those goals may be. He believed that social action is oriented toward one of four ideal types—ideal, not in the sense of being the most desirable, but as a gauge against which actual behavior can be compared. In the case of social action, an ideal type is a deliberate simplification or caricature of what motivates people to act, in that it exaggerates and emphasizes the distinguishing characteristics that make one type of action distinct from another. In reality, social action is not so clear-cut but involves some mixture of the four types:

1. *Traditional*—a goal is pursued because it was pursued in the past (that is, "that is the way it has always been").
2. *Affectional*—a goal is pursued in response to an emotion such as revenge, love, or loyalty (soldiers throw themselves on a grenade out of love and sense of duty for those in their unit).
3. *Value-rational*—a valued goal is pursued with a deep and abiding awareness of the symbolic significance of the actions taken to pursue the goal, meaning that "there can be no compromises or cost-accounting, no rational weighing of one end against another" (Weintraub and Soares 2005). Instead, action is guided by codes of conduct that prohibit certain kinds of behavior and permit others. With value-rational action, the way in which people go about achieving a goal is valued as much as the goal itself—perhaps even more so as, in an effort to stay true to a code of conduct, the goal may not be realized (Weintraub and Soares 2005).
4. *Instrumental-rational*—a valued goal is pursued by the most efficient means, often without considering the appropriateness or consequences of those means. It is result-oriented action. In the context of the Industrial Revolution, the valued goal is profit and the most efficient means are the cost-effective ones taken without regard for their consequences to workers or the environment. In contrast to value-rational action, this type of action does not require or prohibit the manner by which people go about achieving goals—any way of achieving the desired end is allowed as long as the end is achieved. One might equate this type of action with an addiction in the sense that the person will work to acquire a drug or other desired state at any cost to self or to others. There is an inevitable self-destructive quality to this form of action (Henri 2000). In the short run, the instrumental-rational action (with no constraints on behavior) will defeat the value-rationally motivated actors. However, in the long run, the "anything goes" approach will eventually collapse on itself.

Weber maintained that in the presence of industrialization, behavior was less likely to be motivated by tradition or emotion and was more likely to be instrumental-rational. Weber was particularly concerned about instrumental-rational action because he believed that it could lead to disenchantment, a great spiritual void accompanied by a crisis of meaning.

Weber on Mobile Phones Weber believed that sociologists should focus on the broad reasons that people pursue goals, whatever those goals may be. If Weber were alive today, he would consider how people reject or embrace mobile phones as a way of achieving a variety of goals. Specifically, under what conditions do people turn to the mobile phone as a means to preserve a tradition? As means to expressing emotion? As a vehicle for maintaining a valued standard or code of conduct? As a vehicle for achieving a goal, regardless of adverse consequences to self and others?

Tradition-Driven Social Action People can embrace the mobile phone as a tool for preserving some valued tradition. One specific example involves Pope Benedict and the Vatican staff who use mobile phone technology to convey the Catholic message to youth seeking a connection to that religion. On World Youth Day, the Pope texted daily messages to attendees, and organizers erected digital prayer walls even as the Vatican warned that obsessions with such technologies are affecting time needed to engage in spiritual pursuits (Beaumont 2008). The Vatican has also posted religious manuscripts and ancient texts online.

Emotion-Driven Social Action People can choose the mobile phone as a tool that allows them to quickly access social networking sites such as Facebook, Twitter, and dating sites to fulfill the emotional need for social contact with others. People may also choose the mobile phone as a way to connect with and mobilize those who share causes to which they are deeply committed.

Value-Rational Action Here people choose mobile phones as a means for meeting responsibilities or as a way of maintaining contact from a distance. That responsibility may

social actions Actions people take in response to others.

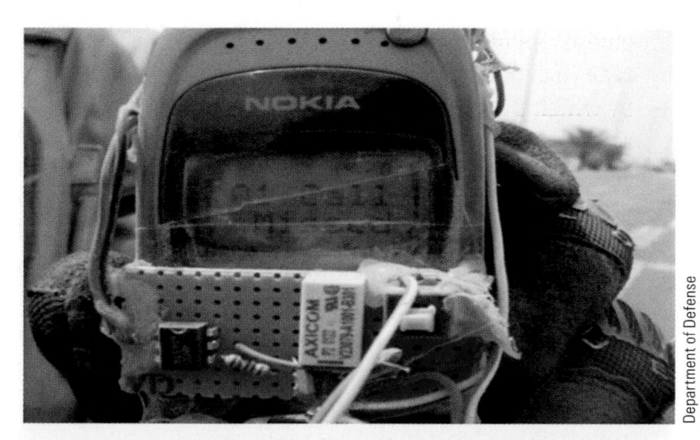

Department of Defense

This mobile phone, rigged as a detonator for an improvised explosive device, represents an example of instrumental rational action in that the mobile phone is used as a tool to achieve a goal by any means necessary.

involve parenting children from a distance, acting as a caregiver to parents or grandparents, monitoring criminal behavior so as catch and record "wrongdoings" of violations of codes, and so on.

Instrumental-Rational Action The mobile phone can be part of a "by any means necessary mentality" for achieving a desired goal. In this situation, the phone is part of an arsenal of tools to achieve a goal. The phone might become a tool thieves use to pinpoint potential victims, as when a criminal watches for bank customers withdrawing large sums of cash and then texts an accomplice to follow and rob that person at the right moment. The phone could be used to detonate a roadside or other homemade bomb from a safe distance.

W.E.B. DuBois (1868–1963)

W.E.B. DuBois wrote about the "strange meaning of being black" and about the color line. In *The Souls of Black Folk* (1903)—a book that has been republished in 119 editions (Gates 2003)—DuBois announced his preoccupation with the "strange meaning of being black here in the dawning of the Twentieth Century." The strange meaning of being black in America includes a **double consciousness** that DuBois defined as "this sense of always looking at one's self through the eyes of others, of measuring one's soul by the tape of a world that looks on in amused contempt and pity." The double consciousness includes a sense of two-ness: "an American, a Negro; two souls, two thoughts, two unreconciled strivings; two warring ideals in one dark body, whose dogged strength alone keeps it from being torn asunder." DuBois's preoccupation with the strange meaning of being black was no doubt affected by the facts that his father was a Haitian of French and African descent and his mother was an American of Dutch and African descent (Lewis 1993). Historically in the United States, a person has been considered "black" even when his or her parents are of different or blended "races." To accept this idea, we must act as if whites and blacks do not marry each other or produce offspring together and as if one parent, the "black" one, contributes a disproportionate amount of genetic material—so large that it negates the genetic contribution of the other parent.

In addition to writing about the "strange meaning of being black" and about racial mixing, DuBois also wrote about the **color line**, a barrier supported by customs and laws separating nonwhites from whites, especially with regard to their place in the division of labor. The color line originated with the colonial expansion that accompanied the Industrial Revolution. That expansion involved rival European powers (Britain, France, Germany, Belgium, Portugal, the Netherlands, and Italy) competing to secure colonies, and by extension, the labor and natural resources within those colonies. The colonies' resources and labor fueled European and American industrialization. DuBois (1919) traced the color line's origin to the scramble for Africa's resources, beginning with the slave trade upon which the British Empire and American republic were built, costing black Africa "no less than 100,000,000 souls" (p. 246). DuBois maintained that the world was able "to endure this horrible tragedy by deliberately stopping its ears and changing the subject in conversation" (p. 246). He further maintained that an honest review of Africa's history could only bring us to conclude that Western governments and corporations coveted Africa for its natural resources and for the cheap labor needed to extract them.

DuBois on the Mobile Phone If W.E.B. DuBois were alive today, he would think about the mobile phone, using the color line as a lens. He would focus on the "scramble" for the resources needed to produce mobile phones and other electronic products. The raw materials needed to produce wireless phones come from all over the world. The resources needed to make circuit boards, LCD screens, and batteries include gold, lead, nickel, zinc, tantalum, oil, limestone, mercury, copper, cadmium, and lithium (U.S. EPA 2010). The phrase *scramble* is an apt one because it suggests that all parties involved in the hunt to secure and control these resources are part of a frantic competition. The history of all previously named resources and many others involved a scramble to exploit labor and resources of non-European

---

double consciousness According to DuBois, "this sense of always looking at one's self through the eyes of others, of measuring one's soul by the tape of a world that looks on in amused contempt and pity." The double consciousness includes a sense of two-ness: "an American, a Negro; two souls, two thoughts, two unreconciled strivings; two warring ideals in one dark body, whose dogged strength alone keeps it from being torn asunder."

color line A barrier supported by customs and laws separating nonwhites from whites, especially with regard to their place in the division of labor.

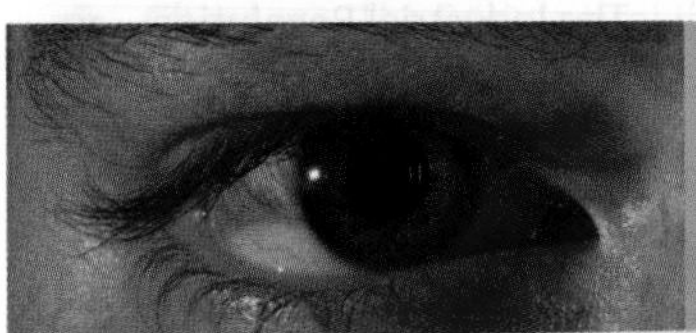

The Sociological Imagination Unplugged

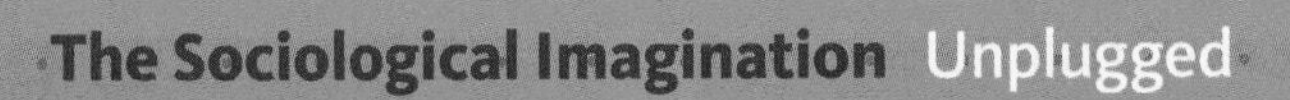

During my junior year of college, I quickly realized that for my resume to stand out I needed to become actively involved on campus. While searching for a group, I came across the student organization Kiksuya, which is dedicated to promoting Native American awareness and raising funds for the reservation of Pine Ridge, South Dakota.

While my knowledge of Native American culture was rather limited in that I knew that in general Native Americans face hardships, I wanted to help—I just didn't know how. So I joined and the following semester, I, along with other Kiksuya members, organized a volunteer trip to Pine Ridge Reservation with the help from our faculty advisor, Dr. Nicole Grant, who provided us with the information and cultural sensitivity needed to make the trip successful. As we prepared for the trip we discussed technology. Since we were traveling to a poverty-stricken area, we all agreed to unplug—to leave our technological devices at home or at least in the van. We knew that those who lived in Pine Ridge couldn't afford such luxuries and so we determined that it really wasn't necessary to take our iPods or mobile phones along.

For the first time since being in my possession, my purple Verizon flip phone stayed at home. That mobile phone usually inhabits my right pocket, nothing else. I need to have a connection with the outside world. What if someone needs me? What if something happens? What if I receive a text message, or heaven forbid, someone actually calls me? No matter the time of day or my location, I am on duty. Always. My phone startles me awake in the morning with shrill beeps. My phone is the first thing I look at upon waking and the last thing I touch before reaching sleep. The mobile phone is my god. Yet on this volunteer trip, my pocket remained empty. I actually didn't bring any electronic devices, even though the others did. I wanted to push away anything that had to do with my culture so I could learn and focus all my energy on the experiences at hand. My mind wanders enough without me texting someone every four seconds to find out what is going on in the "world." I know too well the temptations of technology.

At our fingertips we have a tool that allows us to do just about anything all within our own personal bubbles. For many, no matter if in class or driving, the task at hand never takes priority over checking the mobile phone. Text messages must be read and answered immediately. There were moments where I wished that I had taken my phone on the trip. Or a book. Something. Staring out the window of the van, all I could see was relentless land. Dusty hills sweeping across the plains. The beauty is unsurpassable yet it still seemed as if the drive might never end. As we neared our destination, I did not see mobile phone towers placed awkwardly amidst the hills. Few, if any, electrical lines stretched strangely across plateaus.

While on the reservation, however, one asinine inconvenience stood out: the lack of technology. And not only a lack, but it was nonexistent. Not once did I see a mobile phone or an iPod except in the hands of a volunteer. Not once did I see someone on Pine Ridge updating their status on Facebook or even posing for a photograph. Technology on the reservation seemed to only exist in the minds, and literally in the hands, of the volunteers. Is technology a necessity, something one cannot exist without? Is technology required to live and to thrive as a people? To be a part of a culture that breathes peace and humility is technology a true necessity?

Unhooked and unplugged, I'm glad I left my phone at home. Instead of carrying on meaningless chatter with those who will survive perfectly fine while I am away, I was able to learn so much more. Technology is an amazing convenience, but it can also is a major inconvenience. If I had been texting, I would have missed the young child who asked if she could show me her family's sweat lodge. I would have missed the boy climbing onto his new bunk bed we constructed; the joy on his face for this was his first bed. I would have missed the sweet scents drifting with the wind. I would not have noticed the People. We tend to forget not only the people, but also that those people are not merely faces in faraway places. They are our brothers, our sisters, our friends . . . Mitakuye Oyasin. We are all related. We are all in this fight together. If only we could look past the insignificant, trivial issues that the media promotes, stop bowing down to the Cell Phone God, and remove the blinders that technology can create, then perhaps we might pay attention to issues of social justice.

On that volunteer trip, I was handed an opportunity to try to bridge the ravine that plummets between us and the

such program is SafeLink Wireless; it provides wireless phones free of charge to those who cannot afford them and offers free minutes—typically 68 minutes a month—which are paid for by the Universal Service Fund tax added to all phone bills. The SafeLink Program has over 2 million customers in 32 states. This service allows those who are low-income and homebound, for example, to make calls in an emergency (SafeLink Wireless 2011; Richtel 2009).

The Importance of a Global Perspective

CORE CONCEPT 6 A global perspective assumes that social interactions do not stop at political borders and that the most pressing social problems are part of a larger, global situation.

Native American People. To work directly alongside the Lakota I learned about their customs and to see life from a different perspective—the perspective of the lost and forgotten, the perspective of the unwanted, the bruised and the hurt. To witness firsthand the poverty on Pine Ridge Reservation kindled a fire deep inside volunteers, and deep within myself. The generosity of the Lakota people and the love they carry for each other and their land, brings tears in all eyes. What I tried to give to the reservation by volunteering manual labor is absolutely minute in comparison with the values and new ideas with which I returned. I received a new outlook on life, an appreciation for everything previously taken for granted, and for all the luxuries we don't even consider as luxuries like the ability to walk in the next room and flip on the light switch. Or the air conditioning. Or the heat. I wouldn't have acquired that new outlook if I had been wrapped up in my cell phone, or in an iPod or a laptop. I would have instead heard, "And she said that he said. . . ." Wasted time. I'm glad I was able to smell the clean, unpolluted air, and walk through mountainous terrain, listening to songs of the wind. My pocket's not so lonely now; it can survive some time alone, just as I can unhook myself from mainstream culture, and think about the actual necessities in life.

There is a peaceful, ignorant belief that there is no such thing as poverty in America, and if there is, the individual must be the problem. Never society. Oh, the luxury of complete obliviousness. Before this trip, the Native American seemed to be something in a childhood story, a myth—cowboys and Indians, feathers and headdresses, scalping, warriors on a raid riding to kidnap a young blue-eyed child, the noble savage, primitive and scantily clad. I viewed the Native Americans as something in the past, as a lost culture, a lost group of people.

Now Disney representations no longer pollute my mind. I no longer imagine the painted warrior, or the young Indian princess, or the thriving casino; instead I think of third-world countries. I think of no electricity, no garbage facilities for disposal, little to no running water, limited access to vaccinations against disease, extreme depression, alcoholism, the mother who does without so her child might survive. I think of daunting weather conditions, of tired hands and sore feet, and of men who cannot be breadwinners for there are virtually no jobs on the reservation. But above all, I remember to think that in spite of the harsh life poverty has brought, hope persists—a stubborn streak, clinging to life and praying for change. Amidst the debris the wind carries across the land, a family lives. A people struggle to just see the dawn of a new day. With the average annual income less than $4,000 a year, how can one survive? Sacrifice.

Ken Banks, kiwanja.net

The NKU students who traveled to Pine Ridge helped with a variety of construction projects such as the one pictured.

I have come to learn that amidst broken promises and disregarded treaties, the Lakota culture still flourishes. Core values of honesty, humility, sacrifice, love, peace, and respect still define the People. Technology is not the necessary force needed to usher in the dawn of a new day or to whisper the first words to a newborn. Technology is not necessary in a culture that gives highest priority, not to the individual, but to the entire community. These priorities are the values with which the People live and the values by which the People die. Maybe, if we put down our phone and stopped texting trivial information, we'd realize that to create change and equality, we first must change within ourselves.

Source: Written by Staci Wood, Northern Kentucky University, Class of 2011.

Global interdependence is a situation in which human interactions and relationships transcend national borders and in which social problems within any one country—such as unemployment, drug addiction, water shortages, natural disasters, or the search for national security—are shaped by social forces and events taking place outside the country, indeed in various parts of the globe. Global interdependence is part of a dynamic process known as **globalization**—the

global interdependence A situation in which social activity transcends national borders and in which one country's problems—such as unemployment, drug abuse, water shortages, natural disasters, and the search for national security in the face of terrorism—are part of a larger global situation.

globalization The ever-increasing flow of goods, services, money, people, information, and culture across political borders.

THEORETICAL PERSPECTIVES AND METHODS OF SOCIAL RESEARCH

2

With Emphasis on MEXICO

Sociologists view theory and research as interdependent because both are needed to analyze any issue. Theories are frameworks that help us to think about and describe what we see going on in the world around us, and research involves strategies for making objective observations about those issues. In this chapter, we draw upon sociological theories and methods of research to analyze the 700 miles of newly constructed fence that separate the United States from Mexico. This photo shows one stretch of the fence that runs on the Nogales, Arizona side of the border.

SGT JIM GREENHILL/Department of Defense

Why Focus On MEXICO?

The United States and Mexico share a 2,000-mile border that hundreds of millions of people on both sides cross each week to work, shop, socialize, and vacation. The border includes 700 miles of walls, border fences, and other barriers to prevent undocumented immigrants from crossing into the United States. About 80 miles of these barriers, referred to as the Wall of Shame in Mexico, were built in the mid-1990s. In 2006, Congress passed the Secure Fence Act, authorizing the construction of at least 630 miles of strategically placed barriers, including "two layers of reinforced fencing, the installation of additional physical barriers, roads, lighting, cameras and sensors" between Tecate, California, and Brownsville, Texas (see Figure 2.1). Today about 40 percent of this 2,000-mile border is fortified. The border has also become militarized. In addition to the 20,000-plus border guards and thousands of army reservists, high-tech surveillance devices, drones, and military equipment are also dedicated to preventing undocumented immigrants and illicit drugs from entering the United States. One description of the border barriers follows: "It stops and starts, without rhyme or reason, along the Rio Grande River's levees, leaving miles of gaps. It highlights the city's economic divide: It's the first thing folks in the poorer barrios see when they look out their windows, while richer folks enjoy unaltered views of palm trees and manicured fairways when they tee off on private golf courses. It zigs and zags through residents' backyards, through citrus orchards—an ugly red scar on a green, subtropical landscape" (Del Bosque 2010).

We draw upon the three major sociological perspectives and methods of social research to assess the border barriers. The three perspectives help us to describe and think about the fences and their purposes. The methods of research offer guidelines for collecting and analyzing data related to the barriers' impact on undocumented immigrants, surrounding communities, and other affected parties. Although we are focusing on understanding the border barriers, we can apply the perspectives and methods of research to any issue or situation.

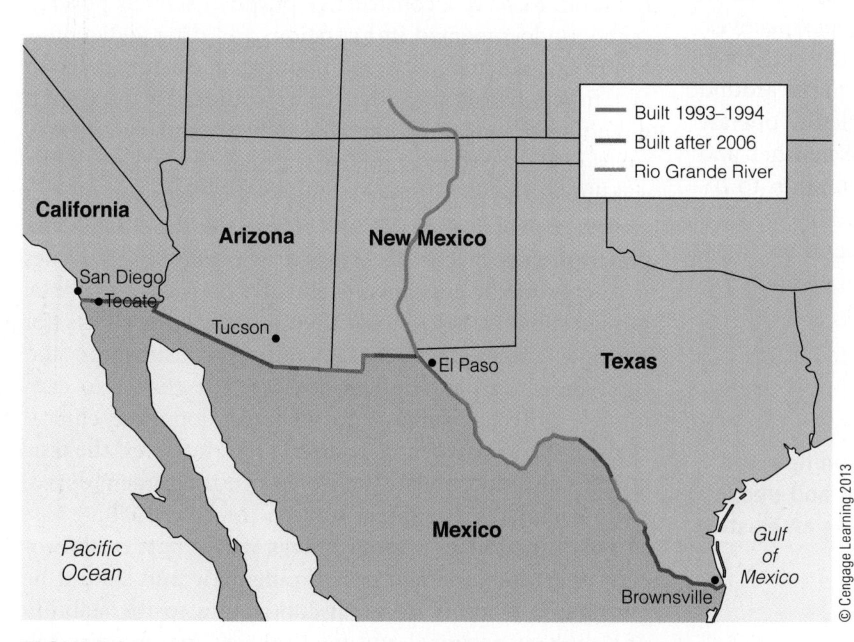

FIGURE 2.1 This map shows the locations of the existing 700 miles of fences. Do you think the fence will be effective in preventing undocumented people from entering the United States? How will we know if it is effective?

Online Poll

Before reading this introduction, did you know that the U.S. Congress had passed a law requiring the construction of border barriers along the U.S.–Mexican border?

○ Yes

○ No

Do you support the construction of 700 miles of walls, fences, and other barriers?

○ Yes

○ No

○ Don't know

To see how other students responded to these questions, go to **www.cengagebrain.com**.

What do we make of the facts that the U.S. government has constructed 700 miles of barriers placed in strategic locations along its border with Mexico, but that it constructed no such barriers to secure U.S. shorelines or its border with Canada? Will these barriers, and focus on the Mexican border, be effective at preventing undocumented workers from crossing into the United States? For answers to these questions, we can draw upon the three theoretical perspectives and the methods of social research.

Theoretical perspectives are frameworks for thinking about what is going on in the world around us. In sociology, there are three broad theoretical perspectives: functionalist, conflict, and symbolic interaction. Each perspective has a distinct focus and is organized around fundamental assumptions about how societies operate and how the people in them relate to one another and their surroundings. Each offers a central question to direct thinking about any issue or situation. Finally, each perspective offers a vocabulary to help answer these questions. We begin an overview of the functionalist perspective and then apply it to the border barriers.

Functionalist Perspective

CORE CONCEPT 1 Functionalists focus on how the "parts" of society contribute in expected and unexpected ways to maintaining and disrupting an existing social order.

Functionalists focus on the existing order and how it is maintained. They define society as a system of interrelated, interdependent parts. To illustrate this vision, functionalists use the human body as an analogy for society. The human body is composed of parts such as bones, cartilage, ligaments, muscles, a brain, a spinal cord, nerves, hormones, blood, blood vessels, a heart, a spleen, kidneys, and lungs. All of these body parts work together in impressive harmony. Each functions in a unique way to maintain the entire body, but it cannot be separated from other body parts that it affects and that in turn help it function.

Society, like the human body, is made up of parts, such as schools, automobiles, sports teams, funeral rites, ways of greeting people, religious rituals, laws, languages, household appliances, and border barriers. Like the various body parts, each of society's parts is interdependent and *functions* to maintain a larger system. Functionalists define a **function** as the contribution a part makes to order and stability within the society.

Consider sports teams—whether they be Little League, grade school, high school, college, city, Olympic, or professional teams. Sports teams function to draw audiences whose members are often extremely different from one another economically, culturally, linguistically, politically, religiously, and in other ways. Loyalty to a sports team transcends individual differences and fosters a sense of belonging to the school, company, city, or country associated with it.

In the most controversial form of this perspective, functionalists argue that all parts of society—even those that do not seem to serve a constructive purpose, such as poverty, crime, undocumented immigration, and drug addiction—contribute in some way to maintaining an existing order. In fact, functionalists argue that a part would cease to exist if it did not serve some function. Thus they strive to identify how parts—even seemingly problematic ones—contribute to the stability of the social order. Consider one function of poverty: Poor people often "volunteer" for over-the-counter and prescription drug tests. Most new drugs, from AIDS vaccines to allergy medicines, must eventually be tried on healthy human subjects to determine their potential side effects (for example, rashes, headaches, vomiting, constipation, and drowsiness) and appropriate dosages. The chance to earn money motivates subjects to volunteer for these clinical trials. Because payment is relatively low, however, the tests attract a disproportionate share of low-income, unemployed, or underemployed people as subjects (Morrow 1996).

This function of poverty shows why a part of the society that everyone agrees is problematic and should be eliminated remains intact: It contributes to the stability of the pharmaceutical and medical systems, the credibility of which depends on human trials. Without poverty, these systems would be seriously strained to find human subjects to test out procedures and new products. As you might imagine, this line of reasoning can lead to charges that functionalists defend the status quo. To address some

theoretical perspectives Frameworks for thinking about what is going on in the world around us.

function The contribution a part of a society makes to the existing social order.

U.S. Air Force photo by Tech. Sgt. Charity Barrett/Released

From a functionalist perspective, sport teams and celebrities function to transcend differences among fans and foster a sense of belonging to the school, city, or country the team or celebrity represents.

of this criticism, sociologist Robert K. Merton (1967) introduced the concepts of *manifest* and *latent functions,* as well as *dysfunctions*. These concepts help us to think, not just about a part's contribution to order and stability, but its other effects, especially unanticipated effects.

Manifest and Latent Functions and Dysfunctions

Merton distinguished between two types of functions that contribute to maintaining an existing social order: manifest functions and latent functions. **Manifest functions** are a part's anticipated or intended effects on that order. **Latent functions** are the unanticipated or unintended disruptive effects to the social order. To illustrate this distinction, consider the manifest and latent functions associated with annual community-wide celebrations, such as fireworks displays on the Fourth of July and concerts in the park. Corporate sponsors often join with city government to mount such events. Three manifest functions readily come to mind: The community celebration functions (1) as a marketing and public relations event for the city and for corporate sponsors, (2) as an occasion to plan activities with family and friends, and (3) as an experience that draws the community together for celebration.

At the same time, several unanticipated, or latent, functions are associated with community celebrations. First, such celebrations put the spotlight on public transportation systems as people take buses or ride trains to avoid traffic jams. Second, such events function to break down barriers across neighborhoods. People who do drive may find that they must park some distance from the event, often in neighborhoods that they would not otherwise visit. Consequently, after they park, people have the opportunity to walk through such neighborhoods and observe life up close instead of at a distance.

Merton also points out that parts can have **dysfunctions**; that is, they can have disruptive consequences to the social order or to some segment of society. Like functions, dysfunctions can be either manifest or latent. **Manifest dysfunctions** are a part's anticipated disruptions to order. Anticipated disruptions that seem to go hand in hand with community-wide celebrations include traffic jams, closed streets, piles of garbage, and a shortage of clean public toilets.

In contrast, **latent dysfunctions** are unanticipated or unintended disruptions to the social order. For instance,

manifest functions Intended or anticipated effects that a part has on the existing social order.

latent functions Unintended or unanticipated effects that a part has on the existing order.

dysfunctions Disruptive consequences of a part to the existing social order.

manifest dysfunctions A part's anticipated disruptions to an existing social order.

latent dysfunctions Unintended, unanticipated disruptions to an existing social order.

community-wide celebrations often have some unanticipated disruptive consequences. Sometimes people celebrate so vigorously that the celebration has the unintended consequence of lowering worker productivity, as people miss class or work the day of the event or the day after.

The Functionalist Perspective on Border Barriers

To see how the functionalist theory can be applied to a specific issue, we will consider how functionalists analyze the U.S.–Mexico border barriers. Functionalists ask, "What are the anticipated and unintended consequences of the border barriers?" Functionalists apply the concepts of manifest and latent functions and dysfunctions to answer this question. The purpose of the analysis that follows is not to generate an exhaustive list of functions and dysfunctions associated with the border barriers, but to demonstrate how functionalists frame any issues.

Manifest Functions

To identify the manifest functions (*anticipated* effects on social order and stability) of the border barriers, we need to understand why the United States constructed them in the first place. In the mid-1990s, three major border cities constructed 80 miles of barriers: San Diego (Operation Gatekeeper), El Paso (Operation Hold the Line), and Nogales (Operation Safeguard). The three operations were a response to the real or imagined belief that the United States was being overrun by undocumented immigrants. In addition, the overall reported crime rate along the border was 30 percent higher than the national average. Newspaper accounts at that time "described large groups of immigrants, serviced by Mexican food and drink vendors in a carnival atmosphere," waiting on the Mexican side of the border for after-dark surges into the United States, overwhelming the border agents (U.S. Department of Justice 2007). In response, the U.S. government shifted its emphasis from apprehension after entry to prevention, erecting barriers, and increasing the number of border agents in areas believed to have the highest numbers of undocumented entries.

In 2006, Congress passed the Secure Fence Act, authorizing construction of more strategically placed border barriers. This act was a response to reports that millions of undocumented immigrants were living in the United States and to post-9/11 priorities of achieving operational control "over the entire international land and maritime borders of the United States." The act gave highest priority to the southwest border, calling for 700 miles of fencing and security improvements between the Pacific Ocean and the Gulf of Mexico (Secure Fence Act 2006). See Figure 2.1.

Eugenio del Bosque

The construction of 700 miles of fence has redirected flows of undocumented immigrants in the direction of the 1,300 miles of unfenced border areas.

The manifest functions associated with constructing the border barriers have included the following:

1. A reported decrease in the number of undocumented immigrants apprehended crossing the newly secured areas of the border
2. Success in forcing undocumented entries away from secured urban areas to less populated unsecured areas and through rough terrain and climates (such as steep mountains, deep canyons, thick brush, the extreme cold of winter, and the searing heat of summer) to give Border Patrol agents a strategic advantage
3. An overall drop in the reported crime rate on the U.S. side of the border from 30 percent higher than the national average to 12 percent

Latent Functions

The construction of barriers along the border has had the following latent functions (*unanticipated* effects on social order and stability):

1. Cooperation between Mexican and U.S. officials in launching the Border Safety Initiative Program to prevent injuries and fatalities of those trekking through the desert and other rough terrain to enter the United States
2. The creation of the Border Patrol Search, Trauma, and Rescue team, which responds to *all* incidents involving people in distress, not just incidents involving undocumented immigrants (The team has rescued tens of thousands of people.)
3. A border barrier that doubles as a volleyball net, allowing U.S. and Mexican volleyball players on each side of the border to face off as part of goodwill festivals and other cross-border celebrations

Manifest Dysfunctions

The construction of the barriers has been associated with several manifest dysfunctions (anticipated disruptions of the existing order), including the following:

1. Increased apprehensions of undocumented immigrants in border counties not secured by barriers
2. A reported crime rate that was higher than the national average in some thinly populated counties with unsecured barriers
3. Increased fatalities as undocumented immigrants now risk their lives to enter the United States through the desert and other inhospitable terrain
4. Increased numbers of undocumented immigrants paying smugglers, or coyotes, to guide them into the United States. Before the barrier construction, human smuggling was a "mom-and-pop" operation with a typical fee of $500. Now human smuggling is highly organized, and the typical fee has increased to more than $2,000. Organized drug cartels are also now in the business of smuggling the undocumented across the border (*The Economist* 2008).
5. An increase in illicit businesses that facilitate undocumented immigrants' entry into the United States (for example, businesses that issue fraudulent documents)

Courtesy of Tomas Castelazo

The border fences have forced undocumented immigrants to enter the United States through desert and other inhospitable terrains, resulting in a manifest dysfunction of increasing the number of fatalities among illegal immigrants. The monument is to those who died crossing the border since 1994, the year the first fence was built dividing Tijuana from San Diego.

Department of Defense

A border fence now separates Nogales, Arizona, from Sonora, Mexico. How do you think the construction of the border fence affected interactions among people in the two cities?

Latent Dysfunctions

Several latent dysfunctions (unanticipated disruptions to the social order) have followed the construction of the barriers:

1. Dramatic disruptions to the grazing, hunting, watering, and migration patterns of wildlife ("If it doesn't fly, it's not getting across." [Pomfret 2006])
2. Some barrier construction sites were not subjected to environmental impact review (Archibold 2009)
3. Decrease in the number of undocumented workers returning home to Mexico after completing seasonal work in the United States for fear that they may be unable to cross back into the United States to work
4. Redirected flows of undocumented immigrants to areas of the United States unaccustomed to this movement, fueling the perception that the United States is being "invaded" by undocumented migrants (Massey 2006)
5. Disruptions to economically and socially interdependent border communities now separated by barriers. About 75 percent of those who legally cross from Mexico into the United States do so to shop. Towns like Nogales, Arizona, with 20,000 inhabitants, depend on Mexican shoppers from Nogales, Sonora (population 190,000). As the number of border patrol officers increased and took more time to check documents and to search vehicles, the time waiting to cross increased. From a business perspective, Mexicans are now spending more time waiting and less time shopping. Because of the longer lines and wait, fewer Mexicans drive into the United States, but instead come in as pedestrians. And pedestrians cannot carry as many purchases as they can with a car (*The Economist* 2008).

Conflict Perspective

CORE CONCEPT 2 The conflict perspective focuses on conflict over scarce and valued resources and the strategies dominant groups use to create and protect the social arrangements and practices that give them an advantage in accessing and controlling those resources.

National Archives and Records Administration

Library of Congress Prints and Photographs Division

For at least 100 years or more, the United States has both invited and deterred workers from the Mexican side of the border. One photo, taken in 1926, shows Border Patrol officers in Laredo, Texas, forming a "wall" with cars and guns to prevent illegal immigrants from crossing. The other photo shows Mexican workers recruited by U.S. Farm Security Program in 1943, traveling by train to Arkansas, Colorado, Nebraska, and Minnesota to harvest beets.

In contrast to functionalists, who focus on social order, conflict theorists focus on conflict as an inevitable fact of social life and as the most important agent for social change. Conflict can take many forms, including physical confrontations, exploitation, disagreement, tension, hostility, and direct competition. In any society, advantaged and disadvantaged groups compete for scarce and valued resources (access to material wealth, education, health care, well-paying jobs, and so on). Those who gain control and access to these resources strive to protect their own interests against the competing interests of others.

facade of legitimacy An explanation that members of dominant groups give to justify the social arrangements that benefit them over others.

Conflict theorists ask this basic question: Who benefits from a particular social arrangement, and at whose expense? In answering this question, they seek to identify the dominant and subordinate groups as well as to describe the social arrangements that the dominant groups have established, consciously or unconsciously, to promote and protect their interests. Exposing these practices helps explain the unequal access to valued and scarce resources. Not surprisingly, the advantaged seek to protect their position, whereas the relatively disadvantaged seek to change their position.

Dominant groups explain their advantages using a **facade of legitimacy**—an explanation that members of dominant groups give to justify the social arrangements that benefit them over others. On close analysis, however, this explanation turns out to be based on "misleading arguments, incomplete analyses, unsupported assertions, and implausible premises" (Carver 1987, pp. 89–90). To illustrate, consider that employers who exploit workers justify such exploitation by pointing out that workers who are dissatisfied with their working conditions, wages, or benefits are free to quit. On close analysis, we see that this justification does not hold if only because the employee "has but wages to live upon, and must therefore take work when, where, and at what terms he can get it. The workman . . . is fearfully handicapped by hunger" (Engels 1886). Another common justification employers give for exploiting labor is to suggest that the exploited really benefit (for example, that a \$2.00-per-hour job—or even a \$0.48-per-hour job—is better than no job). Consider the following interview with an American woman who employs an undocumented immigrant to care for her children. In particular, pay attention to how this woman justifies the demands she makes on this woman:

> *MR. BEARDEN (reporter):* There are some that believe that people who hire undocumented aliens gain an unfair power over them; it gives them influence over them because they're, in a sense, collaborating in something that's against the law. Do you agree with that, or have any thoughts about that?
>
> *DENVER WOMAN:* I guess I would disagree with that. The one thing that you get in undocumented child care or the biggest thing that you probably get, my woman from Mexico was available to me 24 hours a day. I mean, her cost of living in Mexico and quality of life in Mexico compared to what she got in my household were two extremes. When we hired her, she said, "I'll be available all hours of the day, I'll clean the house, I'll cook." They do everything. And if you hire someone from here in the States, all they're going to do is take care of your children. So not only do you have a differentiation in price, you have a differentiation in services in your household. I have to admit that was, at that point, with a newborn infant, wonderful to have someone who was so available. . . .
>
> *MR. BEARDEN:* And it's not like indentured servitude?

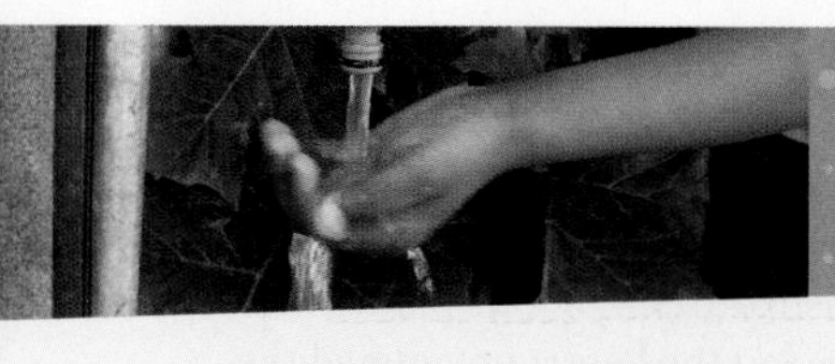

Working for Change Remittance Income from Migrants

International remittances are monies earned by people living or working in one country and sent to someone (usually family and friends) in a home country. An estimated 200 million migrants worldwide send home more than $440 billion to help 500 million people pay for food, medicine, clothing, housing, education, and land. Because senders do not always use official channels (such as banks or Western Union) to send this money, it is likely that the actual amount remitted exceeds the $440 billion estimate by at least another $33 billion (World Bank 2010; Robinson 2003). As one measure of how widespread this practice is, consider that an estimated 20 million people born in Latin American countries live in a foreign country (*Stalker's Guide to International Migration* 2003). Half of these 20 million people send home an estimated $23 billion per year (Van Doorn 2003). An Inter-American Development Bank poll found that almost one in five adult Mexican residents receives money from relatives working in the United States. One in every four Guatemalan and El Salvadoran adults receives such money (Suro 2003b; Thompson 2003).

The sum of $414 billion in remittance represents the "monetary expression of a profound human bond" between migrants and the families they left behind (Suro 2003a, p. 2). Taken together, remittances represent an important source of income for developing countries. For many small island economies, remittances—along with foreign aid and tourism—represent one of the only sources of income. Many critics of foreign aid policies and programs believe that remittance "aid" represents an ideal altruistic self-help model (Kapur 2003, p. 10).

Although remittance aid clearly has many positive effects, it would be naive to think that remittances alone could eliminate poverty, drive economic development, and reduce budget deficits (Lowell, De la Garza, and Hogg 2000). Some of the potential positive effects are reduced by the cost to send money home, a cost that can exceed 23 percent of a remittance when check-cashing fees, money transfer fees, currency conversion fees, and fees on the receiving end are considered. Some critics estimate that reducing these fees by just 5 percent could generate another $16 billion in remittance aid. Banks and other financial institutions have taken notice of this money flow and are competing with Western Union and MoneyGram to offer transfer services. Such competition may work to reduce transfer fees. Nevertheless, many migrants cannot use banks' services because of the migrants' undocumented status and because of high checking account and electronic transfer fees.

Top Five Countries Receiving Annual Remittances (in billions)

India	$55
China	$51
Mexico	$22.6
Philippines	$21.3
France	$15.9

Source: Data from World Bank 2010.

> *DENVER WOMAN:* That crossed my mind, and after she had been here for six months or so, we went to a schedule where she finished at 6 or 7 o'clock at night. And I don't think I ever really took advantage of her. Once a week I'd have her get up with the baby, so I didn't. . . . She was available to me, but I don't feel like I really took advantage of her, other than the fact that I paid her less and she was certainly more available. But she got paid more here than she would have gotten paid if she'd stayed where she was. (Bearden 1993, p. 8)

Conflict theorists take issue with the logic that this Denver woman employs to justify the low pay and high demand she makes. An honest assessment would suggest that this Denver woman is protecting and promoting her interests (having someone available at all hours of the day to cook, clean, and provide child care) at the expense of the undocumented worker's freedom to have a life apart from work.

The Conflict Perspective on the Border Barriers

In analyzing the construction of 700 miles of border barriers along the U.S.–Mexican border, conflict theorists ask "Who benefits from the barriers and at whose expense?" In answering this question, they would point out an obvious fact: The barriers secure a border that separates a high-wage and low-wage economy. In this regard, (1) the barriers are just one of many measures that the United States has put in place over time to control the flow of low-wage undocumented labor from Mexico, but not eliminate it; and (2) the barriers serve as a potent political symbol used to convey the illusion that United States is in control of its borders during a time of almost intolerable economic uncertainty.

Controlling the Flow of Undocumented Labor The barriers can be viewed as part of a long series of efforts by U.S. government to strategically draw from a very large low-wage labor pool in Mexico while also protecting the country from an "invasion" of Mexico's poor who have historically been viewed as unfair competitors taking jobs away from Americans because employers play them less (Zolberg 1999). Although it is certainly true that many undocumented immigrants risk life and limb to escape an economy where they are paid about $4.50 per day to enter one that pays $60 to $80 per day, we must remember that they are enticed by employers offering them jobs, albeit jobs with wages and conditions (for example, seasonal, long hours) that are unacceptable to many Americans. Since 1880, it has been the employers (those who purchase labor) on the U.S. side who have ultimately determined the size and destination of migration flows to the United States from Mexico. In fact, the U.S. economy has always depended on documented and undocumented Mexican workers to fill low-status, physically demanding, entry-level, seasonal, and cyclical jobs—jobs like dishwashers, farm workers, meat packers, maids, day laborers, roofers, and caretakers of children and elderly—whether or not the American public or politicians liked it (Massey, Durand, and Malone 2002, Cornelius 1981, Judis 2006).

In light of this longstanding and mutual dependence, the barriers are simply one part of the "odd bundle of legal measures and administrative practices that have accumulated over the decades" to control the sea of low-wage laborers "willing" to come to the United States to work (Zolberg 1999, p. 76).

The Barriers as a Political Symbol From a conflict point of view, the construction of barriers and the accompanying militarization of the border is less about deterring undocumented border crossings from Mexico and more about crafting an image that gives the appearance of control and security during a time of economic and national insecurity (Andreas 2009). In essence, constructing barriers and militarizing the southwestern border focuses everyone's attention on the border as *the* source of the United States' problems. In the meantime, the real forces threatening economic and national security—which are global in scale and difficult to control—are left largely unchallenged and unaddressed.

One irony of the 2006 Secure Fence Act relates to its stated aim of gaining control over all borders. Yet the act called for 700 miles of strategically placed fencing and security improvements along the 2,000 miles of U.S.–Mexico border. Also left largely unsecured were the 3,987 miles with Canada, the 1,538-mile Alaska–Canada border, and the coastlines of the United States (Beaver 2006). A 2010 Homeland Security report estimated that only about 900 miles of all U.S. borders can be classified as secure (U.S. Customs and Border Patrol 2010).

Another irony of the Secure Fence Act aimed at securing borders as a response to 9/11 was the fact that the 16 hijackers who flew and successfully crashed three of four planes into the World Trade Center towers and the Pentagon entered the United States legally (with documents or visas). Still, the Secure Fence Act gave highest priority to securing its southwest border. In addition, the most credible research studies estimate that 40 to 50 percent of 11.6 million undocumented immigrants living in the United States entered legally and overstayed their visas (Andreas 2009). In evaluating these statistics, simply consider that, in 2009, 162.3 million nonimmigrant visas were issued to foreign nationals who legally came to the United States to work, vacation, and study.

In addition, when we consider the fact that more than 15 million jobs have been outsourced from the United States to India, China, and elsewhere since 2000, and that the Internet allows workers around the world to compete for information-based jobs, we can see that undocumented labor from Mexico is one small player in what is a global competition for jobs. In the midst of economic insecurity, politicians can point to the Mexican border as evidence that concrete steps are being taken to addres the situation—for example, building barriers, militarizing the border by sending the National Guard and increasing the number of Border Patrol agents, and issuing lucrative contracts to such corporations as Boeing, Lockheed Martin Raytheon, and Northrop Grumman to construct barriers and provide equipment needed to secure the border. For example, Boeing was awarded a multibillion-dollar contract to supply small unmanned aerial surveillance vehicles that can be launched from Border Patrol truck beds, and equip as many as 1,800 watchtowers with cameras, heat and motion detectors, and other sensors (Witte 2006).

In light of this information, conflict theorists have no trouble answering the question, "Who benefits from the construction of the 700 miles of barriers, and at whose expense?" The answer is clear: American politicians, employers, and defense contractors are among the clear beneficiaries.

Symbolic-Interactionist Perspective

CORE CONCEPT 3 Symbolic interactionists focus on social interaction and related concepts of self-awareness/reflexive thinking, symbols, and negotiated order.

Sociologist Herbert Blumer coined the term *symbolic interaction* and outlined its essential principles. Symbolic interactionists focus on **social interaction**, everyday encounters in which people communicate, interpret, and respond to one another's words and actions. These theorists ask, when involved in interaction, how do people involved in interaction "take account of what each other is doing or is about to do" and then direct their own conduct accordingly

social interaction Everyday encounters in which people communicate, interpret, and respond to each other's words and actions.

(Blumer 1969)? The process depends on (1) self-awareness, (2) shared symbols, and (3) negotiated order.

Self-Awareness

Self-awareness occurs when a person is able to observe and evaluate the self from another's viewpoint. People are self-aware when they imagine how others are viewing, evaluating, and interpreting their words and actions. Through this imaginative process, people become objects to themselves; they come to recognize that others see them, for instance, "as being a man, young in age, a student, in debt, trying to become a doctor, coming from an undistinguished family and so forth" (Blumer 1969, p. 172). In imagining others' reactions, people respond and make adjustments (apologize, change facial expressions, lash out, and so on).

Shared Symbols

A **symbol** is any kind of object to which people assign a name, meaning, or value (Blumer 1969). Objects can be classified as physical (cell phones, cars, a color, the facial expression), social (a friend, a parent, a celebrity, a bus driver), or as abstract (freedom, greed, justice, empathy). In the context of driving, for example, the color green has come to symbolize "go or proceed." Objects can take on different meanings depending on audience and context: A tree can have different meanings to an urban dweller, a farmer, a poet, a home builder, an environmentalist, or a lumberjack (Blumer 1969). People learn meanings that their culture attaches to objects from observing others. By observing, people learn such things as that a wave of the hand means good-bye or that one should not send text messages during a funeral service.

Courtesy of Jonathan McIntosh

To illustrate the difference in symbolic meanings that Americans and Mexicans attach to a map of the United States, consider that when most Americans view a map of their country, they do not think that one-third of the territory once belonged to Mexico. In contrast, when people in Mexico view the map of the United States, they likely think the Southwest portion was once part of their country.

Negotiated Order

When we enter into interaction with others, we take for granted that a system of expected behaviors and shared meanings is already in place to guide the interaction. Although expectations are in place, symbolic interactionists emphasize that established meanings and ways of behaving can be reinforced and affirmed during interaction, but that they can also be ignored, challenged, or changed (Blumer 1969). In most interactions, room for negotiation exists; that is, the parties involved have the option of negotiating other expectations and meanings. The **negotiated order**, then, is the sum of existing expectations and newly negotiated ones (Strauss 1978).

To illustrate, college students know, for example, that when they enter a classroom on the first day of class, they should not walk to the front of the room and give instructions to the class. Likewise, professors know that on the first day of class, students expect them to give an overview of the course and lay out expectations such as no texting during class. Already established expectations are in place, guiding interaction. Usually, however, some room for negotiation exists; that is, the parties involved have the option of negotiating a social order. So on the first day of class, a professor may negotiate with students, indicating it is okay to check text messages when the professor is passing out assignments but not during lecture or class discussions. However, professors know that they cannot "negotiate" a social order in which students pay money to receive a desired grade.

The Symbolic-Interactionist Perspective on United States–Mexico Barriers

As we have learned, symbolic interactionists study people as they engage in social interaction. Thus, with regard to the border barrier, the following would be of great interest to symbolic interactionists who would seek to immerse themselves in the border world:

- interactions between border control agents and those crossing legally and illegally;
- the way Border Patrol agents are recruited and trained;

symbol Any kind of physical phenomenon to which people assign a name, meaning, or value.

negotiated order The sum of existing expectations and newly negotiated ones.

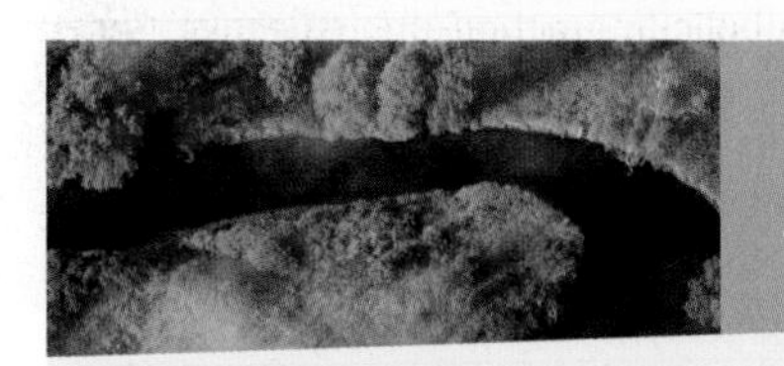

No Borders No Boundaries Interaction That Straddles the U.S.–Mexican Border

The border between Mexico and the United States stretches for almost 2,000 miles. The border region extends 60 miles south into Mexico and 60 miles north into the United States and includes 12 million people (Brown 2004). "From the perspective of the border, borderlines are not lines of sharp demarcation, but broad scenes of intense interactions in which people from both sides work out everyday accommodations based on face-to-face relationships. Each crosser seeks something that exists on the other side of the border. Each needs to make herself or himself understood on the other side of the border, to get food or gas or a job" (Thelen 1992, p. 437).

This map shows the tremendous amount of two-way traffic across the border, as well as attempts by the United States to control that traffic.

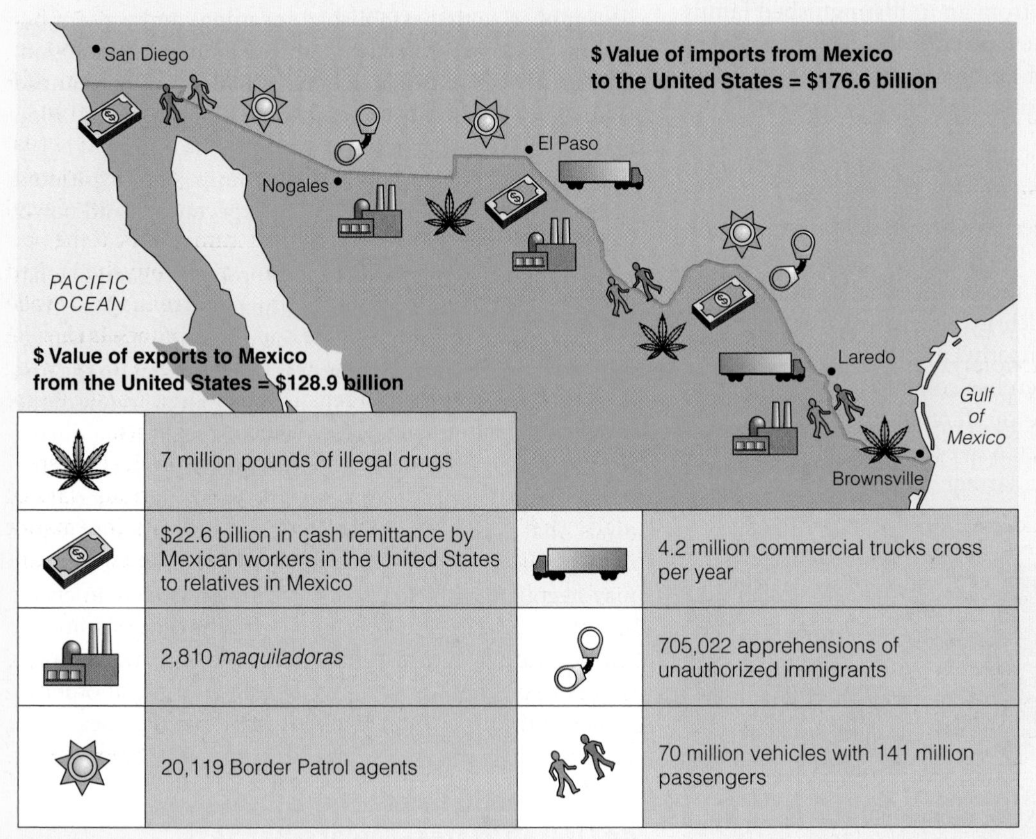

FIGURE 2.2
Selected Inventory of Border Activity Per Year
Sources: Data from U.S. Department of Homeland Security (2009, 2010), Migration Immigration Source (2006), U.S. Department of Transportation (2010), Ellingwood (2009), Haddal (2010)

- the interactions between undocumented immigrants and contacts already living in the United States;
- the way employers knowingly or unknowingly hire undocumented immigrants;
- the strategies undocumented immigrants use to blend into American society upon entry; and
- the strategies undocumented immigrants use to escape detection when passing though official border crossings.

As one example of the symbolic-interactionist emphasis on the close-up and personal, consider the following description of border agents inspecting cars and their passengers entering one of the world's busiest border crossings in California. This description helps us understand how many unauthorized immigrants manage to blend in with the crowds passing through official ports of entry (see No Borders, No Boundaries: "Interaction That Straddles the U.S.–Mexican Border"):

> Primary inspectors had to dispose of entering vehicles at an average of one per 45 seconds. In this time, the officer had to enter the license plate number into the Customs computer system, read the results (to see if there was any previous record of smuggling or illegal entry, or indeed any "lookout" rumors in the system), verbally ask about nationality and, if occupants had brought anything back from Mexico, inspect any immigration documents provided and examine the occupants of the car for their comportment (to see, for example, if they were nervous or sitting rigidly). Using this set of clues, some clear and others quite vague, the officer either cleared the car into

the U.S. without further inspection, or sent the vehicle to the side for an appropriate secondary inspection. If an admitted car was later stopped and found to have some narcotics or undocumented immigrants, the computer records would reveal who allowed the vehicle to enter. On the other hand, if the inspector took too long making decisions, it would back up traffic, with consequences including exacerbating air pollution and slowing down the interchange of people and goods on the Tijuana–San Diego corridor, the most important passage in the Mexico–U.S. free trade system. (Heyman 1999, p. 626)

Based on this description of border-crossing activity, it is easy to apply the core concepts that drive the symbolic-interactionist analysis of interactions: self-awareness/reflexive thinking, symbols, and negotiated order. In such situations, both drivers/passengers and agents are imagining how the other party is viewing their actions, evaluating their appearance, attaching meaning to their motives, and interpreting their words. Both parties are aware of the existing expectations, rules, and policies for crossing the border, but do on-the-spot "negotiations" to speed up inspection.

Critique of Three Sociological Theories

To this point, we have seen how each of the three sociological perspectives guide analysis of the border barriers through central questions and key concepts. Keep in mind that, taken alone, no single perspective can offer a complete picture of a situation. But we can acquire a more complete picture by applying all three. Of course, each theory has its strengths and weaknesses.

Concerning the border barriers, one strength of the functionalist theory is that it gives a balanced overview of the barriers' intended and unintended consequences to the existing social order. One weakness of that perspective is that it leaves us wondering about the border barriers' *overall* effects. That is, do the manifest and latent functions of the border barriers outweigh the manifest and latent dysfunctions?

One strength of the conflict theory is that it forces us to look beyond popular justifications for border barriers—to control undocumented migration—and to explore questions about whose interests are being protected and promoted and at whose expense. A weakness of the conflict theory is that it presents a simplistic view of the relationship between advantaged and disadvantaged groups: Dominant groups are portrayed as all-powerful and capable of imposing their will without resistance from subordinate groups, who are portrayed as exploited victims.

One strength of the symbolic-interactionist theory is that it encourages firsthand, extensive knowledge about how the border barriers shape interactions between Border Patrol agents and legal and undocumented immigrants. If the border barriers stop unauthorized movement into the United States, how do undocumented immigrants manage to evade the Border Patrol? One weakness of the approach is that we cannot be sure if the observations are unique to those being observed or apply to all interactions between Border Patrol agents and legal and undocumented immigrants.

We have reviewed the three major theoretical perspectives (see Figure 2.3). Now we turn to the various ways of data gathering and analysis that sociologists (no matter what their perspective) use to formulate and answer meaningful research questions.

	Functionalist Perspective	Conflict Perspective	Symbolic Interactionist Perspective
Focus:	Order and stability	Conflict over scarce and valued resources	Social interaction
Key Terms:	Function, dysfunction, manifest, latent	Conflict facade of legitimacy	Self-awareness, shared symbols, negotiated order
Vision of Society:	System of interrelated parts	Dominant and subordinate groups in conflicts over scarce and valued resources	Web of social interactions
Central Question(s):	Why does a part exist? What are the anticipated and unintended consequences of a part?	Who benefits from a particular pattern or arrangement, and at whose expense?	How do involved parties experience, interpret, influence, and respond to what they and others are doing in the course of interacting?
Strength:	Balanced analysis of positive and negative effects	Encourages analysis beyond popular explanations	Encourages direct, firsthand, and extensive analysis
Weakness:	Defends existing social arrangements Difficult to determine overall effect	Presents simplistic view of dominant-subordinate groups or relationships	Generalizability of observation is difficult to determine

FIGURE 2.3 Overview of the Three Theoretical Perspectives

Methods of Social Research

CORE CONCEPT 4 Sociologists adhere to the scientific method; that is, they acquire data through observation and leave it open to verification by others.

Research is a data-gathering and data-explaining enterprise governed by strict rules (Hagan 1989). **Research methods** are the various techniques that sociologists and other investigators use to formulate or answer meaningful research questions and to collect, analyze, and interpret data in ways that allow other researchers to check the results.

Theory and research are interdependent, because (1) theories inspire research, whose results can be used to support, disprove, or modify those theories; (2) the results of social research can inspire theories; and (3) theories are used to interpret facts generated through research. All sociologists are guided by the scientific method when they investigate human activities. The **scientific method** is an approach to data collection that relies on two assumptions: (1) knowledge about the world is acquired through observation, and (2) the truth of that knowledge is confirmed by verification—that is, by others making the same observations.

Researchers collect data that they and others can see, hear, taste, touch, and smell; that is, they focus on what they observe through the senses. They report the process they used to make their observations so that interested parties can duplicate, or at least critique, that process. If observations cannot be duplicated, or if repeating the study yields results that differ substantially from those of the original study, we consider the study suspect. Findings endure as long as they can withstand continued reexamination and duplication by the scientific community. "Duplication is the heart of good research" (Dye 1995, p. D5). No finding can be taken seriously unless other researchers can repeat the process and obtain the same results.

When researchers know that others are critiquing and checking their work, the process serves to reinforce careful, thoughtful, honest, and conscientious behavior. Moreover, this "checking" encourages researchers to maintain **objectivity**; that is, it encourages them not to let their personal, or subjective, views about the topic influence their observations or the outcome of the research.

This description of the scientific method is an ideal one, because it outlines how researchers and reviewers should behave. Ideally, researchers should be guided by the core values of honesty, skepticism, fairness, collegiality, and openness (National Academy of Sciences 1995). In practice, though, some research is dismissed as unimportant and unworthy of examination simply because the topic or the researcher is controversial or because the findings depart from mainstream thinking. Moreover, some researchers fabricate data to support a personal, an economic, or a political agenda. The extent to which researchers actually adhere to core values remains unknown.

Research should be carefully planned; the enterprise of gathering and explaining facts involves a number of interdependent steps (Rossi 1988):

1. Choosing the topic for investigation or deciding on the research question
2. Reviewing the literature
3. Identifying core concepts
4. Choosing a research design, forming hypotheses, and collecting data
5. Analyzing the data
6. Drawing conclusions

Researchers do not always follow these six steps in sequence, however. They may not define the topic (step 1) until they have familiarized themselves with the literature (step 2). Sometimes an opportunity arises to gather information (step 4), and a project is defined to fit that opportunity (step 1).

Although the six steps need not be followed exactly in sequence, all must be completed at some point to ensure the quality of the project. In the sections that follow, we will examine each stage. Along the way, we will refer to sociological and other research that focuses on unauthorized immigration and the border barriers' effect on deterring entry. Where appropriate, we will emphasize two sociological studies: *Patrolling Chaos: The U.S. Border Patrol in Deep South Texas* by Robert Lee Maril (2004) and "The Social Process of Undocumented Border Crossing among Mexican Migrants" by Audrey Singer and Douglas S. Massey (1998). Singer and Massey's research is part of a larger project known as the Mexican Migration Project, a multidisciplinary effort involving researchers in the United States and Mexico (Mexican Migration Project 2007).

Step 1: Choosing the Topic for Investigation

The first step of a research project involves choosing a topic or deciding on a research question. It would be impossible to compile a comprehensive list of the topics that

research A data-gathering and data-explaining enterprise governed by strict rules.

research methods Techniques that sociologists and other investigators use to formulate or answer meaningful research questions and to collect, analyze, and interpret data in ways that allow other researchers to verify the results.

scientific method An approach to data collection in which knowledge is gained through observation and its truth is confirmed through verification.

objectivity A stance in which researchers' personal, or subjective, views do not influence their observations or the outcomes of their research.

sociologists study, because almost any subject involving humans represents a potential target for investigation. Sociology is distinguished from other disciplines not by the topics sociologists investigate, but by the perspectives needed to study topics. Researchers choose their topics for a number of reasons. Personal interest is a common and often understated motive. It is perhaps the most significant reason that someone picks a specific topic to study, especially if we consider how researchers eventually choose one topic from a virtually infinite set of possibilities.

Good researchers explain to their readers why their topic or research question is significant. This explanation is vital because it clarifies the purpose and importance of the project and the researcher's motivation for doing the work. Sociologists Audrey Singer and Douglas S. Massey (1998) studied "the social process of undocumented border crossings among Mexican migrants" (p. 561). The two researchers chose this topic for several reasons: (1) It is a politically divisive topic in the United States; (2) little is known about how migrants evade borders guarded by agents, maneuver around or bypass border barriers (some triple-deep), and escape detection by surveillance equipment; and (3) knowing the extent of undocumented entries helps us judge whether border barriers or other barriers are effective deterrents.

In this step, researchers often announce the perspective guiding their investigation. Sometimes they announce it directly by indicating that they are writing from one or more theoretical traditions or perspectives. Sometimes they announce their guiding perspective(s) indirectly; that is, readers surmise the perspective from the way researchers frame the question or analysis. That Singer and Massey (1998) focused on undocumented immigrants' social ties to others who have crossed the border successfully without authorization suggests that the two researchers are drawing upon the symbolic interaction perspective to frame their analysis. One also notices that Singer and Massey take a conflict perspective when they suggest that constructing border barriers and implementing other border control strategies "sit well with the public," because the U.S. government "appears to be defending the United States against alien invaders while not antagonizing U.S. business interests" (pp. 563–564).

Step 2: Reviewing the Literature

All good researchers consider existing research. They read what knowledgeable authorities have written on the chosen topic, if only to avoid repeating earlier work. More importantly, reading the relevant literature can generate insights that researchers may not have considered. Even if researchers believe that they have a revolutionary idea, they must still consider the works of other thinkers and show how their new research verifies, advances, and corrects past research.

At the end of most research papers, authors cite the literature that has influenced that work. This list can include dozens to hundreds of citations. For their research on undocumented border crossings, Singer and Massey (1998) cited 44 references. They used the existing literature to identify factors that help them predict how many times an undocumented migrant will be apprehended trying to evade detection. Those factors: (1) the nature and intensity of U.S. enforcement efforts, (2) characteristics of the migrant that enhance the chance of success (such as previous success at crossing borders), and (3) ties to other undocumented migrants (such as a parent, other relative, or friend) who succeeded in unauthorized entry.

Step 3: Identifying and Defining Core Concepts

After deciding on a topic and reading the relevant literature (albeit not necessarily in that order), researchers typically state their core concepts. **Concepts** are powerful thinking and communication tools that enable researchers to give and receive complex information efficiently. The mention of a concept triggers in the minds of people who know its meaning a definition and a range of important associations that help frame and focus observations. One core concept for sociologists studying undocumented immigration is the *unauthorized immigrant* (also referred to as an undocumented or illegal immigrant or migrant), defined as a noncitizen residing in the United States whom the American government has not admitted for permanent residence or for specific authorized temporary work or stays. Another core concept is *interpersonal ties*, or the network of connections that include parents, siblings, or other relatives or friends and acquaintances who guide or otherwise help (or fail to help) migrants enter the United States undetected.

Step 4: Choosing a Research Design and Data-Gathering Strategies

Once researchers have clarified core concepts, they decide on a **research design**, a plan for gathering data on the topic they have chosen. A research design specifies the population to be studied and the **methods of data collection**, or the procedures used to gather relevant data. One research design is not inherently better than another; researchers

concepts Thinking and communication tools used to give and receive complex information efficiently and to frame and focus observations.

research design A plan for gathering data that specifies who or what will be studied and the methods of data collection.

methods of data collection The procedures a researcher follows to gather relevant data.

choose the design that best enables them to address the research question at hand (Smith 1991).

Researchers must decide whom or what they are going to study. The most common "thing" sociologists study is individuals, but they may also decide to study traces, documents, territories, households, small groups, or individuals (Rossi 1988):

- **Traces** are materials or other evidence that yield information about human activity, such as the items that people throw away, the number of lights turned on in a house, or changes in water pressure. Researchers who study undocumented border crossings might learn about paths undocumented immigrants take into the United States by observing the litter they leave behind, including "one-gallon plastic bottles, jeans, T-shirts, candy wrappers, socks, underwear, discarded purses, and inexpensive tennis shoes" (Maril 2004, p. 166).
- **Documents** are written or printed materials, such as magazines, advertisements, graffiti, birth certificates, death certificates, prescription forms, and traffic tickets. Researchers interested in the process by which Homeland Security chose and secured land for the 700 miles of barriers might look at U.S. Army Corp of Engineers' letters proclaiming the land was needed to build barriers and specifying the process by and condition under which land would be relinquished (Del Bosque 2010).
- **Territories** are places that have known boundaries or that are set aside for particular activities. They include countries, states, counties, cities, neighborhoods, streets, buildings, and classrooms. Sociologists who study territories focus on activity within the territory they select. Sociologists may choose to study communities divided by the barriers.
- **Households** include all related and unrelated people who share the same dwelling. When studying a household, researchers collect information about the household itself. They might want to determine the number of people living in the household and the household income (that is, the combined income of all people living in the same dwelling). For their research on undocumented border crossings, Singer and Massey (1998) drew upon interviews of households in 34 Mexican communities. Among other kinds of information, the researchers collected data on household size and on the number of household members who have made successful unauthorized entries into the United States.
- **Small groups** are defined as 2 to about 20 people who interact with one another in meaningful ways (Shotola 1992). Examples include father–child pairs, doctor–patient pairs, families, sports teams, circles of friends, and committees. Concerning a doctor–patient interaction, a sociologist might study the length of time the doctor spends with each patient. Keep in mind that the focus is on the doctor–patient *relationship*, rather than on the doctor or the patient *per se*. For his book *Patrolling Chaos*, Robert Maril studied a small group. Specifically, he accompanied 12 Border Patrol agents on 60 ten-hour shifts along the border.

Researchers who study unauthorized border crossings may look for artifacts left behind, dropped, or discarded as a way to document paths migrants take as they trek toward the United States. Here we see a pair of pants left behind in a remote area on the Mexican side of the border.

Sociologists who study activity along a border station such as that pictured here have chosen to study territories. They may choose to determine the average time it takes a vehicle, once it pulls into a line, to cross from the Mexican side to the U.S. side.

traces Materials or other forms of physical evidence that yield information about human activity.

documents Written or printed materials used in research.

territories Settings that have borders or that are set aside for particular activities.

households All related and unrelated people who share the same dwelling.

small groups Groups of two to about 20 people who interact with one another in meaningful ways.

Because of time constraints, researchers cannot study entire **populations**—the total number of individuals, traces, documents, territories, households, or groups that exist. Instead, they study **samples,** or portions of the cases from a larger population.

Sampling Ideally, sociologists should study a **random sample,** in which every case in the population has an equal chance of being selected. The classic, if inefficient, way of selecting a random sample is to assign a number to every case, place cards on which the numbers are written into a container, thoroughly mix the cards, and pull out one card at a time until the desired sample size is achieved. Rather than employ this tedious system to generate their samples, most of today's researchers use computer programs. If every case has an equal chance of becoming part of a sample, then theoretically the sample should be a **representative sample**—that is, one with the same distribution of characteristics (such as age, sex, and ethnic composition) as the population from which it is selected. For example, if 56.4 percent of the population from which a sample is drawn is at least 30 years old, then approximately 56.4 percent of a representative sample should be that age. In theory, if the sample is representative, then whatever holds true for the sample should also hold true for the larger population.

Obtaining a random sample is not as easy as it might appear. For one thing, researchers must begin with a **sampling frame**—a complete list of every case in the population—and each member of the population must have an equal chance of being selected. Securing such a complete list can be difficult. Campus and city telephone directories are easy to acquire, but lists of, say, U.S. citizens, adopted children in the United States, or American-owned companies with operations in Mexico are more difficult to obtain. Almost all lists omit some people (such as individuals with unlisted telephone numbers, members too new to be listed, or between-semester transfer students) and include some people who no longer belong (such as individuals who have moved, died, or dropped out). What is important is that researchers consider the extent to which the list is incomplete and update it before drawing a sample. Even if the list is complete, researchers must also think of the cost and time required to take random samples and consider the problems of inducing all sampled people to participate.

Researchers sometimes select nonrandom samples to study people who they know are not representative of the larger population but who are easily accessible such as high school and college students. Researchers may choose unrepresentative samples for other important reasons: (1) little is known about members of the sample, (2) they have some special quality or characteristic, or (3) their experiences clarify important social issues.

Singer and Massey (1998) used data collected from a random sample of 6,341 households in 34 communities in Mexico. This data was supplemented by a nonrandom sample of 484 U.S. households in which undocumented immigrants from each of the 34 communities in Mexico were now living. For his study *Patrolling Chaos*, Maril chose a nonrandom sample that consisted of one Border Patrol station among nine in the McAllen, Texas, sector of the southwest border. The border station, which he observed for two years, employed 300 men and women to guard a 45-mile stretch of the Rio Grande, which marks the Texas–Mexico border. In particular, Maril observed 12 agents as they worked ten-hour patrol shifts.

Methods of Data Collection

Besides identifying whom or what to study, the design must include a plan for collecting information. Researchers can choose from a variety of data-gathering methods, including self-administered questionnaires, interviews, observation, and secondary sources.

Self-Administered Questionnaire A **self-administered questionnaire** is a set of questions given to respondents, who read the instructions and fill in the answers themselves. The questions may require respondents to write out answers (open ended) or to select one that best reflects their answer from a list of responses (forced choice). Self-administered questionnaires are one of the most common methods of data collection. The questionnaires found in magazines or books, displayed on tables or racks in service-oriented establishments (such as hospitals, garages, restaurants, grocery stores, and physicians' offices), and mailed to households are all self-administered questionnaires.

This method of data collection offers a number of advantages. No interviewers are needed to ask respondents questions, and the questionnaires can be given to large numbers of people at one time. Also, an interviewer's facial expressions or body language cannot influence respondents, so the respondents feel freer than they otherwise might to give unpopular or controversial responses.

populations The total number of individuals, traces, documents, territories, households, or groups that could be studied.

samples Portions of the cases from a larger population.

random sample A type of sample in which every case in the population has an equal chance of being selected.

representative sample A type of sample in which those selected for study have the same distribution of characteristics as the population from which it is selected.

sampling frame A complete list of every case in a population.

self-administered questionnaire A set of questions given to respondents who read the instructions and fill in the answers themselves.

Self-administered questionnaires pose some problems, too. Respondents can misunderstand or skip over questions. When questionnaires are mailed, set out on a table, or published in a magazine or newspaper, researchers must wonder whether the people who choose to fill them out have opinions that differ from those of people who ignore the questionnaires. The results of a questionnaire depend not only on respondents' decisions to fill it out, answer questions conscientiously and honestly, and return it, but also on the quality of the questions and on a host of other considerations.

Interviews Compared with self-administered questionnaires, **interviews** are more personal. In these face-to-face or telephone conversations between an interviewer and a respondent, the interviewer asks questions and records the respondent's answers. As respondents give answers, interviewers must avoid pauses, expressions of surprise, or body language that reflects value judgments. Refraining from such conduct helps respondents feel comfortable and encourages them to give honest answers.

Interviews can be structured or unstructured, or some combination of the two. In a **structured interview**, the wording and sequence of questions are set in advance and cannot be altered during the interview. In one kind of structured interview, respondents choose answers from a response list that the interviewer reads to them. In another kind of structured interview, respondents are free to answer the questions as they see fit, although the interviewer may ask them to clarify answers or explain them in more detail. For their study on undocumented border crossings, Singer and Massey (1998) relied on data collected from structured interviews with heads of households in Mexico and the United States. Among other questions, interviewers asked the following: How many people have ever lived in the household, including those who no longer live there? How many people from the household have experience in migrating to the United States? For each person who has such experience, how many trips have they made to the United States, including the year of the first and last trip? How long did they stay in the United States? What occupation did they hold, and what did it pay? How many of those trips were made without documentation? How many trips resulted in deportation? What was the place of crossing and with whom was the crossing made? Did they use a coyote, and if yes, how much was the coyote paid? Interviewers also asked about each immigrant's ability to speak and understand English.

interviews Face-to-face or telephone conversations between an interviewer and a respondent, in which the interviewer asks questions and records the respondent's answers.

structured interview An interview in which the wording and sequence of questions are set in advance and cannot be changed during the interview.

unstructured interview An interview in which the question-and-answer sequence is spontaneous, open-ended, and flexible.

observation A research technique in which the researcher watches, listens to, and records behavior and conversations as they happen.

United States Census Bureau

Interviews are face-to-face sessions where the interviewer asks questions and records the respondent's answers. The interview can be structured or unstructured.

In contrast to the structured interview, an **unstructured interview** is flexible and open ended. The question-and-answer sequence is spontaneous and resembles a normal conversation in that the questions are not worded in advance and are not asked in a set order. The interviewer allows respondents to take the conversation in directions they define as crucial. The interviewer's role is to give focus to the interview, ask for further explanation or clarification, and probe and follow up on interesting ideas expressed by respondents. The interviewer appraises the meaning of respondents' answers and uses the information learned to ask follow-up questions. Talk show hosts, for instance, often use an unstructured format to interview their guests. Sociologists, however, have goals much different from those of talk show hosts. For one thing, sociologists do not formulate questions with the goal of entertaining an audience. In addition, they strive to ask questions in a neutral way, and no audience reaction influences how respondents answer. Robert Maril (2004) used unstructured interviews as he rode with border agents during their ten-hour shifts and at other times. As he describes it, "under the scorching sun and in the dead of night along the banks of the Rio Grande, I asked these men and women what they knew, what they had seen, and what they thought. In no uncertain terms and with direct, sometimes alarming honesty, they told me" (p. 16).

Observation As the term implies, **observation** involves watching, listening to, and recording behavior and

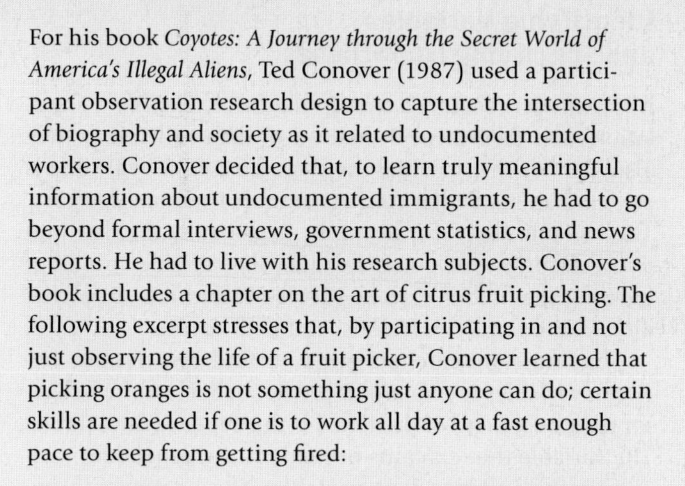

The Sociological Imagination The Life of a Citus Picker

For his book *Coyotes: A Journey through the Secret World of America's Illegal Aliens,* Ted Conover (1987) used a participant observation research design to capture the intersection of biography and society as it related to undocumented workers. Conover decided that, to learn truly meaningful information about undocumented immigrants, he had to go beyond formal interviews, government statistics, and news reports. He had to live with his research subjects. Conover's book includes a chapter on the art of citrus fruit picking. The following excerpt stresses that, by participating in and not just observing the life of a fruit picker, Conover learned that picking oranges is not something just anyone can do; certain skills are needed if one is to work all day at a fast enough pace to keep from getting fired:

> It required a vast store of special knowledge and dexterity. Not only did you need to know the optimum position of the ladder against a given tree, for example, you also had to be able to get it there fast and deftly. Men half my size could manipulate the twenty-foot ladders as though they were balsa wood, but in my hands the ladder was a heavy, deadly weapon. I had bruised a friend's arm with my ladder one day and had nearly broken my own when, with sixty pounds of oranges in the bag around my neck, I had slipped from its fifth rung and been grazed as it followed me to the muddy ground. . . . Handling the fruit itself was a further challenge. To be successful, you had to pick with both hands simultaneously. This meant your balance had to be good—you couldn't hold on to a branch or the ladder for support. And you had to twist the fruit off: pull hard on a Valencia without twisting and you're likely to get the whole branch in your hand. Planning was also important: Ideally, you would start with your sack empty at the top of the tree and work your way down. Oranges would just be peeking out the top of the sack when your feet touched the soil, if you did it right; you topped off the bag by grabbing low-hanging fruit on your way to the tractor. (pp. 42–43)

Source: *Coyotes: A Journey through the Secret World of America's Illegal Aliens,* Ted Conover (1987), pp. 42–43.

conversations as they happen. This research technique may sound easy, but it entails more than just watching and listening. The challenge of observation lies in knowing what to look for while remaining open to other considerations; success results from identifying what is worth observing. "It is a crucial choice, often determining the success or failure of months of work, often differentiating the brilliant observer from the . . . plodder" (Gregg 1989, p. 53). Good observation techniques must be developed through practice; observers must learn to recognize what is worth observing, be alert to unusual features, take detailed notes, and make associations between observed behaviors.

If observers come from a culture different from the one under study, they must be careful not to misinterpret or misrepresent what is happening. Imagine for a moment how an uninformed, naive observer might describe a sumo wrestling match: "One big, fat guy tries to ground another big, fat guy or force him out of the ring in a match that can last as little as three seconds" (Schonberg 1981, p. B9). Actually, for those who understand it, sumo wrestling is "a sport rich with tradition, pageantry, and elegance and filled with action, excitement, and heroes dedicated to an almost impossible standard of excellence down to the last detail" (Thayer 1983, p. 271).

Observational techniques are especially useful for three purposes: (1) studying behavior as it occurs, (2) learning information that cannot be surveyed easily, and (3) acquiring the viewpoint of the people under observation. Observation can take two forms: participant and nonparticipant. **Nonparticipant observation** consists of detached watching and listening; researchers merely observe but do not interact with the study subjects or become involved in their daily life. In contrast, researchers engage in **participant observation** when they join a group and assume the role of a group member, interact directly with individuals whom they are studying, assume a position critical to the outcome of the study, or live in a community under study (see Sociological Imagination: "The Life of a Citrus Picker"). Maril's research qualifies as participant observation, because, for all practical purposes, he became involved in the daily life of the people he studied, even "actively participated in some of the policing techniques and procedures [he] was observing" (p. 18).

In both participant and nonparticipant observation, researchers must decide whether to hide their identity and purpose or to announce them. One major reason for choosing concealment is to avoid the **Hawthorne effect**,

nonparticipant observation A research technique in which the researcher observes study participants without interacting with them.

participant observation A research technique in which the researcher observes study participants while directly interacting with them.

Hawthorne effect A phenomenon in which research subjects alter their behavior when they learn they are being observed.

a phenomenon in which research subjects alter their behavior when they learn they are being observed. The term *Hawthorne effect* originated from a series of worker productivity studies involving female employees of the Hawthorne, Illinois, plant of Western Electric, conducted in the 1920s and 1930s. Researchers found that no matter how they varied working conditions—bright versus dim lighting, long versus short breaks, frequent versus no breaks, piecework pay versus fixed salary—workers' productivity increased. One explanation for this finding was that workers were responding positively to having been singled out for study (Roethlisberger and Dickson 1939).

If researchers choose to announce their identity and purpose, they must give participants adequate time to adjust to their presence. Usually, if researchers are present for a long enough time, their subjects will eventually display natural, uninhibited behaviors. Maril (2004) chose to announce his identity, and agents often referred to him as the "professor from the university." He noted that the majority of agents who worked at the station seemed to pay little attention to him after a couple of months. Maril believed that the agents were, with a few exceptions, "open and honest with me not just because they came to trust me, but because few, if any, had ever been asked about their work as agents. . . . They were anxious to talk and show me what they knew" (p. 16).

Secondary Sources (Archival Data) Another data-gathering strategy relies on **secondary sources (archival data)**—that is, data that have been collected by other researchers for some other purpose. Government researchers, for example, collect and publish data on many areas of life, including births, deaths, marriages, divorces, crime, education, travel, and trade. Any researcher who takes some existing data set and then applies it to address a different research question is using a secondary source (Horan 1995).

Another kind of secondary data source consists of materials that people have written, recorded, or created for reasons other than research (Singleton, Straits, and Straits 1993). Examples include television commercials and other advertisements, letters, diaries, home videos, poems, photographs, artwork, graffiti, movies, and song lyrics.

Identifying Variables and Specifying Hypotheses

As researchers acquire a conceptual focus, identify a population, and determine a method of data collection, they also identify the variables they want to study. A **variable** is any characteristic that consists of more than one category. The variable "sex," for example, is generally divided into two categories: male and female. The variable "mode of crossing the border" can be divided into three categories: alone, with family or friends, and with coyote.

Sometimes researchers strive to find associations between variables to explain or predict behavior. The behavior to be explained or predicted is the **dependent variable**. The variable that explains or predicts the dependent variable is the **independent variable**. The relationship between independent and dependent variables is described in a **hypothesis**, or trial explanation put forward as the focus of research, which predicts the relationship between independent and dependent variables. Specifically, it predicts how a change in an independent variable brings about a change in a dependent variable. Hypotheses that could be tested using Singer and Massey's data include the following:

- **Hypothesis 1.** The more proficient undocumented immigrants are in English, the less likely they are to be apprehended by Border Patrol.
- **Hypothesis 2.** The more times an undocumented immigrant crosses the border into the United States without being apprehended, the more likely the immigrant will be to cross alone.

In hypothesis 1, the independent variable is English-language proficiency and the dependent variable is

Eugenio del Bosque

Sociologists Singer and Massey worked to identify the variables that help us predict the probability that undocumented immigrants will be apprehended.

secondary sources (archival data) Data that have been collected by other researchers for some other purpose.

variable Any trait or characteristic that can change under different conditions or that consists of more than one category.

dependent variable The variable to be explained or predicted.

independent variable The variable that explains or predicts the dependent variable.

hypothesis A trial explanation put forward as the focus of research; it predicts how independent and dependent variables are related and how a dependent variable will change when an independent variable changes.

likelihood of apprehension. In hypothesis 2, the independent variable is number of successful border crossings and the dependent variable is likelihood of crossing alone.

CORE CONCEPT 5 If findings are to matter, researchers must create meaningful operational definitions.

A major reason that researchers collect data is to test and propose hypotheses. If their findings are to matter, other researchers must be able to replicate their study. So, researchers need to give clear, precise definitions and instructions about how to observe and/or measure the variables under study. In the language of research, such definitions and accompanying instructions are called **operational definitions**.

An analogy can be drawn between an operational definition and a recipe. Just as people with basic cooking skills should be able to follow a recipe to achieve a desired end, people with basic research skills should be able to replicate a researcher's observations if they know the operational definitions (Katzer, Cook, and Crouch 1991). Operational definitions include clear, precise definitions and instructions about how to observe or measure variables. They help researchers determine whether a behavior of interest has occurred.

Suppose a researcher is interested in the question of who washes hands after using a public toilet. An operational definition of hand washing would include an account of what must take place for a researcher to count someone as a hand washer. If people simply run water over their fingertips or rinse their hands quickly without using soap, should the behavior count as hand washing? What if people use soap but wash only their fingertips? Should a behavior count as hand washing only if it satisfies the guidelines issued by the American Society for Microbiology? Those guidelines specify using warm or hot running water and soap while washing for 10 to 15 seconds "all surfaces thoroughly, including wrists, palms, back of hands, fingers and under fingernails" (American Society for Microbiology 1996). Researchers must address these kinds of questions in creating operational definitions.

Consider the operational definition of English-language proficiency used in the Singer and Massey study:

Do you speak and understand English?
___ Does not speak or understand.
___ Does not speak but understands a little.
___ Does not speak but understands well.
___ Both speaks and understands a little.
___ Both speaks and understands well.

Other questions related to English-language proficiency include asking the immigrants in which settings they speak English when living in the United States. The settings of interest include at home, at work, with friends, and while shopping; the frequency of using English in each setting is recorded as never, sometimes, often, or always.

Operational definitions do not have to take the form of questions. They may be precise accounts or descriptions of what a researcher observed and what the context of the observations was. If operational definitions are not clear or do not indicate accurately the behaviors they were designed to represent, they have questionable value. Good operational definitions are both reliable and valid. **Reliability** is the extent to which an operational definition gives consistent results. For example, Singer and Massey's operational definition of English-language proficiency many not yield reliable answers when the person answering questions has weak English-language proficiency. Thus, if you asked a respondent the question at two different times, he or she might give two different answers. One way to increase the reliability of this question could be to have a Spanish- and English-proficient interviewer rate the person's level of proficiency.

Validity is the degree to which an operational definition measures what it claims to measure. Professors, for example, give tests to measure students' knowledge of a particular subject as covered in class lectures, discussions, reading assignments, and other activities. Students may question the validity of this measure if the questions on a test reflect only the material covered in lectures. They may argue that the test does not measure knowledge of all the material covered or assigned. Remember, when assessing validity, always ask, *Is the operational definition really measuring what it claims to measure?* For example, is the question *Do you speak and understand English?* likely to yield valid responses about English-language proficiency? One problem with this operational definition is disagreement as to what constitutes speaking *a little* English and speaking English *well*. What if a person's proficiency falls somewhere between *a little* and *well*? All people who answer this question probably have their own understanding of what speaking *a little* or *well* means.

Likewise, is the number of apprehensions a valid measure of the effectiveness of a border control initiative such as barrier construction? Should effectiveness be measured by increased apprehensions or decreased apprehensions? Increased apprehensions could mean barriers are effective in forcing undocumented immigrants into territories where the Border Patrol has a strategic advantage; they could also mean that the barriers are ineffective, because for every one undocumented immigrant caught, four manage to cross (see Figure 2.4).

operational definitions Clear, precise definitions and instructions about how to observe and/or measure the variables under study.

reliability The extent to which an operational definition gives consistent results.

validity The degree to which an operational definition measures what it claims to measure.

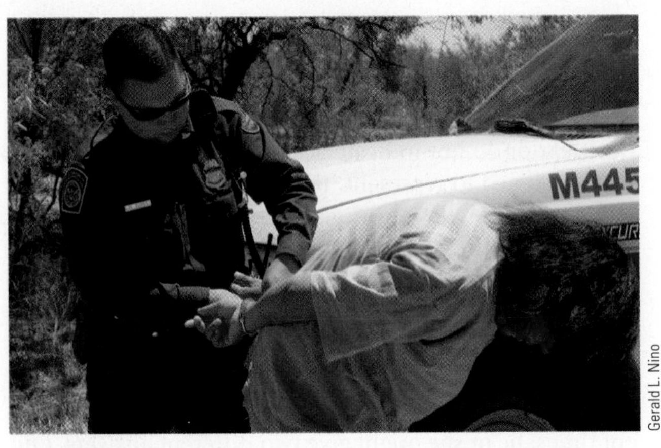

Gerald L. Nino

Is number of apprehensions the best way to measure the effectiveness of border fences? Increased apprehensions could mean one of two things: the fence is effective in forcing undocumented immigrants into territories where border patrol have a strategic advantage, or the fence is ineffective because, for every one undocumented immigrant caught, four manage to cross.

Steps 5 and 6: Analyzing the Data and Drawing Conclusions

When researchers reach the stage of analyzing collected data, they search for common themes, meaningful patterns, and links. Researchers must "pick and choose among the available numbers [and observations] and then fashion a format" (Hacker 1997, p. 478). In presenting their findings, researchers may use graphs, frequency tables, photos, statistical data, and so on. The choice of presentation depends on which results are significant and how they might be best shown (see Table 2.1).

Generalizability Besides choosing a format in which to present data, sociologists comment on the **generalizability** of findings, the extent to which the findings can be applied to the larger population from which the sample was drawn. Both the sample used and the response rate are important in determining generalizability. If a sample is

TABLE 2.1 Basic Statistics Researchers Use to Convey Findings

The following table shows the number of unauthorized immigrants apprehended each year from 1992 through 2010. One can see that the number of apprehensions has dropped with construction of the border barriers. But are apprehensions the best measure of effectiveness? Could it be that the undocumented are simply staying in the United States rather than returning home after seasonal employment ends? Could it be that the undocumented have found new ways to enter undetected? Massey thinks that this might be the case.

Year	Undocumented Immigrants Apprehended on U.S–Mexico Border (in millions)
2010	.45
2009	.69
2008	.73
2007	.88
2006	1.11
2005	1.17
2004	1.14
2003	.91
2002	.93
2001	1.24
2000	1.64
1999	1.54
1998	1.52
1997	1.37
1996	1.51
1995	1.27
1994	.98
1993	1.20
1992	1.18

Source: **Data from Homeland Security, Office of Immigration Statistics (2010)**

n: number of cases (19 years of data)

Mean: The sum of all apprehensions from 1992 through 2010, divided by the total number of years considered (19). Over the 18-year period, the average number of apprehensions was 1.13 million.

Standard deviation (s.d.): A measure that shows how far the data are spread from the mean. About two-thirds of data falls one standard deviation from the mean. Most of data (at least 95 percent) falls within two standard deviations of the mean, and almost all of data (99.7 percent) will fall within three standard deviations of the mean. For the 19-year period specified in the table, the standard deviation is .32 million or 320,000. We will not present the formula here but the standard deviation can be easily calculated in Microsoft Excel or some other spreadsheet. To calculate the spread of data around the mean, add *and* subtract the standard deviation from the mean once, twice, or three times. To calculate the range of apprehensions that are one standard deviation from the mean, simply add .32 to the mean (1.13 million) and subtract .32 from the mean. The results tell you that 66 percent of the data falls between .81 and 1.45 million. To calculate the range of apprehensions over the past 19 years that are two standard deviations, add and subtract the standard deviation multiplied by two from the mean.

Minimum: The lowest number of apprehensions (450,000 apprehensions in 2010).

Maximum: The highest number of apprehensions (1.64 million apprehensions in 2005).

Range: The difference between the highest value and the lowest value. In this case, the range of apprehensions falls between .45 million in 2010 and to 1.64 million in 2005, a difference of 1.19 million.

Mode: The value that occurs most often. In this case, there is no mode as the 19 values are all different.

n: A symbol that stands for the number of respondents or cases. In this case, n is 19.

Median: The number that 50 percent of the cases fall above and 50 percent fall below. For the 19 years of apprehension data, half of the years listed had fewer than 1.17 million apprehensions, and half had more than 1.17 million.

Singer and Massey (1998) calculated the probability that an undocumented migrant would be caught for the years 1964 through 1994. In 1964 the probability of getting caught was about 33 percent. In 1994 that probability dropped to about 15 percent. This drop in probability suggests that a decline in the number of apprehensions is only one measure of success at deterring undocumented migrants. As we will see in the next section, the more crucial measure involves knowing the number that attempted to enter and the percentage who were eventually caught.

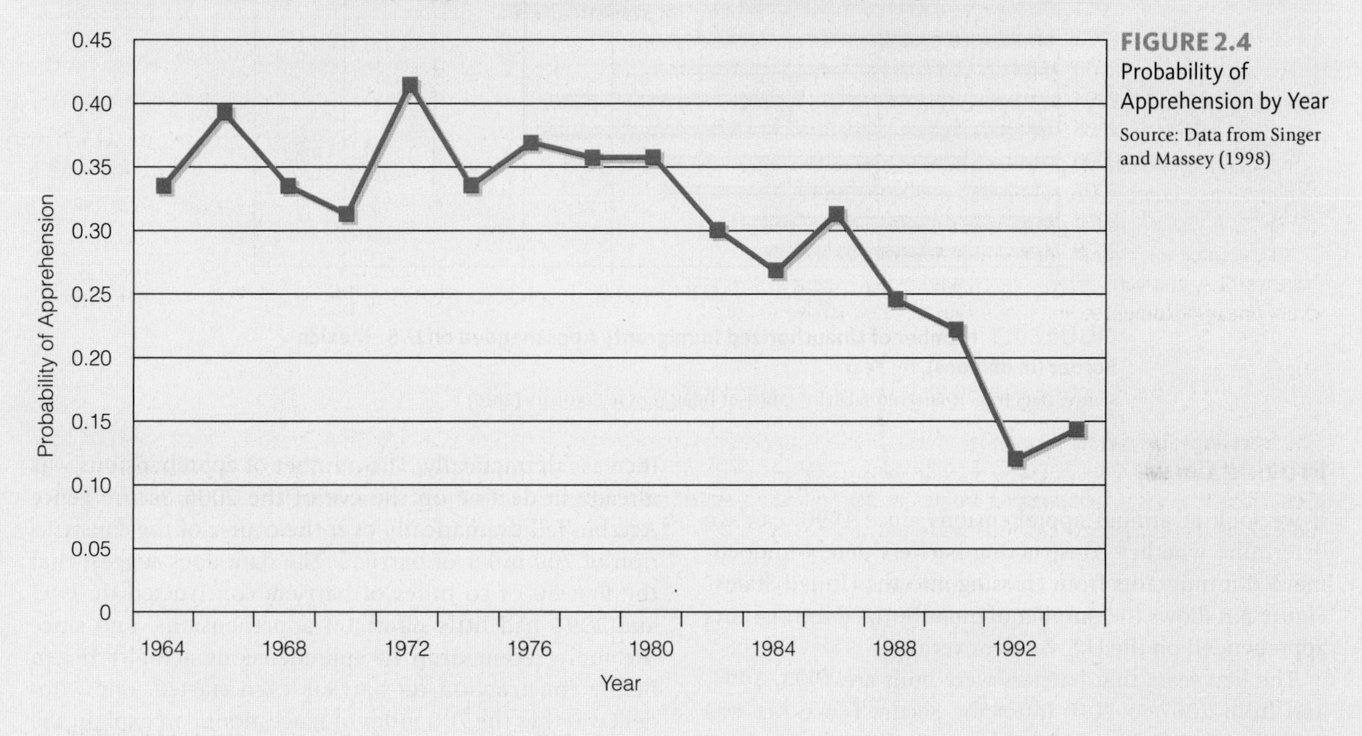

FIGURE 2.4 Probability of Apprehension by Year
Source: Data from Singer and Massey (1998)

randomly selected, if almost all subjects agree to participate, and if the response rate for every question is high, we can say that the sample is representative of the population and that the findings are theoretically generalizable to that population. Recall that Singer and Massey (1998) used data derived from structured interviews with a sample of 6,341 randomly selected households in 34 Mexican communities. The response rate varied by community (from a 100 percent response rate in two rural communities to an 84 percent response rate in a large urban community); the average response rate was 94 percent. Such a high response rate suggests that Singer and Massey's findings are generalizable to the undocumented population. In particular, Singer and Massey found that, despite the apparent buildup of enforcement resources and the implementation of Operations Hold the Line and Gatekeeper, the probability of apprehensions fell in the late 1980s through early to mid-1990s. Singer and Massey's conclusions can be extended to analyzing the success of the border barriers. Although the border barriers may indeed be successful, their success cannot be determined from by simply showing a decline in apprehensions. To really determine success, we must also know how many undocumented attempted to enter, what percentage were caught, and how many have just stayed put in the United States.

Also keep in mind that even when samples are random, the generalizations that follow can never be presented as statements of certainty that apply to everyone. In the case of Maril's (2006) research, the findings probably cannot be generalized to *all* Border Patrol stations. Still his research on the border from the perspective of the Border Patrol agents does lend support to Singer and Massey's findings. Specifically, Maril agued that highly visible border control strategies—border barriers, increases in the number of agents, and increased surveillance—are "a grand pretense" because the majority of undocumented immigration and opportunities to enter illegally exist "far from the public eye" (p. 287).

generalizability The extent to which findings can be applied to the larger population from which a sample is drawn.

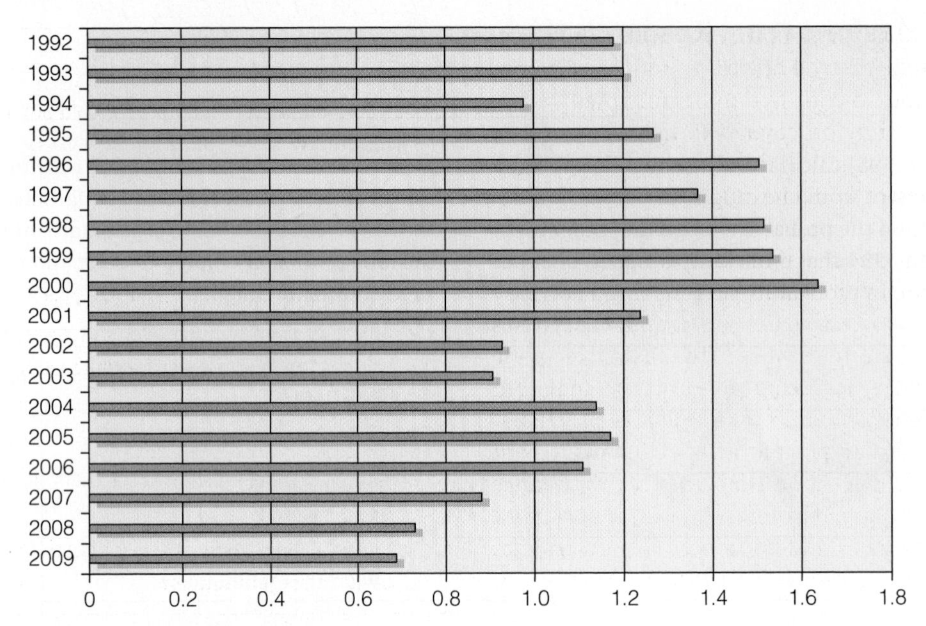

FIGURE 2.5 Number of Unauthorized Immigrants Apprehended on U.S.-Mexico Border (in millions), by Year

Source: Data from Homeland Security, Office of Immigration Statistics (2010)

Proving Cause

If we look at annual apprehensions since 1990, can we determine whether constructing barriers deters undocumented immigrants from crossing into the United States? Figure 2.5 shows the number of unauthorized immigrants apprehended on the U.S. border over time.

The key years that barriers were built are 1993, 1994, and from 2007 to 2010 (after the Secure Fence Act was passed in 2006). Figure 2.5 shows a decline in apprehensions around 1993 and 1994, but then apprehensions rise again to pre-barrier levels. The year before the barriers were constructed apprehensions began to decline. Should we consider the barriers a success? That is, are decreases in apprehensions the result of the barriers? To "prove" this, we must demonstrate three things: First, that barrier construction occurred before apprehensions began their decline. Figure 2.5 shows that apprehensions increased in the years following construction in 1993 and 1994. The bar chart also shows that the decline in apprehensions begin in 2006, the year the Secure Fence Act was passed and before construction of an additional 700 miles of fence began.

Second, there must be a clear association between barrier construction and drop in apprehensions as barriers were put in place. Note that apprehensions decline between 1993 and 1994 but then in subsequent years increase dramatically. The number of apprehensions was already in decline on the eve of the 2006 Secure Fence Act, but fell dramatically over the course of the construction of 700 miles of barriers. The data does suggest that the first 80 or so miles of barriers constructed in 1993 and 1994 had little effect on apprehensions, and since the more recent drop in apprehensions actually began before construction on barriers even started, one wonders whether the 700 miles of fence alone can explain the sharp decline.

Third, to further prove that the border barriers are responsible for the decline must eliminate the possibility of a **spurious correlation**. That means we must show that the association between barrier construction and the decline in apprehensions is not the result of coincidence. In other words, the barrier construction may not actually be the "cause" of the decline. Instead some third factor makes it appear as if the barriers "caused" the decline. In this regard, sociologists consider other reasons that might explain the decline. One possibility is that the ongoing economic recession starting in 2007 and 2008, accompanied by high unemployment, was the real reason apprehensions declined as job opportunities in the United State were less plentiful. Consider that many undocumented migrants work in construction. Might the decline in apprehensions really be explained by the collapse of the housing market? Why immigrate to the United States when there is no work? Second, undocumented workers might have changed the way they enter the United States and the length of time they stay after arriving, giving the appearance that the flow of undocumented workers has subsided. That is, undocumented workers return home

spurious correlation A correlation that is coincidental or accidental because the independent and dependent variables are not actually related; rather, some third variable related to both of them makes it seem as though they are.

less often for fear of getting apprehended; they hire coyotes; they travel alone instead of with a group; and so on. By one estimate, four out of five undocumented immigrants have come to rely on coyotes to help them evade the border patrol or to guide them through unsecured areas or remote desert and mountainous areas (Cornelius 2008).

Third, there is a possibility that border control agencies manipulate data on apprehensions so that it appears they are meeting performance goals. A Government Accountability Office (2009) study showed many instances where Border Patrol reported achieving performance goals but did not have documentation to back up their proclamations. In fact, the study reported "a lack of management oversight, and unclear checkpoint data collection guidance resulted in the overstatement of checkpoint performance results in fiscal year 2007 and 2008 agency performance reports, as well as inconsistent data collection practices at checkpoints" (p. i).

Online Poll

Do you think the United States should build a wall across the entire 2,000-mile border with Mexico?

- ○ Yes
- ○ No
- ○ Don't know

Do you think the dramatic decline in border apprehensions is a good measure of the border barriers' effectiveness at reducing the flow of undocumented workers into the United States?

- ○ Yes
- ○ No, some other factor such as economic recession must be considered.

To see how other students responded to these questions, go to **www.cengagebrain.com**.

Summary of CORE CONCEPTS

In this chapter, we learned that sociological perspectives offer a set of guiding questions and key concepts to explain social issues. Three major perspectives inform the discipline of sociology: functionalist, conflict, and symbolic-interactionist. We used these perspectives to analyze the barriers, walls, and fences along the U.S.–Mexico border. We also used the methods of sociological research to assess whether the border barriers deter undocumented immigration.

CORE CONCEPT 1 Functionalists focus on how the "parts" of society contribute in expected and unexpected ways to maintaining and disrupting an existing social order.

We used the functionalist perspective to understand the border barriers as a "part" of society and to ask what are its anticipated (manifest) and unintended (latent) consequences to the existing social order? The barriers manifest functions include deterring and controlling the flow of undocumented immigration; the barriers' latent functions include providing a volleyball "net" for residents on either side to use during community gatherings. The barriers also have disruptive functions that are manifest and latent. One manifest dysfunction is that the increased border security pushes undocumented immigrants to seek help crossing from coyotes. A latent dysfunction is that many communities have been severed by a barrier.

CORE CONCEPT 2 The conflict perspective focuses on conflict over scarce and valued resources and the strategies dominant groups use to create and protect the social arrangements and practices that give them an advantage in accessing and controlling those resources.

Conflict theorists ask this basic question: Who benefits from a particular social pattern or arrangement? In the case of the barriers, conflict theorists argue that an honest analysis would suggest that American politicians, employers, and military contractors benefit the most at the expense of undocumented immigrants from Mexico. The focus on the Mexican border draws attention away from a global trend to outsource jobs to the lowest-wage markets and makes it appear that something is being done to protect U.S. workers.

CORE CONCEPT 3 Symbolic interactionists focus on social interaction and related concepts of self-awareness/reflexive thinking, symbols, and negotiated order.

Symbolic interactionists focus on social interaction and ask, How do involved parties experience, interpret, influence, and respond to what they and others are doing while interacting? They draw upon the following concepts: (1) reflexive thinking, (2) symbol, and (3) negotiated order to think about any social situation. This perspective encourages researchers to personally immerse themselves in the situation they are studying. Symbolic interactionists observe interactions between border guards and those crossing with an eye toward understanding how undocumented immigrants cross through checkpoints into the United States.

CORE CONCEPT 4 Sociologists adhere to the scientific method; that is, they acquire data through observation and leave it open to verification by others.

When doing research, sociologists explain why their research topic is important, tie their research in with existing research, and specify the core concepts guiding analysis. Sociologists decide on a plan for gathering data, identifying whom or what they will study and how they will select (sample) subjects for study. Sociologists use a variety of data-collection methods, including self-administered questionnaires, interviews, observation, and secondary sources. In addition, they propose and test hypotheses that specify the relationship between independent and dependent variables. In their study of undocumented migration into the United States, Audrey Singer and Douglas Massey followed the scientific method. In his study of the border patrol, so did Robert Lee Maril.

CORE CONCEPT 5 If findings are to matter, researchers must create meaningful operational definitions.

Operational definitions are clear, precise definitions and instructions about how to observe and/or measure the variables under study. They must be both reliable and valid. The operational definition must yield consistent measures and it must measure what it claims to measure. The Department of Homeland Security, for example, is using the number of border apprehensions as one important operational definition of the barriers' effectiveness, but we have learned this measure may not be the best indicator of effectiveness because undocumented immigrants may be simply changing the way they enter the United States. Also keep in mind that applying this measure to the southwest border barriers may take our eye off other legal and undocumented ways of entering the country. Recall that as many as half of the estimated 11.6 million undocumented immigrants in the United States may simply have overstayed their visas, of which 162 million are issued each year.

Resources on the Internet

Login to CengageBrain.com to access the resources your instructor requires. For this book, you can access:

 **Sociology CourseMate**

Access an integrated eBook, chapter-specific interactive learning tools, including flash cards, quizzes, videos, and more in your Sociology CourseMate.

Take a pretest for this chapter and receive a personalized study plan based on your results that will identify the topics you need to review and direct you to online resources to help you master those topics. Then take a posttest to help you determine the concepts you have mastered and what you will need to work on.

CourseReader

CourseReader for Sociology is an online reader providing access to readings, and audio and video selections to accompany your course materials.

Visit **www.cengagebrain.com** to access your account and purchase materials.

Key Terms

concepts 41
dependent variable 46
documents 42
dysfunctions 31
facade of legitimacy 34
function 30
generalizability 48
Hawthorne effect 45
households 42
hypothesis 46
independent variable 46
interviews 44
latent dysfunctions 31
latent functions 31
manifest dysfunctions 31
manifest functions 31
methods of data collection 41
negotiated order 37
nonparticipant observation 45
objectivity 40
observation 44
operational definitions 47
participant observation 45
populations 43
random sample 43
reliability 47
representative sample 43
research 40
research design 41
research methods 40
samples 43
sampling frame 43
scientific method 40
secondary sources (archival data) 46
self-administered questionnaire 43
small groups 42
social interaction 36
spurious correlations 50
structured interview 44
symbol 37
territories 42
traces 42
theoretical perspectives 30
unstructured interview 44
validity 47
variable 46

CULTURE

3 With Emphasis on NORTH AND SOUTH KOREA

Sociologists see people as shaped by their cultural experiences. In the broadest sense of the word, *culture* is the human-created strategies for adapting and responding to one's surroundings. In this chapter, we will consider how the 1953 construction of a strict border zone separating North from South Korea has shaped and still shapes the lives of the Koreans on both sides. This photo offers some insights about how the Korean people's lives have been affected. Here, North Koreans are shown waving good-bye to their South Korean relatives who crossed the border to visit them at the Diamond Mountain Resort in North Korea. That visit marked the first time, since 1953, that they have seen or otherwise communicated with one another. In that year, the Korean War ended and the peninsula was divided into North and South Korea. The moment the division took effect, an estimated five million Koreans were working, living, visiting, and shopping on the "wrong" side and found themselves stuck, never to return home or see relatives stuck on the other side.

Why Focus On NORTH AND SOUTH KOREA?

Today, about 27,000 U.S. military personnel are stationed in South Korea. U.S. military involvement on the Korean Peninsula dates back to the end of World War II, when Premier Joseph Stalin of the Soviet Union, Prime Minister Winston Churchill of Great Britain, and President Franklin Roosevelt of the United States met in 1945 and, "without consulting even one Korean," agreed to chop Korea in half (Kang 1995, p. 75). The Korean War began in 1950, after the North Korean government invaded South Korea. Both sides endured heavy casualties as they fought to control the peninsula. In 1953, the war ended in a stalemate, with the 1945-drawn boundary still in place. A 2.5-mile-wide border of barbed wire and land mines known as the demilitarized zone (DMZ) separates North from South Korea. To the north of the border are 1.2 million North Korean soldiers. To the south are 740,000 South Korean troops, in addition to the U.S. troops.

Over the course of the past 60 years, more than 7.5 million U.S. servicemen and women have fought, died, and otherwise served to maintain the division between the two countries. Today, North Korea possesses a communist-style government and has one of the most isolated and centrally planned economies in the world. South Korea, on the other hand, is a republic, and its economy ranks among the top 15 in the world (U.S. Central Intelligence Agency 2011).

In this chapter, we consider how sociologists think about and describe any culture. We apply the sociological framework to North and South Korea. More specifically, we use this framework to think about the profound effects the Korean War and the subsequent division of the Korean Peninsula into north and south has had on Korean culture, on the meaning of being Korean, and on the relationship between the United States and the two Koreas. This division has affected the life of every North and South Korean resident who lived through the event and who has been born since. It has also affected the lives of millions of U.S. servicemen and servicewomen, their families, and significant others.

FIGURE 3.1

AP Photo/ Korea Pool, Jeon Yeon-ho

Online Poll

Do you have a connection with South Korea? (Check all that apply.)

- ○ No
- ○ Yes, served in military, or a family member served
- ○ Yes, I am/my immediate family is from Korea
- ○ Yes, I claim Korean ancestry
- ○ Yes, I know someone who adopted a Korean baby
- ○ Yes, a person of Korean ancestry married into our family
- ○ Yes, a friend who is Korean
- ○ Yes, other

To see how other students responded to these questions, go to **www.cengagebrain.com.**

The Challenge of Defining Culture

CORE CONCEPT 1 In the most general sense, culture is the way of life of a people.

More specifically, sociologists define **culture** as the human-created strategies for adjusting to their surroundings and to those creatures (including humans) that are part of those surroundings. As we will learn, the list of human-created strategies is endless. These strategies include the automobile as a strategy for transporting people (and their possessions) from one point to another; language as a strategy for communicating with others; and the online community Facebook.com as a strategy for presenting the self to others, learning about others, and building social networks. Culture cannot exist without a **society,** a group of interacting people who share, perpetuate, and create culture.

We use the word *culture* in ways that emphasize differences: "The cultures of X and Y are very different"; "There is a culture gap between X and Y"; "It is culture shock to come from X and live in Y." Our use of the word suggests that we think of culture as having clear boundaries, as an explanation for differences, and as a source of misunderstandings. In light of the seemingly clear way in which we use the word, we may be surprised to learn that sociologists face several conceptual challenges in thinking about culture:

- How do you describe a culture? To put it another way, is it possible to find words to define something so vast as the way of life of a people? What exactly is American or Korean culture?
- How do we know who belongs to a culture? Does a person who appears Korean and who has lived in the United States most of his or her life belong to Korean or American culture? Is everyone who lives in South Korea or North Korea, Korean?
- What are the distinguishing characteristics that set one culture apart from others? Is eating rice for breakfast a behavior that makes someone Korean? Is an ability to speak Korean a characteristic that makes someone Korean? Are Koreans who live in Mexico and speak Spanish Korean?

This chapter offers a framework for thinking about culture that considers both its elusiveness and its importance in shaping human life. A few words of caution are in order before delving into this subject. Although this chapter considers the cultures of South Korea, North Korea, and the United States, we can apply the concepts discussed here to understand any culture and frame other cross-cultural comparisons. As you read about culture, remember that North and South Korea are broadly referred to as countries possessing an Eastern (or Asian) culture and that the United States is regarded as a country possessing a Western culture. Therefore, many of the patterns described here are not necessarily unique to the Koreas or the United States; rather, they are shared with other Eastern or Western societies. At the same time, do not overstate the similarities among countries that share a broad cultural tradition.

Do not assume, for example, that South and North Korea share the same culture. Although they undoubtedly have many similarities, the two nations have very different economic and political systems. "North Korea is a place that is shrouded in mystery and conjecture; . . . for so long it has chosen to close itself off from the rest of the world that little information flows in or out of the place" (Sharp 2005a). Likewise, do not assume that South Korea is just like Japan, for example. As we will see, much of South Korean identity is intricately linked with the idea of being "not Japanese" (Fallows 1988). To assume that South Korea is like Japan is equivalent to assuming that the United States is just like a Western European country, such as Germany. As we know, however, the United States is a country that celebrates its independence from European influence.

Courtesy of Prince Brown, Jr.

Spc. Jason C. Adolphson

Since 1945, millions of American soldiers of all races and ethnic groups, including Korean Americans, have served in South Korea. The top photo shows soldiers serving in South Korea in the 1960s, the bottom photo shows soldiers serving in South Korea today.

culture The way of life of a people; more specifically, the human-created strategies for adjusting to their surroundings and to those creatures (including humans) that are part of those surroundings.

society A group of interacting people who share, perpetuate, and create culture.

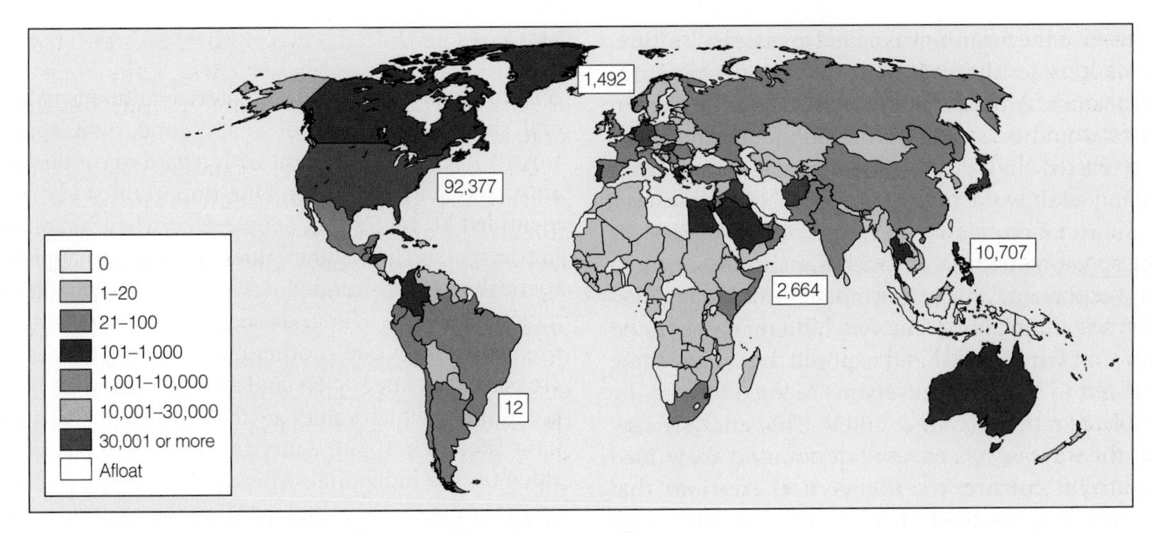

FIGURE 3.2 Number of U.S. Military Personnel by Country
The U.S. military presence in South Korea has shaped the cultures of North and South Korea. The Korean Peninsula is not the only place where the United States has such a presence. Figure 3.2 shows U.S. military personnel are stationed in 140 countries. How might the presence of the U.S. military affect a country's culture?
Source: Data from Department of Defense (2009)

Material and Nonmaterial Components

Culture consists of material and nonmaterial components. **Material culture** consists of all the natural and human-created objects to which people have attached meaning. Material culture includes plants, trees, minerals or ores, dogs, cars, trucks, microwave ovens, computers, video cameras, and iPods. When sociologists think about material culture, they consider the uses to which an item is put and the meanings assigned by the people who use it (Rohner 1984).

Learning the meanings that people assign to material culture helps sociologists grasp the significance of those objects in people's lives. Faucets, showers, tubs, soap, and towels are examples of material culture that people use to cleanse their bodies. Most Americans associate these items with bathrooms, relatively small rooms that offer a private space for washing the body. Although private bathrooms exist in Korea, most Koreans also associate these items with public bathhouses. An American woman visiting Korea described her experience with a bathhouse:

> Looking around, I noticed that all the women were completely naked—at a Korean bath, you check your modesty at the door, and the towel is for scrubbing, not drying or draping. After stripping down, I tentatively stepped through steamy glass doors, into the world of the baths—a large, noisy, cheerful area where about 100 women of all ages and small children of both sexes were scrubbing, chatting, and soaking. To one side were rows of washing stations, with faucets, hand showers, and mirrors set low to the ground. (Koreans, like Japanese, sit while washing.) (McClane 2000)

Two European women visiting South Korea offer similar observations: "The most amazing thing is the range of ages here, from grandmother to babies, all enjoying the same place. . . . It takes a few trips here to get used to walking around naked. . . . And you never see your own grandmother naked in [Great Britain]" (ABC News/Travel 2010).

Department of Defense

The Korean War Memorial in Washington, D.C. is an example of material culture. It includes 19 statues, approximately 7 feet, 3 inches tall: 14 Army members, three Marines, one Navy member, and one Air Force member. The statues represent an ethnic cross section of the men who fought in the Korean War: 12 Caucasians, three African Americans, two Hispanic Americans, one Asian American, and one Native American.

material culture All the natural and human-created objects to which people have attached meaning.

In addition to the meanings assigned to material culture, sociologists consider the ways material culture shapes social relationships. American sociologists studying Korean bathhouses would be struck by the public nature of the bath, the relaxed and casual relationships among nude children and adult women, the lack of self-consciousness, and acceptance of one's own body and others' bodies. As one Western woman who went with her sister-in-law to a bathhouse explained, "She just stripped . . . and doing likewise to her son, didn't notice my very hesitant moves to do the same. . . . I felt so weird and exposed, but at the same time tried not to show it, as everyone seemed to be quite comfortable like that" (Chung 2003). This analysis suggests that the ways towels are used or not used are guided by **nonmaterial culture**, the nonphysical creations that people cannot hold or see. In this case, the Korean women do not define a towel (material culture) as something used to cover themselves because they are influenced by the nonmaterial component of Korean culture, which encompasses beliefs, values, norms, symbols, and language.

Beliefs

Beliefs are conceptions that people accept as true, concerning how the world operates and where the individual fits in relationship to others. Beliefs can be rooted in blind faith, experience, tradition, or in science. Whatever their accuracy or origins, beliefs can exert powerful influences on actions as they are used to justify behavior, ranging from the most generous to the most violent. Some beliefs follow:

- Anyone who wants to can grow up to become president of the United States.
- Ongoing conversation, rather than silence, validates a relationship.
- Athletic talent is something you are born with.
- Anyone can develop athletic talent if they work hard, practice, and persist.
- It is fine for young children of both sexes to bathe with their mothers, grandmothers, and other women in a public bathhouse.

nonmaterial culture Intangible human creations, which we cannot identify directly through the senses.

beliefs Conceptions that people accept as true, concerning how the world operates and where the individual fits in relationship to others.

values General, shared conceptions of what is good, right, appropriate, worthwhile, and important with regard to conduct, appearance, and states of being.

norms Written and unwritten rules that specify behaviors appropriate and inappropriate to a particular social situation.

folkways Norms that apply to the mundane aspects or details of daily life.

Values

A second component of nonmaterial culture is **values**: general, shared conceptions of what is good, right, appropriate, worthwhile, and important with regard to conduct, appearance, and states of being. One important study on values identified 36 values that people everywhere share to differing degrees, including the values of freedom, happiness, true friendship, broad-mindedness, cleanliness, obedience, and national security. The study suggested that societies are distinguished from one another not according to which values are present in one society and absent in another, but rather, according to which values are the most cherished and dominant (Rokeach 1973). Americans, for example, place high value on the individual, whereas Koreans place high value on the group. These values manifest themselves in the American preference to bathe alone and the Korean preference to share the experience with others in public bathhouses.

Sports offer some insights about a culture's values. The national sport of South Korea is tae kwon do. That sport values physical power when it is used in self-defense and only in an amount necessary to gain control over an aggressor. Tae kwon do athletes also value freedom, justice, and using power to build a better world. By contrast, football—arguably the national sport of the United States—places a high value on aggression. Hard hits to opponents are highly valued and replayed as game highlights. The object of the American football game is to advance the ball into "enemy" territory and score by invading an opponent's end zone.

Norms

A third component of nonmaterial culture is **norms**, written and unwritten rules that specify behaviors appropriate and inappropriate to a particular social situation. Examples of written norms are rules that appear in college student handbooks, on signs in restaurants (No Smoking Section), and on garage doors of automobile repair centers (Honk Horn to Open). Unwritten norms exist for virtually every kind of situation: wash your hands before preparing food; do not hold hands with a friend of the same sex in public; leave at least a 20 percent tip for waiters; remove your shoes before entering the house. Some norms are considered more important than others, and so the penalties for their violation are more severe. Depending on the importance of a norm, punishment can range from a frown to death. In this regard, we can distinguish between folkways and morés.

Folkways are norms that apply to the mundane aspects or details of daily life: when and what to eat, how to greet someone, how long the workday should be, how many times caregivers should change babies' diapers each day. As sociologist William Graham Sumner (1907) noted, "Folkways give us discipline and support of routine and habit"; if we were forced constantly to make decisions

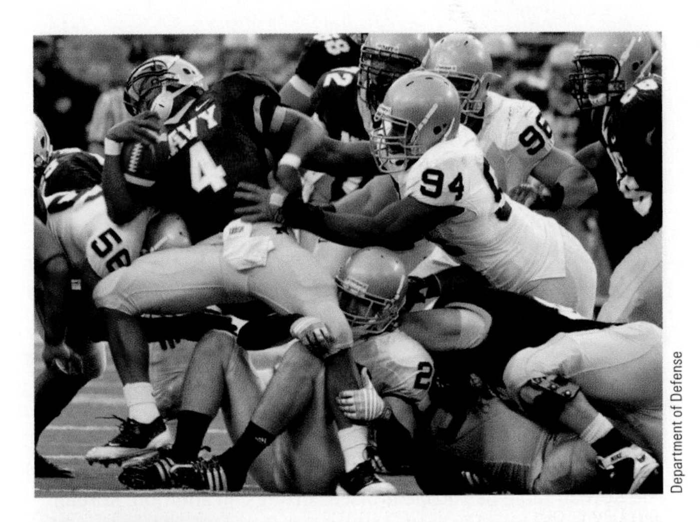

We can gain insights about what values dominate in a society when we examine its most popular, celebrated sports. The photo from a U.S. Naval Academy football game suggests aggressive and hard-hitting play is valued. The photo of South Korean marines demonstrating tae kwon do skills suggests an appreciation for skills that can be used to defend oneself from an aggressor.

about these details, "the burden would be unbearable" (p. 92). Generally, we go about everyday life without asking why until something reminds us or forces us to see that other ways are possible.

Consider the folkways that govern how a meal is typically eaten at Korean and American dinner tables. In Korea, diners do not pass items to one another, except to small children. Instead, they reach and stretch across one another and use their chopsticks to lift small portions from serving bowls to individual rice bowls or directly to their mouths. The Korean norms of table etiquette—reaching across instead of passing, having no clear place settings, and using the same utensils to eat and serve oneself food from platters and bowls—deemphasize the individual and reinforce the greater importance of the group.

Americans follow different dining folkways. They have individual place settings, marked clearly by place mats or blocked off by eating utensils. It is considered impolite to reach across another person's space and to use personal utensils to take food from the communal serving bowls. Instead, diners pass items around the table and use special serving utensils. That Americans have clearly marked eating spaces, do not typically trespass into other diners' spaces, and use separate utensils to take food reinforces values about the importance of the individual.

Often, travel guides and cultural guides list folkways that foreign travelers should follow when visiting a particular country. The U.S. military introduced a new housing policy in February 2010 that allows military families to live off base in housing complexes with Korean neighbors. In preparation, the military created culturally oriented videos to educate Americans about Korean folkways, pointing out that "we are guests" in Korea and not to expect Koreans to make concessions to American ways of doing things. The videos titled "Being a Good Neighbor" describe a number of Korean folkways:

- Dogs in Korea are likely tiny things that you can carry in your pocket so big dogs are a new phenomenon, especially for children. Keep your dogs on a leash.
- Barbeques are a no-no. Koreans see them as fire hazards, and the smoke is a nuisance.
- Recycling is very important in Korea. It is a way of life enforced by law, and citizens pay for those items that cannot be recycled. (2nd Infantry Division 2010, CNN 2010)

Morés are norms that people define as essential to the well-being of a group. People who violate morés are usually punished severely: They may be ostracized, institutionalized in prisons or mental hospitals, sentenced to physical punishment, or condemned to die. In contrast to folkways, morés are regarded as "the only way" or "the truth" and as thus are unchangeable. Most Americans, for example, have strong morés against public nudity, especially when adults

morés Norms that people define as critical to the well-being of a group. Violation of morés can result in severe forms of punishment.

are in the presence of children. They believe that children should be shielded from seeing adults without clothes and that children are more vulnerable when naked in the presence of adults who are also nude. Koreans, on the other hand, do not view public nakedness in the right context to be morally wrong or as a danger to children. Instead, they view the body as something to be accepted for what it is. Americans who visit Korean bathhouses report, to their surprise, that they adjust quickly to social nudity and come to see being naked among others as unremarkable.

Symbols

We learned in Chapter 2 that symbols are any kind of physical or conceptual phenomenon—a word, an object, a sound, a feeling, an odor, a gesture or bodily movement, or a concept of time—to which people assign a name and a meaning. The meaning assigned is not evident from the physical phenomenon or idea alone. For example, what do the numbers 2-0-1-5 and 1-0-0, taken on their own, mean? Many societies, Christian and non-Christian, locate themselves in time by referencing the birth of Jesus Christ. Thus, "AD 2015" symbolizes 2,015 years since the traditionally recognized birth of Christ. AD is the abbreviation for *anno Domini* (Latin for "in the year of the Lord"). In North Korea, people locate themselves in time by referencing the year Kim Il Sung, the country's founding and "eternal" president, was born. At the time of this writing, the numbers 1-0-0 symbolize 100 years since his birth in 1912, with 1912 being year 1.

Language

In the broadest sense of the word, **language** is a symbol system involving the use of sounds, gestures (signing), and/or characters (such as letters or pictures) to convey meaning. When people learn language, they learn a symbol system. Those learning spoken languages must learn the agreed-upon sounds that convey words, and they must learn the rules that specify relationships among the chosen words. That is, we cannot convey ideas by vocalizing the relevant words in any order we choose. As the author of this textbook, I often ask my students, "Are you reading the book?" I cannot expect students to understand this question if I say the words in some random order like "Are book reading you?" English-language speakers follow a subject-verb-object order (We are reading the book). Koreans, on the other hand, follow a subject-object-verb format (We book are reading). As another example, rules governing word order apply to stating first and last names. Consider that Koreans tend to identify themselves by stating their family name first and then their given name. In effect, the family name is given precedence over the individual's first name.

language A symbol system involving the use of sounds, gestures (signing), and/or characters (such as letters or pictures) to convey meaning.

C. Todd Lopez

The Korean table has no clear place settings. In addition, Korean diners use the same utensils for serving and eating. Most people living in the United States prefer individual place settings marked clearly by place mats or blocked off by eating utensils.

Learning a language and its rules is the key to human development and interaction. The level and complexity of human language sets people apart from the other animals. In addition, language is among the most important social institutions humans have created. That is, language is a predictable social arrangement among people that has emerged over time to facilitate human interaction and communication.

The Role of Geographic and Historical Forces

CORE CONCEPT 2 Geographic and historical forces shape culture.

Sociologists operate under the assumption that culture acts as a mediator between people and their surroundings (Herskovits 1948). Thus, material and nonmaterial culture represent responses to historical and geographic challenges. Note that cultural responses are not always constructive and may have unintended consequences. The 1953 division of the Korean Peninsula into North and South Korea was a response to a historical event that affected the personal lives and culture of Koreans on both sides of the DMZ. It was a geographic event in that the division confined some Koreans to the north of the line and others to the south. For the most part, people in the two Koreas have not communicated or interacted with each other since 1953, effectively evolving into two separate cultures. To date about 20,000 North and South Koreans have reunited briefly, and an estimated 80,000 Koreans, now in their mid-80s, are on waiting lists to see relatives. Some South Koreans travel to newly opened resorts in North Korea that are isolated from that county's population

This map on a highway memorial sign in South Dakota shows the Korean Peninsula and the boundary known as the DMZ (demilitarized zone) that has separated the two societies since 1953. It is considered the most militarized border in the world.

(Korean Overseas Information Service 2006). About 100 South Korean firms, part of a special industrial zone, employ tens of thousands of North Koreans workers (CNBC 2010).

Although the division of the Korean peninsula was one key event that shaped North and South Korean cultures, one important historical event that has shaped American culture—especially the way Americans think about and use energy—was the Spindletop, a Texas oil gusher of 1901. That discovery made the United States the largest producer of oil at the time and the most powerful nation in the world. "Oil was the new currency of the industrialized world, and America was rich. . . . Few Americans realized that their country was different or particularly fortunate. . . . They soon began to take their subterranean wealth for granted. . . . People in other industrialized nations were more aware of America's blessing. Being less sure of their sources of energy, they were warier about its dispensation. America quickly turned its industrial plant and its electrical grids over to oil" (Halberstam 1986, p. 87).

In contrast to the United States, North and South Korea produce no oil and consequently must import all their oil. Until the mid-1970s, the United States produced all the oil it consumed. It still produces 50 percent of the oil it consumes (about 9 million barrels a day). This history of abundance shaped the way Americans use appliances. For example, South Koreans, North Koreans, and Americans use appliances such as refrigerators differently. Consider that South Koreans try to minimize the amount of electricity the refrigerator uses to keep food cold by opening the refrigerator

North Korea

Size of military. 1.17 million (active); 9.7 million (reserves)

Total Annual military expenditures. $5.5 billion

Per capita military spending. $230.9

Military expenditures (percentage of GDP). 33.9%

South Korea

Size of military. 1.1 million (active); 8.2 million (reserves)

Total Annual military expenditures. $27.1 billion

Per capita military spending. $557

Military expenditures (percentage of GDP). 2.8%

United States

Size of forces stationed in South Korea. 26,339

Total Annual military expenditures for Korean operation (est.). $3 + billion

FIGURE 3.3 The most heavily armed border in the world is the demilitarized zone that divides North and South Korea. Consider as one indicator of the costs of maintaining this division is that North Korea spends about 33.9 percent of its gross domestic product (GDP) on its military, whereas South Korea spends $27.1 billion, representing 2.8 percent of GDP. The United States spends about $3 billion each year to station 26,339 troops in South Korea.

Sources: Data from U.S. Central Intelligence Agency (2010), Global Issues (2010), Department of Defense (2008)

Thomas the Train, the creation of a British clergyman for his son, first appeared in a 1946 book, *Thomas the Tank Engine*. In 1979, a British writer and producer turned the book into a successful TV series, *Thomas and Friends*, that eventually became a worldwide hit (Just Thomas 2011). This Thomas the Train pictured is part of the Lotus Lantern Festival in South Korea. Notice that the Koreans borrowed Thomas and turned him into a lantern. They also added a child conductor holding a lotus flower. The festival celebrates the birth of Buddha.

learn that male circumcision in South Korea can be traced to contact with the U.S. military during the Korean War. Koreans, however, depart from the American practice of circumcising male babies at birth. In fact, only one percent of South Korean babies are believed to be circumcised at birth; most circumcisions occur during the elementary and middle school years (Ku et al. 2003).

In contrast to South Korea, the North Korean government limits cultural diffusion opportunities by restricting access to information from the outside world. With rare exceptions, the 22.5 million people of North Korea cannot receive mail or telephone calls from outside the country. Nor can they travel beyond their country's borders (Brooke 2003a). One North Korean defector claims that "all the tape recorders and radios have to be registered. At registration, they cut off and solder the tuning dial to make sure you don't have a 'free' radio" (Brooke 2003b, p. A8). In addition, North Korean officials confiscate mobile phones from visitors while they are in the country, assign official escorts, and inspect and delete some photographs from cameras upon departure (Branigan 2010). Although the North Korean government restricts cultural diffusion, North Korean people, especially those living along the border with China, find ways to acquire illicit radios, mobile phones, CD players, stereos, and televisions, which offer access to a world beyond North Korea (Mac-Kinnon 2005; Caryl and Lee 2006). In the past five years, North Korea has issued visas to U.S. and other foreign peoples to attend important festivals and celebrations (BBC News 2005). Foreign reporters covering these events learned, to their surprise, that the *Da Vinci Code* was a hit and Celine Dion songs are popular (Branigan 2010).

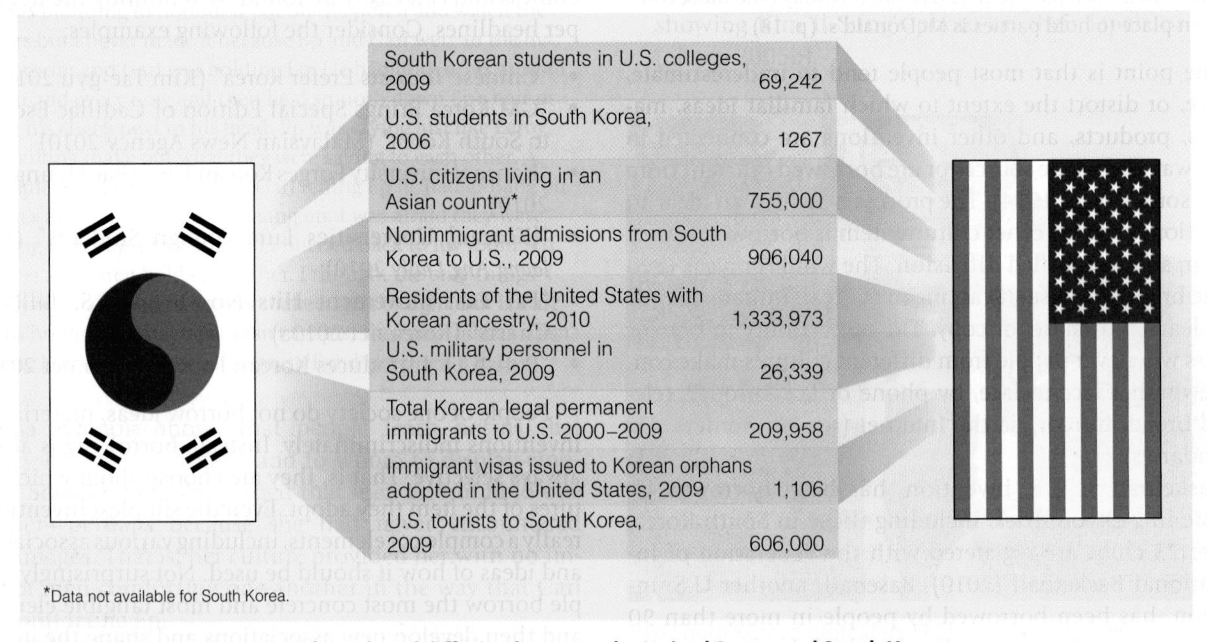

South Korean students in U.S. colleges, 2009 69,242
U.S. students in South Korea, 2006 1267
U.S. citizens living in an Asian country* 755,000
Nonimmigrant admissions from South Korea to U.S., 2009 906,040
Residents of the United States with Korean ancestry, 2010 1,333,973
U.S. military personnel in South Korea, 2009 26,339
Total Korean legal permanent immigrants to U.S. 2000–2009 209,958
Immigrant visas issued to Korean orphans adopted in the United States, 2009 1,106
U.S. tourists to South Korea, 2009 606,000
*Data not available for South Korea.

FIGURE 3.4 Opportunities for cultural diffusion between the United States and South Korea
There are many opportunities for Americans in Korea to learn and borrow Korean ways as evidence by the presence of U.S. troops in South Korea, international trade, and the departure and return of many Koreans who have attended colleges and universities in the United States. Notice, for example, that almost 70,000 South Koreans are enrolled in U.S. colleges and universities.

Source: Data from U.S. Bureau of the Census (2010), U.S. Central Intelligence Agency (2011); U.S. Department of Commerce (2010); U.S. Department of Defense (2008); U.S. Department of Homeland Security (2010); Association of American Residents Overseas (2011)

No Borders No Boundaries K-Pop/Korean Pop Music

Sociologists define *popular culture* as any aspect of culture that is embraced by the masses within and outside of that society from which it is believed to originate. Examples include a sandwich (such as the Big Mac), a doll (such as Barbie), a television show (such as *Desperate Housewives*), a book (*Seven Habits of Highly Effective People*), a movie (such as *Harry Potter and the Deathly Hallows*), an item of clothing (such as blue jeans), and a phrase (such as "She's so fly"). Any analysis of popular culture must consider the industries that sell and market it, including the ways by which it reaches the masses: whether it be via commercials, television programs, radio, You-Tube or newspapers—to name a few. K-pop or Korean pop music is making a sustained debut on the global market. At the time of this writing, some of the best -known South Korean groups were Big Bang, TVXQ, 2PM, and Girls' Generation; a Korean American group Far East Movement (for example, "Like a G6") was also popular. In response to the growing popularity of Korean pop music, two Korean Americans—Johnny Noh and Paul Han—launched a celebrity news website Allkpop.com in 2007, to "help spread this now global phenomenon." Their website reportedly attracts 2 million visitors each month. About 40 percent of the visits are from K-pop fans based in the United States, with 10 percent from fans based in Canada and 10 percent from Singapore. Noh and Han believe that "Korea has the idol formula down pat; they are very polished in their mannerisms on stage and in society. Fans are able to fall in love not only with artists' music, but their personalities as well" (Garcia 2010).

Girls' Generation, a nine-member female pop group, can introduce themselves in four languages—Korean, English, Japanese, and Chinese. The group believes that learning foreign languages is the key to reaching audiences outside of South Korea (Wall Street Journal 2010). "We've been really focusing on the language these days, especially since we want to be able to connect with our foreign fans. It takes time, patience, and a lot of practice but we really wanna be able to express ourselves to our foreign fans." These nine females seek to circumvent the U.S. and European music markets, by unifying the Asian market and building it into "the biggest market in the world" (Allkpop 2010).

Kyodo/Newscom

The Home Culture as the Standard

CORE CONCEPT 6 The home culture is usually the standard that people use to make judgments about another culture.

Most people come to learn and accept the ways of their culture as natural. When they encounter foreign cultures, therefore, they can experience mental and physical strain that comes with reorienting to new ways of thinking and behaving. Sociologists use the term **culture shock** to describe that strain. In particular, they must adjust to a new language and to the idea that the behaviors and responses they learned in their home culture, and have come to take for granted, do not apply in the foreign setting. The intensity of culture shock depends on several factors: (1) the extent to which the home and foreign cultures differ, (2) the level of preparation for or knowledge about the new culture, and (3) the circumstances (such as vacation, job transfer, or war) surrounding the encounter. Some cases of culture shock are so intense and unsettling that people become ill. Among the symptoms are "obsessive concern with cleanliness, depression, compulsive eating and drinking, excessive sleeping, irritability, lack of self-confidence, fits of weeping, nausea" (Lamb 1987, p. 270).

Minor league outfielder for the Charlotte Stone Crabs, Kyeong Kang experienced culture shock when he moved with his family from South Korea to Norcross, Georgia, when he was 14 years old. His family moved to the United States so he could play baseball in high school. Kang experienced culture shock because when he first came to the

culture shock The strain that people from one culture experience when they must reorient themselves to the ways of a new culture.

More specifically, it is the process by which humans (1) acquire a sense of self or social identity, (2) learn about the social groups to which they belong and do not belong, (3) develop their human capacities, and (4) learn to negotiate the social and physical environment they have inherited. Socialization is a lifelong process, beginning at birth and ending at death. It takes hold through **internalization,** the process in which people take as their own and accept as binding the norms, values, beliefs, and language that their socializers are attempting to pass on.

No discussion of socialization can ignore the importance of two factors: nature and nurture. **Nature** comprises one's human genetic makeup or biological inheritance. **Nurture** refers to social experiences that make up every individual's life. Some scientists debate the relative importance of genes and the social experiences, arguing that one is ultimately more important than the other. Such a debate is futile, because it is impossible to separate the influence of the two factors or to say that one is more forceful. Both nature and nurture are essential to socialization (Ornstein and Thompson 1984).

The relationship between the cerebral cortex and spoken language illustrates rather dramatically the inseparable qualities of nature and nurture. As part of our human genetic makeup (nature), we possess a cerebral cortex, which allows us to organize, remember, communicate, understand, and create. Scientists believe that humans inherit a cerebral cortex "setup" to learn any of the more than 6,000 known human languages. In the first months of life, all babies are biologically capable of babbling the essential sounds needed to speak any language. As children grow, this enormous potential is reduced, however, by their social experiences (nurture) with language. Most obviously, babies eventually learn the language (or languages) they hear spoken. For the most part, Palestinian babies hear standard Arabic spoken at home and learn Hebrew in school. Israeli babies, for the most part, hear modern Hebrew spoken at home. Because Israel is a land of immigrants, many Jewish children are exposed to Russian, Yiddish, Ladino, or Romanian as babies, but they all must eventually learn Hebrew. Most Israeli Jews do not learn Arabic.

Although humans have a biological makeup that allows them to speak a language, the language itself is learned through interactions with others. If babies are not exposed to language, they will not acquire that communication tool. Though social interaction is essential to language development, it is also essential to human development in general. If babies are deprived of social contact with others, they cannot become normally functioning human beings.

Father Rob Waller, Courtesy of Joan Ferrante

Cherie A. Thurlby

Millions of social experiences culminate to create a sense of self. These Israeli children's sense of self develops through interaction experiences that may include celebrations of Jewish cultural tradition. Social experiences can also include military training that leads one to self-identify as an Israeli soldier.

The Importance of Social Contact

CORE CONCEPT 2 Socialization depends on meaningful interaction experiences with others.

Cases of children raised in extreme isolation or in restrictive and sterile environments show the importance of

socialization The process by which people develop a sense of self and learn the ways of the society in which they live.

internalization The process in which people take as their own and accept as binding the norms, values, beliefs, and language that their socializers are attempting to pass on.

nature Human genetic makeup or biological inheritance.

nurture The social environment, or the interaction experiences that make up every individual's life.

social contact (nurture) to normal development. Some of the earliest and most systematic work in this area was done by sociologist Kingsley Davis. His work on the consequences of extreme isolation demonstrates how neglect and lack of social contact influence emotional, mental, and even physical development.

Davis (1940, 1947) documented and compared the separate yet similar lives of two girls: Anna and Isabelle. During the first six years of their lives, the girls received only the minimum of human care. Both children, living in the United States in the 1940s, were classified as illegitimate. Because of that status, both were rejected and forced into seclusion; each was living in a dark, attic-like room. Anna was shut off from her family and their daily activities. Isabelle was shut off in a dark room with her deaf-mute mother. Both girls were 6 years old when authorities intervened. At that time, they exhibited behavior comparable to that of 6-month-old children.

Anna "had no glimmering of speech, absolutely no ability to walk, no sense of gesture, not the least capacity to feed herself even when food was put in front of her, and no comprehension of cleanliness. She was so apathetic that it was hard to tell whether or not she could hear" (Davis 1947, p. 434). Anna was placed in a private home for mentally disabled children until she died four years later. At the time of her death, she behaved and thought at the level of a 2-year-old child.

Like Anna, Isabelle had not developed speech, but she did use gestures and croaks to communicate. Because of a lack of sunshine and a poor diet, she had developed rickets: "Her legs in particular were affected; they 'were so bowed that as she stood erect the soles of her shoes came nearly flat together, and she got about with a skittering gait'" (Davis 1947, p. 436). She also exhibited extreme fear of and hostility toward strangers. Isabelle entered into an intensive and systematic program designed to help her master speech, reading, and other important skills. After two years in the program, she had achieved a level of thought and behavior normal for someone her age. Isabelle's success may be partly attributed to her establishing an important bond with her deaf-mute mother, who taught her how to communicate through gestures and croaks. Although the bond was less than ideal, it still gave her an advantage over Anna.

Cases of Less Extreme Isolation

Other evidence of the importance of social contact comes from less extreme cases of neglect. Psychiatrist Rene Spitz (1951) studied 91 infants who were raised by their parents during their first three to four months of life but who were later placed in orphanages. When they were admitted to the orphanages, the infants were physically and emotionally normal. Orphanage staff provided adequate care for their bodily needs—good food, clothing, diaper changes, clean nurseries—but gave the children little personal attention. Because only one nurse was available for every 8 to 12 children, the children were starved emotionally. The emotional starvation caused by the lack of social contact resulted in such rapid physical and developmental deterioration that a significant number of the children died. Others became completely passive, lying on their backs in their cots. Many were unable to stand, walk, or talk (Spitz 1951).

Shannon K. Cassidy

A person's overall well-being depends on meaningful interaction experiences with others. Social interaction is essential to developing and maintaining a sense of self.

Children of the Holocaust

Anna Freud and Sophie Dann (1958) studied six German Jewish children whose parents had been killed in the gas chambers of Nazi Germany. The children were shuttled from one foster home to another for a year before being sent to the ward for motherless children at the Terezin concentration camp. The ward was staffed by malnourished and overworked nurses, who were themselves concentration camp inmates. After the war, the six children were housed in three different institution-like environments. Eventually they were sent to a country cottage, where they received intensive social and emotional care.

During their short lives, these six children had been deprived of stable emotional ties and relationships with caring adults. Freud and Dann found that the children were ignorant of the meaning of family and grew excessively upset when they were separated from one another, even for a few seconds. In addition, they "behaved in a wild, restless, and uncontrollably noisy manner":

> During the first days after their arrival, they destroyed all the toys and damaged much of the furniture. Toward the staff they behaved either with cold indifference or with active hostility, making no exception for the young assistant Maureen who had accompanied them from Windermere and was their only link with the immediate past. At times, they

The Palestinian-Israeli conflict has involved six wars (1948, 1956, 1967, 1968–1971, 1982, and 2006), two major intifadas (uprisings), countless clashes and has spilled over into Lebanon on many occasions. Here we consider some of the major issues that have yet to be addressed and resolved.

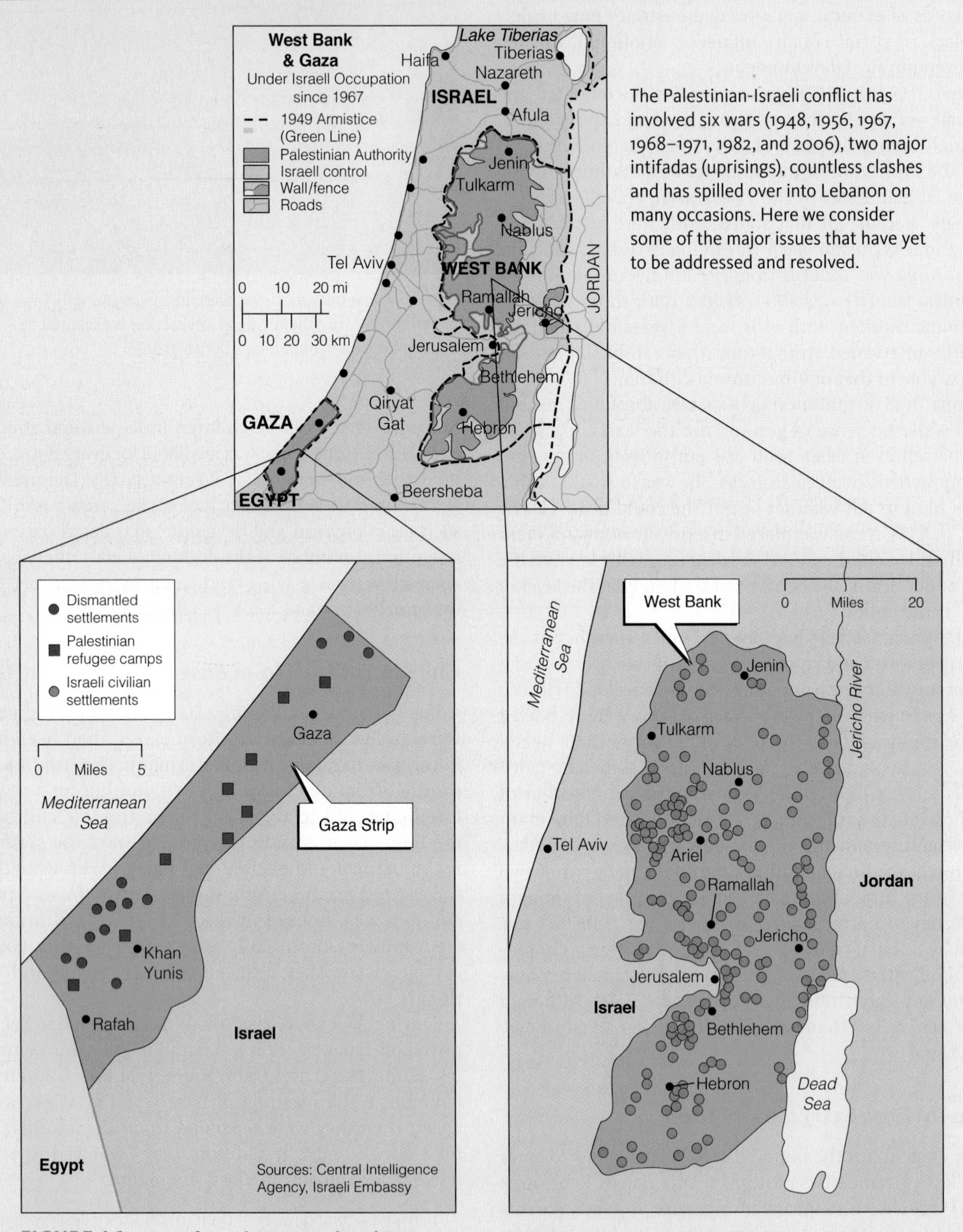

FIGURE 4.1 Maps of Israel, West Bank and Gaza
The top map show the state of Israel and the location of the Gaza Strip and West Bank. The lower maps show where Israeli settlements are located in the West Bank and the settlements that have been dismantled in the Gaza Strip.
Sources: Central Intelligence Agency, Israeli Embassy

The United States and other parties involved in the peace process have proposed a two-state solution to end the conflict. A two-state solution means that Israel remains a Jewish state and a second for Palestinians would be established and recognized. Specifically, the Palestinian territories of the West Bank and Gaza—two geographically disconnected lands—would become the new Palestinian state. The Palestinians living in Israel could opt to remain in Israel as citizens or become citizens of Palestine.

Before the two-state solution can be implemented, a number of critical issues must be resolved:

1. Israeli settlements in the Palestinian territories. Settlements are Jewish-populated communities in what are now the Palestinian Territories. These settlements are diverse in structure, ranging from outposts composed of trailers, campers, and tents to self-contained towns and cities with populations of 10,000 or more. An estimated 325 such settlements house 500,000 Jewish residents (Financial Times 2010). In 2005, the Israeli government evacuated settlers from 21 settlements in the Gaza Strip and from four West Bank settlements. Notwithstanding the recent evacuation, critics argue that the settlements are attempts to establish a significant Jewish presence on Palestinian land so that a permanent solution that gives Palestinian control over this land can never be achieved.
2. Safe passage between Gaza and the West Bank. If Gaza and the West Bank are eventually to be regarded as one state, the pressing question becomes who can control the access roads from one territory to the other? Palestinians? Israelis? A joint force? Currently, Israelis control every access route for moving goods and people into and out of the Palestinian territories.
3. Right of return. The creation of the state of Israel and the subsequent 1948 and 1967 Arab–Israeli wars resulted in Palestinian diasporas, forced scatterings of an ethnic population to various locations around the world. Several million Palestinians immigrated to surrounding countries and now live with their descendants in refugee camps in Jordan (estimated at 1.98 million); the Persian Gulf countries of Kuwait, Saudi Arabia, United Arab Emirates, Qatar (3,711,000), and Iraq (est. 450,000); Lebanon (est. 350,000); Syria (est. 340,000); and elsewhere. Approximately 1.4 million of 4.8 million Palestinians are registered with the UN as refugees. Refugees seek the right to return to the land within Israel from which they fled or were evicted (U.S. Central Intelligence Agency 2010; Bennet 2003b). From the Israeli point of view, if refugees were allowed to return, they would be overwhelmed by the numbers and cease to exist.
4. Status of Jerusalem. Both Palestinians and Israeli Jews claim Jerusalem, which is divided into East and West Jerusalem, as their capital. Under a peace agreement, the Israeli government is to be awarded West Jerusalem and the Palestinians East Jerusalem. But the Israeli government insists that the entire city is its capital. An estimated 200,000 Israelis live in settlements located in East Jerusalem.
5. The Wall. In June 2002, the Israeli government, with the support of 83 percent of Israelis, began construction on the West Bank barrier—a 350-kilometer-long obstacle comprising electrified fencing, razor wire, trenches, concrete walls, and guard towers that wind through the West Bank. The wall puts 14 percent of the West Bank on the Israeli side; at one point, the wall extends some 13 miles into the West Bank (Farnsworth 2004). The wall separates the West Bank from Israel and channels Palestinian movement from the West Bank into Israel through checkpoints. Israel claims that the wall is not a political border but rather a security border designed to keep suicide bombers and other would-be attackers out of Israel and Jewish settlements. The UN estimates that the barrier has disrupted 600,000 Palestinian lives. For example, the wall is constructed so that it completely surrounds 12 Palestinian communities, allowing residents to leave only through Israeli-controlled checkpoints (Myre 2003).
6. The Fatah-Hamas divide. The political party Fatah, founded in 1958, renounced terrorism against Israel in 1988, and acknowledged Israel's right to exist in 1993. Hamas, founded in 1987, is labeled a terrorist organization by Israel, the United States, and the European Union. In 2006, Hamas won 74 of 132 elected seats in democratically held elections, giving it the majority in Palestinian Legislative Council. Hamas (backed by Syria and Iran) took control of the Gaza Strip after five days of civil war. Fatah (backed by the United States and Jordan) remains in control of the West Bank. Israelis refused to recognize Hamas as the legitimate leader of the Palestinian people. After a series of military exchanges between Palestinian hard-liners in Gaza and Israeli forces, Israel sealed off its border with Gaza and prohibited all but the most basic necessities from entering. Egypt also sealed off its border with Gaza, leaving the residents of Gaza imprisoned within. This Fatah-Hamas division has jeopardized the two-state solution and further restricted Palestinian movement between the West Bank and Gaza.

Note, as we read about the Israeli–Palestinian conflict, we will refer to Palestinians who live in Israel as Israeli Palestinians, Jews who live in Israel as Israeli Jews, and Israelis who live in West Bank and East Jerusalem as Israeli settlers.

When exposed to difficult situations, the family can buffer or exacerbate the impact on family members. This family's home has been destroyed by floods. A parent's calm and determined response can offset the potential traumatic effects.

A military unit is a primary group. During military training, soldiers become so close that they fight for one another, rather than for the victory per se, in the heat of the battle.

depression induced by long-term exposure to attacks (Science Daily 2008).

As these cases illustrate, even under extremely stressful circumstances, such as war, the family can teach responses that increase or decrease that stress. Clearly, children in families that emphasize constructive responses to stressful events have an advantage over children whose parents respond in destructive ways.

Like the family, a military unit is a primary group. A unit's success in battle depends on the existence of strong ties among its members. Soldiers in this primary group become so close that they fight for one another, rather than for victory per se, in the heat of battle (Dyer 1985). In Israel, the military represents a place where immigrants from almost 100 national and cultural backgrounds become Israelis, bonding with one another and with native Israelis (Rowley 1998). Military units train their recruits always to think of the group before themselves. In fact, the paramount goal of military training is to make individuals feel inseparable from their unit. Some common strategies employed to achieve this goal include ordering recruits to wear uniforms, to shave their heads, to march in unison, to sleep and eat together, to live in isolation from the larger society, and to perform tasks that require the successful participation of all unit members. If one member fails, the entire unit fails. Another key strategy is to focus the unit's attention on fighting together against a common enemy. An external enemy gives a group a singular direction, thereby increasing its internal cohesiveness.

The Israeli military is an important agent of socialization. Almost every Israeli can claim membership in this type of primary group because virtually every Israeli citizen—male and female—serves in the military (three years for men and about two years for women). Men must serve on active duty for at least one month every year until they are 51 years old (U.S. Central Intelligence Agency 2011).

Similarly, military training is an important experience for many Palestinians. Palestinian youth—especially those living in Syrian, Lebanese, Egyptian, and Jordanian refugee camps (which are outside Israeli control)—join youth clubs and train to protect the camps from attack. The focus on defeating a common enemy helps establish and maintain the boundaries of the military unit. All types of primary groups have boundaries—a sense of who is in the group and who is outside it. The concepts of in-group and out-group help us understand these dynamics.

In-groups and Out-groups

A group distinguishes itself by the symbolic and physical boundaries its members establish to set it apart from nonmembers. Examples of physical boundaries may be gated communities, walls to separate one neighborhood or country from another, special buildings such as prisons or churches, or other distinct geographic settings. Symbolic boundaries include membership cards, colors, or dress codes such as a uniform. A group also distinguishes itself by establishing criteria for membership including a certain ancestry, a particular physical trait, or some accomplishment (for example, a specified level of education, passing a test, qualifying for a team, paying dues, or buying certain attire). The boundaries and membership criteria are ways groups distinguish "us" from "them."

Sociologists use the terms in-group and out-group in reference to intergroup dynamics. An **in-group** is the group to which a person belongs, identifies, admires, and/or feels loyalty. An **out-group** is any group to which a person does not belong. Obviously, one person's in-group is another person's out-group. In-group formation is built on establishing boundaries and membership criteria. In-group members think of themselves as "us" in relation to some specific or even amorphous "them" (Brewer 1999). Examples of specific in-group/out-group pairings include rival gangs such as the Bloods and the Crips or the teams formed for a reality television show such as *Survivor.* More amorphous in-group/out-group pairings include Christians and Muslims, Jews and Palestinians, Fatah and Hamas, Steelers versus Bengals fans, and the far left and far right. Amorphous means that members are very different. In-group and out-group dynamics are such that very different people are bound together because they share something very important to their identity such as a political agenda or religious affiliation.

When sociologists study in-group and out-group dynamics, they ask, "Under what circumstances does the presence of an out-group unify an in-group and create an 'us' versus 'them' dynamic?" Three circumstances are described.

1. An in-group assumes a position of moral superiority over an out-group. Moral superiority is the belief that an in-group's standards represent the only way. In fact, there is no room for negotiation and no tolerance for other ways (Brewer 1999). Moral superiority can express itself in a variety of ways, including refusing to interact with or show interest in anyone in an out-group, establishing laws segregating an in-group from an out-group, or engaging in violence toward an out-group.
2. An in-group perceives an out-group as a threat. In this situation, in-group members believe (rightly or wrongly) that an out-group threatens its way of life. The in-group holds real or imagined fear that the out-group is seeking
    - political power,
    - control of scarce and valued resources,
    - a larger share of the "pie,"
    - retribution for past wrongs, or
    - return of lost power.
3. In-group/out-group tensions may be evoked for political gain. Those with political ambitions may deliberately evoke in-group/out-group tensions as a strategy for mobilizing support for some political purpose (Brewer 1999). Thus, a candidate running for elected office may declare an out-group such as union-busting conservatives, undocumented workers, or gays a threat to the American way of life as a strategy for rallying support.

One can argue that the presence of Palestinians functions to unify Israeli Jews, who are culturally, linguistically, religiously, and politically diverse. Since Israel's founding, Jews from 102 different countries, speaking 80 different languages, have settled there (Peres 1998). Twenty-two percent were born in Europe or the United States. Almost 10 percent were born in an African or Asian country (U.S. Central Intelligence Agency 2011). These diverse Israeli Jews share a common language (Hebrew), the desire for a homeland free of persecution, and an ongoing conflict with the following: Palestinians in the West Bank and Gaza, Palestinians in Israel, and Arabs in surrounding states. "Israeli Arab" is the label the Israeli government applies to the Palestinians who did not leave in 1948 or 1967 and to their descendants. These Palestinians number about 1.3 million people—about 24 percent of Israel's population. They live in 116 "Arab-only" communities and seven so-called "mixed" cities, in which Palestinians live in separate communities adjacent to Jewish communities (Nathan 2006). The Palestinians who live in Israel prefer the label "Israeli Palestinians."

Similarly, the presence of Jews acts to unite an equally diverse Palestinian society, which includes West Bank Palestinians, Gaza Palestinians, and Israeli Palestinians, who also come from different ethnic and religious groups, clans, and political orientations.

Because little interaction occurs between in-group and out-group members, they know little about one another. This lack of firsthand experience deepens and reinforces misrepresentations, mistrust, and misunderstandings between members of the two groups. Thus, members of one group tend to view members of the other in the most stereotypical of terms. Yoram Bilu at Hebrew University of Jerusalem designed and conducted a particularly creative study to examine the consequences of in-group/out-group relations in the West Bank. Bilu and two of his students asked youths aged 11 to 13 from Palestinian refugee camps and Israeli settlements in the West Bank to keep journals of their dreams over a specified period. Seventeen percent of Israeli youths wrote that they dreamed about encounters with Arabs, whereas 30 percent of the Palestinian youths dreamed about meeting Jews. Among the 32 dreams of meetings between Jews and Arabs, not one character was identified by name. Not a single figure was defined by a personal, individual appearance. All the descriptions were completely stereotyped—the characters defined only by their ethnic identification (such as "Jew," "Arab," or "Zionist") or by value-laden terms with negative connotations (such

in-group A group to which a person belongs, identifies, admires, and/or feels loyalty.

out-group Any group to which a person does not belong.

King Leopold's Soliliquy: A Defense of His Congo Rule, By Mark Twain, Boston: The P. R. Warren Co., 1905, Second Edition.

In 1885, Leopold II, king of Belgium, calculated that his "yearly income from the Congo is millions." The Congo Free State, from which Leopold acquired wealth, is now named the Democratic Republic of Congo. Leopold's wealth was derived from forcing African people to extract rubber, diamonds, and ivory. The methods Leopold employed to exploit labor and acquire resources offer important clues to the origin of HIV and AIDS.

Urbain Ureel, Courtesy of Joan Ferrante

A Leopoldville shipyard in the 1930s. Shipyard workers are pushing barges into the Congo River.

Brian Smithson

In a 24-hours-a-day, seven-days-a-week operation, European and Asian logging companies remove an estimated 10 million cubic meters of wood each year from forests in Central African countries, including the DRC (Peterson and Ammann 2003).

to make steel and aluminum dry cell batteries), coltan (a heat-resistant mineral used in cell phones, laptops, and PlayStations), and uranium (needed to generate atomic energy and fuel the atomic bomb) (Oliver 2006). This relentless demand continues to fuel the ongoing conflict in the DRC, even after the independence from Belgium in 1965, when Mobutu Sese Seko assumed power of the country (called Zaire when it gained independence) with the help of the United States. Mobutu and those he appointed exploited the country's resources for their own personal gain. In 1997, Mobutu was forced out of office by Laurent Kabila, who was backed by the governments of Rwanda and Uganda that sent troops into the DRC. From 1997 to 2002, the DRC was the site of what some have called Africa's World War. Armies from Rwanda, Uganda, Burundi, Zimbabwe, Chad, Angola, and Namibia fought with and against Congolese government forces and rebel groups in an effort to control the DRC's resources, especially resources in the northeast and eastern part of the country (French 2009). Today various military forces and rebel groups still fight to control the country's resources (Turner 2009).

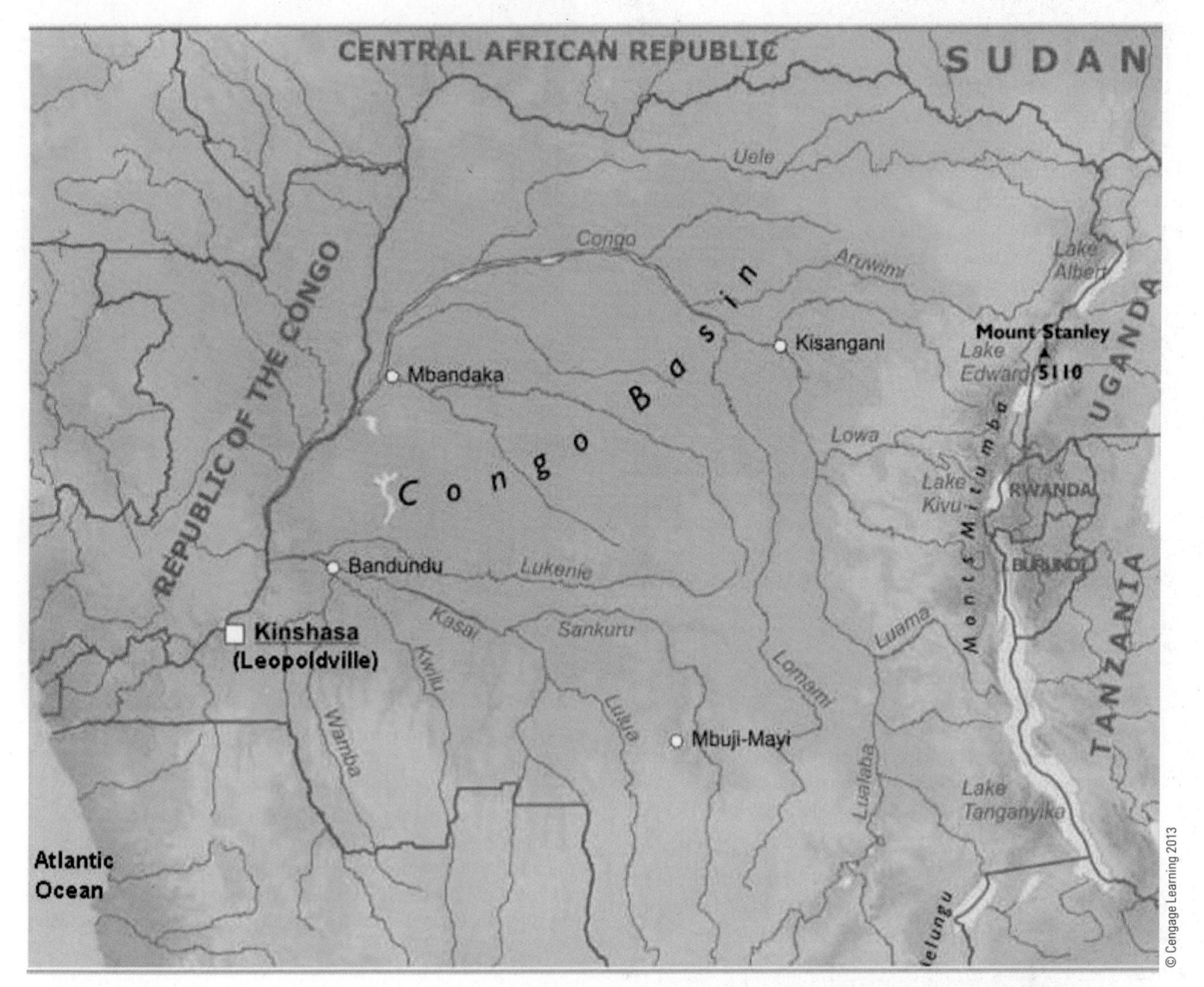

FIGURE 5.2 The map shows Leopoldville, the capital of the Belgian Congo. It was located on the Congo River—the "river that swallows all rivers" running through what is now the DRC, Zambia, Angola, Tanzania, Cameroon, and Gabon. The capital city attracted Europeans and Africans, who traded along the 8,000 miles of waterways that passed through the city and then out to sea.

Solidarity: The Ties That Bind

To this point, we have used the DRC as an example of how industrialization and colonization were inextricably interconnected and as an illustration of how the division of labor expanded to encompass workers and resources from around the world. Durkheim noted that, as the division of labor becomes more specialized and as the sources of materials to make products become more geographically diverse, a new kind of solidarity emerges. He used the term **solidarity** to describe the system of social ties that acts as a glue or cement connecting people to one another and the wider society. Durkheim observed that *mechanical* solidarity is characteristic of preindustrial societies and that a new kind of solidarity, *organic*, is characteristic of industrial societies.

Mechanical Solidarity

Mechanical solidarity characterizes a social order based on common ways of thinking, behaving, and seeing the world. In this situation, a person's "first duty is to resemble everybody else"—that is, "not to have anything personal about one's [core] beliefs and actions" (Durkheim 1933, p. 396). Durkheim believed that this similarity derived from the simple division of labor. In other words, a simple

solidarity The system of social ties that acts as a glue or cement connecting people to one another and the wider society.

mechanical solidarity Social order and cohesion based on a common conscience, or uniform thinking and behavior.

division of labor causes people to be more alike than different, because they do the same kind of tasks to maintain their livelihood. This similarity gives rise to common experiences, skills, core beliefs, attitudes, and thoughts. In societies characterized by mechanical solidarity, the ties that bind individuals to one another are based primarily on kinship, religion, and a shared way of life. To understand how society is organized around a shared way of life, consider the lifestyle of the Mbuti pygmies, a hunting-and-gathering people who, before colonization, lived in the Ituri Forest (an equatorial rain forest) in what is now the northeastern DRC. Their society represents one of the many ways of life that colonization forced people to abandon. Keep in mind that the Mbuti represent but one culture disrupted by colonization.

> The Mbuti share a forest-oriented value system. Their common conscience derived from the fact that the forest gives them food, firewood, and materials for shelter. It is not surprising that the Mbuti recognize their dependence upon the forest and refer to it as "Father" or "Mother" because as they say it gives them food, warmth, shelter, and clothing just like their parents. Mbuti say that the forest also, like their parents, gives them affection. . . . The forest is more than mere environment to the Mbuti. It is a living, conscious thing, both natural and supernatural, something that has to be depended upon, respected, trusted, obeyed, and loved. The love demanded of the Mbuti is no romanticism, and perhaps it might be better included under "respect." It is their world, and in return for their affection and trust it supplies them with all their needs. (1965, p. 19)

Mbuti are aware of the ongoing destruction of the rain forest by companies that push them farther into the forest's interior. The pygmies agree that if they leave the forest and become part of the modern world, their way of life will die: "The forest is our home; when we leave the forest, or when the forest dies, we shall die. We are the people of the forest" (Turnball 1961, p. 260).

Organic Solidarity

A society with a complex division of labor is characterized by **organic solidarity**—a social order or system of social ties based on interdependence and cooperation among people performing a wide range of diverse and specialized tasks. Specialized means that the tasks needed to make a product or to deliver a service performed by workers, often in different locations, who have been trained to do a specified task or tasks in the overall production process. The final product depends on contributions or many occupational categories. Relationships in the larger society reflect this specialization in that people relate to one another in terms of their specialized roles in the division of labor. Thus we buy tires from a dealer; we interact with a customer service representative by telephone or computer; we travel in an airplane flown by pilots and are served by flight attendants; we pay a supermarket cashier for coffee; and we deal with a lab technician when we give or sell blood. We do not need to know these people personally to interact with them. Likewise, we do not need to know the people behind the scenes: the rubber gatherer, the ivory hunter, the coffee grower, or the logger working in the DRC who contributed to making the products we purchase. In sum, most day-to-day interactions are fleeting, limited, impersonal, and instrumental (that is, we interact with people for a specific reason, not to get to know them).

When the division of labor is complex and when the materials for products are geographically scattered, few individuals possess the knowledge, skills, and materials to be self-sufficient. Consequently, people find that they must depend on others for the goods they buy and services they need. A complex division of labor increases differences

Tom Stoddart/Getty Images

In societies characterized by organic solidarity, people's lives depend on those in distant places. Cell phones, laptops, and other electrical equipment upon which we depend to stay connected to others cannot work without minerals such as coltan. Most of us fail to realize that thousands in the Congo mine coltan and minerals with their bare hands and shovels. In addition, not only are their lives at risk from ongoing conflict over resources but miners also are exposed to toxic and radioactive substances (UNEP/GRID—Arendal 2010).

organic solidarity Social order or system of social ties based on specialization, interdependence, and cooperation among people performing a wide range of diverse and specialized tasks.

among people, in turn leading to a decrease in common ways of thinking and behaving. Nevertheless, Durkheim argued, the ties that bind people to one another, based on specialization and interdependence, can be very strong, because people need one another to survive.

Disruptions to the Division of Labor

Durkheim hypothesized that societies become more vulnerable as the division of labor becomes more complex and jobs more specialized. He was particularly concerned with the kinds of events that break down individuals' ability to meaningfully connect with society and others through their work. Such events include (1) industrial and commercial crises caused by plant closings, massive layoffs, epidemics, technological revolutions, or war; (2) workers' strikes; (3) job specialization, insofar as workers are so isolated that few people grasp the workings and consequences of the overall enterprise; (4) forced labor, such that people have no choice in the work they do; and (5) inefficient management and development of workers' talents and abilities, so that work for them is nonexistent, irregular, intermittent, or subject to high turnover.

To illustrate how disruptions to the division of labor break the social ties that bind, consider the case of a factory worker, Joel Goddard, after he was laid off by Ford Motor Company. Goddard was married with two children. After he lost his job, his daily interactions shifted from colleagues at work to contacts with people at the unemployment office. Moreover, he lost the structure in his life that came with the routine of his job. Instead of working, he watched TV, fished, or read want ads. His ties were further disrupted when friends from work moved out of state to find employment. Goddard eventually took a job selling insurance and found himself selling to his acquaintances. He later quit, and then his wife decided to go to work. However, her success at work strained their marriage because it reminded Goddard of his failures.

Since at least 1884, the DRC has experienced these disruptions to the division of labor. In that year, King Leopold II took control of the Congo Free State, and for decades after, the people in that region were subjected to forced labor. Armies of black conscripts under the command of white officers would march into villages and take women—mothers, daughters, sisters, wives—hostage and force men—sons, husbands, fathers, brothers—to go into the rain forest to harvest rubber or to mine resources (Hochschild 2009). As we will see, under this brutal system of forced labor, the conditions that facilitated the emergence of HIV/AIDS arose.

Recall that scientists believe that around 1930 the HIV virus "jumped" form a chimp to a human host. But how did the host come in contact with a chimpanzee? We know that working conditions were so harsh that Congolese people were left vulnerable to disease. "Forced labor camps of thousands had poor sanitation, poor diet, and exhausting labor demands. It is hard to imagine better conditions for the establishment of an immune-deficiency disease" (Moore 2004, p. 545). Biological anthropologist Jim Moore (2004) describes a scenario under which that jump probably occurred:

> [The] fisherman flees his small village to escape a colonial patrol demanding its rubber quota [which he cannot meet]; as he runs, he grabs one of the unfamiliar shotguns that recently arrived in the area. While hiding for several days, he shoots a chimpanzee [for food] and, unfamiliar with the process of butchering it, is infected with simian immunodeficiency virus (SIV), a retrovirus that is found in primates. On return to the village he finds his family massacred and the village disbanded. He wanders for miles, dodging patrols, until arriving at a distant village. The next day he is seized by a railroad press gang and marched for days to the labor site, where he (along with several hundred others) receives several injections for reasons he does not understand. During his months working on the railroad, he has little to eat and is continually stressed, susceptible to any infection. He finds some solace in one of the camp prostitutes (themselves imported by those in charge), but eventually dies of an undiagnosed wasting—the fate of hundreds in that camp alone.

To escape the kind of brutality Moore described, tens of thousands of Congolese fled into the unfamiliar jungle and forests in search of food and safety. There many were bitten by infected tsetse flies and contracted sleeping sickness, an infectious disease with symptoms that include severe headache, fever, weakness, and uncontrollable sleepiness. To care for this condition, "well-meaning, but undersupplied doctors" treated it with serial injections of medicines. Workers were also routinely inoculated to prevent small pox and dysentery. "The problem was that multiple injections given to arriving gangs of tens or hundreds were administered with only a handful of syringes. The importance of sterile technique was known but not regularly practiced. Transfer of pathogens would have been inevitable. And to appease the laborers, in some of the camps sex workers were officially encouraged" (Moore 2006, p. 545).

In the context of HIV/AIDS, it is highly significant that millions of laborers were also tested for sleeping sickness with spinal taps, a medical procedure also involving needles that, at that time in history, were not disposable or subjected to sterilization. A combination of factors set the stage for origin and the eventual transmission of AIDS and HIV across a global stage:

- An exhausted labor force vulnerable to diseases, most notably sleeping sickness and HIV/AIDS;
- Medical testing and treatment that involved unsterilized needles; and
- People from all over the world extracting resources and doing business.

Global Comparisons Percentage of Adult Population with HIV/AIDS by Country

The map shows the prevalence of HIV/AIDS by country. Which areas of the world have the highest prevalence of HIV/AIDS? Notice on the map, which gives country-by-country breakdowns, that between 5 and 15 percent of the population in the DRC has HIV or AIDS. At a minimum, that represents 1 in every 20 people. How does the history of colonization inform your understanding of the AIDS prevalence in Africa?

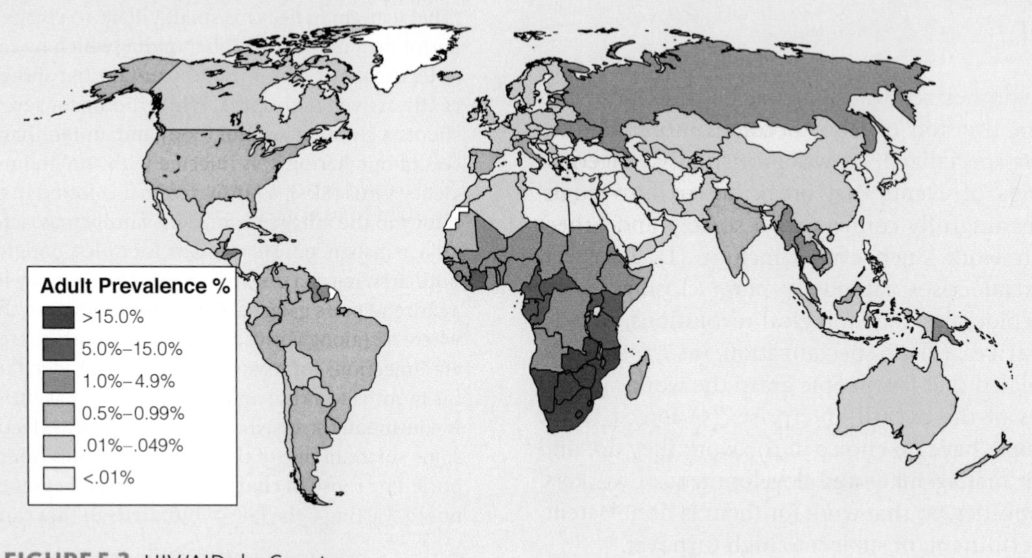

FIGURE 5.3 HIV/AIDs by Country
Source: U.S. Central Intelligence Agency (2007)

Industrialization and colonization drew people from the most remote regions of the world into a process that produced unprecedented quantities of material goods, primarily for the benefit of the colonizing powers. The demand for the raw materials to fuel industrialization pulled people from not just Belgium, but from all over the world into the Congo region and throughout Africa. This was accompanied by concerted efforts to increase the volume of goods and services produced, the reach of markets, and the speed at which goods, services, and labor were delivered. Railroads, freighters, and airplanes (by 1920, there was a Belgium–Congo flight) became part of an infrastructure that increased opportunities for more and more people to enter and leave the DRC and to interact (including sexually) with people they would otherwise never have met (Wrong 2002). In this context, HIV/AIDS cannot be viewed simply as a disease, but as a disease that grew out of millions of social interactions (see Figure 5.4).

To this point, we have considered the context under which HIV/AIDS emerged. In the second half of this chapter, we focus on face-to-face interaction, by which interacting parties communicate and respond through language and symbolic gestures to affect another's behavior and thinking. While interacting, the parties define, interpret, and attach meaning to all that goes on around them and between them. That interaction is embedded in a social structure.

SPC Preston E. Cheeks, USA

As one indicator of the massive movement of humans in general, consider that approximately 703 million people take international trips each year (World Tourism Organization 2007). Approximately 42 million foreign visitors enter the United States (excluding day-trippers), and 25.2 million U.S. residents travel to foreign countries (U.S. International Trade Administration 2007).

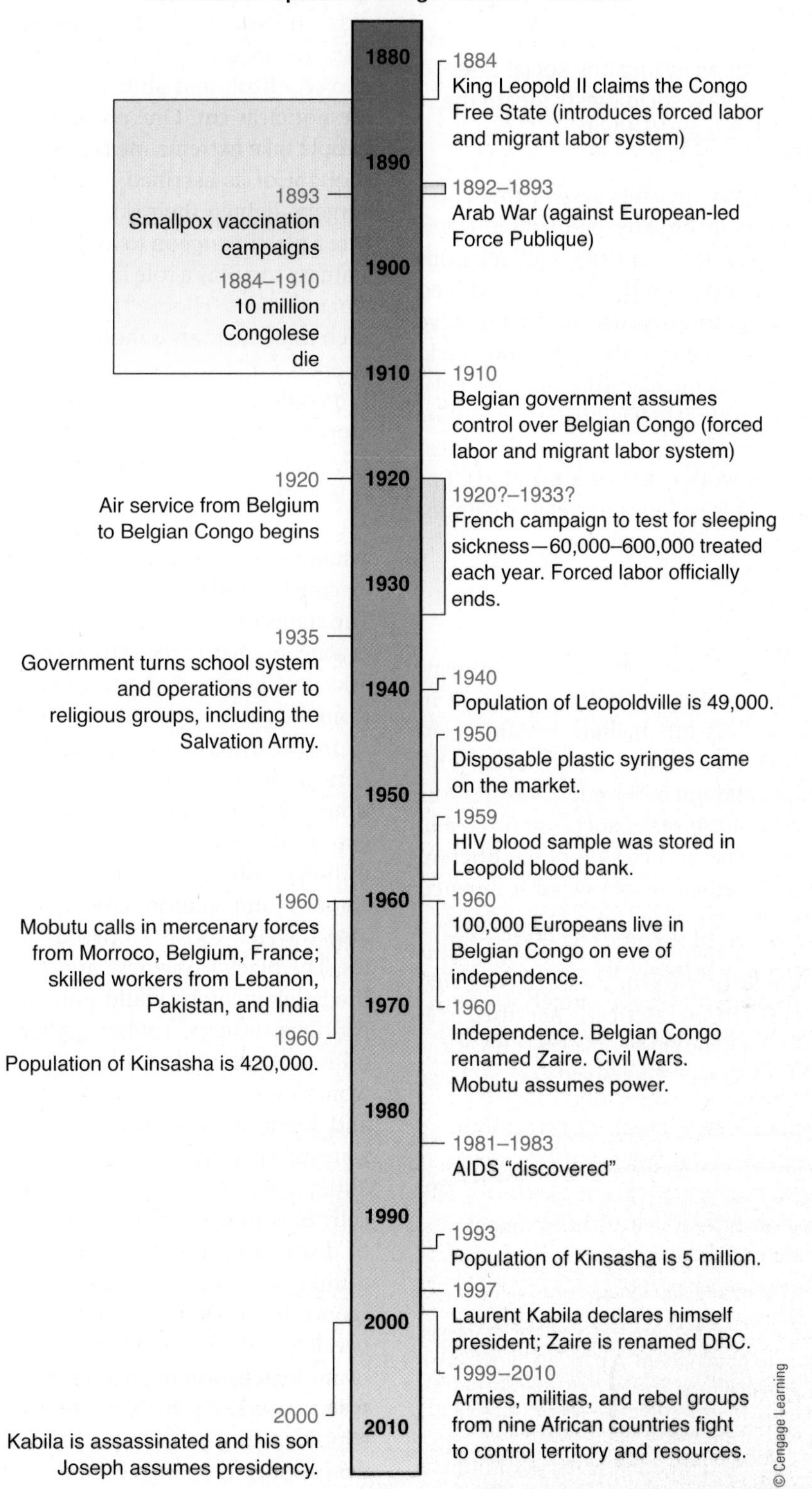
Democratic Republic of Congo Timeline 1880–2010
1880
1890
1900
1910
1920
1930
1940
1950
1960
1970
1980
1990
2000
2010
1884
King Leopold II claims the Congo Free State (introduces forced labor and migrant labor system)
1892–1893
Arab War (against European-led Force Publique)
1893
Smallpox vaccination campaigns
1884–1910
10 million Congolese die
1910
Belgian government assumes control over Belgian Congo (forced labor and migrant labor system)
1920
Air service from Belgium to Belgian Congo begins
1920?–1933?
French campaign to test for sleeping sickness—60,000–600,000 treated each year. Forced labor officially ends.
1935
Government turns school system and operations over to religious groups, including the Salvation Army.
1940
Population of Leopoldville is 49,000.
1950
Disposable plastic syringes came on the market.
1959
HIV blood sample was stored in Leopold blood bank.
1960
Mobutu calls in mercenary forces from Morroco, Belgium, France; skilled workers from Lebanon, Pakistan, and India
1960
Population of Kinsasha is 420,000.
1960
100,000 Europeans live in Belgian Congo on eve of independence.
1960
Independence. Belgian Congo renamed Zaire. Civil Wars. Mobutu assumes power.
1981–1983
AIDS "discovered"
1993
Population of Kinsasha is 5 million.
1997
Laurent Kabila declares himself president; Zaire is renamed DRC.
1999–2010
Armies, militias, and rebel groups from nine African countries fight to control territory and resources.
2000
Kabila is assassinated and his son Joseph assumes presidency.
© Cengage Learning

FIGURE 5.4 Time Line of Major Events in the DRC. This time line begins when King Leopold II claims the Congo Free State as his personal property. With the help of an army, he forces locals to extract rubber, diamonds, ivory, and other resources that fuel industrialization in the West. The time line ends with the current state of affairs in which armies, militia, and rebel groups from as many as nine surrounding countries and from within the Congo fighting to gain control of DRC resources. Ironically, even today armed groups regularly force villagers (as did King Leopold and the Belgium government) to carry supplies, ammunition, and water. One survey found that 50 percent of people living in eastern Congo had been forced to do such work (Hochschild 2009).

Social Structure

CORE CONCEPT 3 When analyzing any social interaction, or any human activity, sociologists locate the social structure in which it is embedded.

A **social structure** is a largely invisible system that coordinates human interaction in broadly predictable ways. Social structures shape people's identities and relationships to others and their opportunities to access valued resources. Sociologists study social structures that involve as few as two people (doctor–patient, shopper–store clerk, parent–child), that are global in scale (the pharmaceutical industry), national in scale (the Democratic Republic of the Congo), and that are of just about any size (a local bar, an extended family, a work camp or a hospital). Social structures encompass at least four interrelated components: statuses, roles, groups, and institutions.

Status

Sociologists use the term *social status* in a very broad way. **Social status** is a human-created and defined position in society. Examples are endless but include female, teenager, doctor, patient, sister, homosexual, heterosexual, employer, employee, soldier, and unemployed. A social status has meaning only in relation to other social statuses. For instance, the status of a physician takes on quite different meanings depending on whether the physician is interacting with another physician, a patient, or a nurse. Thus, a physician's behavior varies depending on the social status of the person with whom he or she is interacting. Note that some statuses, such as sister, are ascribed. **Ascribed statuses** are the result of chance in that people exert no effort to obtain them. Birth order, race, biological sex, and age qualify as ascribed statuses. Other statuses, such as nurse's aide and college student, are **achieved statuses**; that is, they are acquired through some combination of personal choice, effort, and ability. Ascribed and achieved statuses are not clear-cut. One can always think of cases in which people take extreme measures to achieve a status typically thought of as ascribed; they undergo sex transformation surgery, lighten their skin to appear to be another race, or hire a plastic surgeon to look younger. In addition, ascribed statuses can play a role in determining achieved statuses, as when females "choose" to enter a female-dominated career such as elementary-school teacher, or when men "choose" to enter a male-dominated career such as car repairman.

People usually occupy more than one social status. Sociologists use the term **status set** to capture all the statuses any one person assumes (see Figure 5.5). Sometimes one status in a status set overshadows the others such that it shapes every aspect of life and dominates social interactions. Such a status is known as a **master status**. Unemployed, retired, ex-convict, and HIV-infected can qualify as master statuses. The status of physician can be a master status as well, if everyone, no matter the setting (party, church, fitness center), asks health-related questions or seeks health-related advice from the person occupying that status.

In the DRC between 1884 and 1960, European and African nationalities were master statuses. During this time, white European males (predominantly Belgian) occupied the following kinds of positions: commissioned military officers, steamboat captains, district commissioners, and station chiefs. Belgian men and women also traveled to the Congo as physicians and missionaries. The black Congolese and other Africans who lived in the Belgian colony could only find work as porters, canoe paddlers, miners, rubber gatherers, soldiers and low- to mid-level administrative and technical staff. Congolese women worked as maids, house servants, and prostitutes, and because there were always labor shortages, women were also pressed to work on roads and plantations (Lyons 2002). Most Congolese were forced to take these physically demanding positions out of desperation or the threat of death from the Force Publique, a white-led repressive military force comprising African males from many ethnic groups both inside and outside of the Congo. Until independence in 1960, the Congo maintained a rigid system of racial segregation in all areas of life. The segregated school system was such that the Congolese were denied access to education beyond primary school. At the time of independence, only 17 Congolese had earned a university degree (Wrong 2000, Edgerton 2002).

Role

Sociologists use the term **role** to describe the behavior expected of a status in relation to another status—for example, the role of physician in relation to a patient.

social structure A largely invisible system that coordinates human interaction in broadly predictable ways.

social status A human-created and defined position in society.

ascribed statuses Social statuses that are the result of chance in that people exert no effort to obtain them. A person's birth order, race, biological sex, and age qualify as ascribed characteristics.

achieved statuses Social statuses acquired through some combination of personal choice, effort, and ability. A person's marital status, occupation, and educational attainment are considered examples of achieved statuses.

status set All the statuses an individual assumes.

master status One status in a status set that overshadows the others such that it shapes every aspect of life and dominates social interactions.

role The behavior expected of a status in relation to another status.

FIGURE 5.5 Status Set and Master Status

Jillian occupies many statuses including sister, weight lifter, and assistant to a college basketball coach. Her master status, however, is that of a person with Down syndrome. Having Down syndrome is considered a master status because that status overshadows her other statuses. In other words, it is often hard for people to see beyond Jillian's master status and think of her as having a full and active life. Other statuses are, at best treated as secondary to her master status. That is, Jillian is not just seen as a weight lifter; she is a weight lifter with Down syndrome.

Patients can expect doctors to establish a diagnosis, to not overtreat, to prescribe a treatment plan, and to not make sexual advances (Hippocratic Oath 1943). Physicians expect their patients to answer questions honestly and to cooperate with a treatment plan. The distinction between role and status is this: People occupy statuses and enact their statuses through roles.

Any given social status is associated with an array of roles, called a **role set**, (Merton 1957). The social status of school principal is associated with a role set that is composed of relationships with students, parents, teachers, and other school staff. Quite often **role performances**, the actual behaviors of the person occupying a role, do not meet expectations—as when some physicians knowingly perform unnecessary surgery or when some patients fail to comply with treatment plans. The concept of role performance reminds us that, in their actual behavior, people carry out their roles in unique ways. Still, there is a predictability to role performances because when role performances deviate too far from expectations,

role set The array of roles associated with a given social status.

role performance The actual behavior of the person occupying a role.

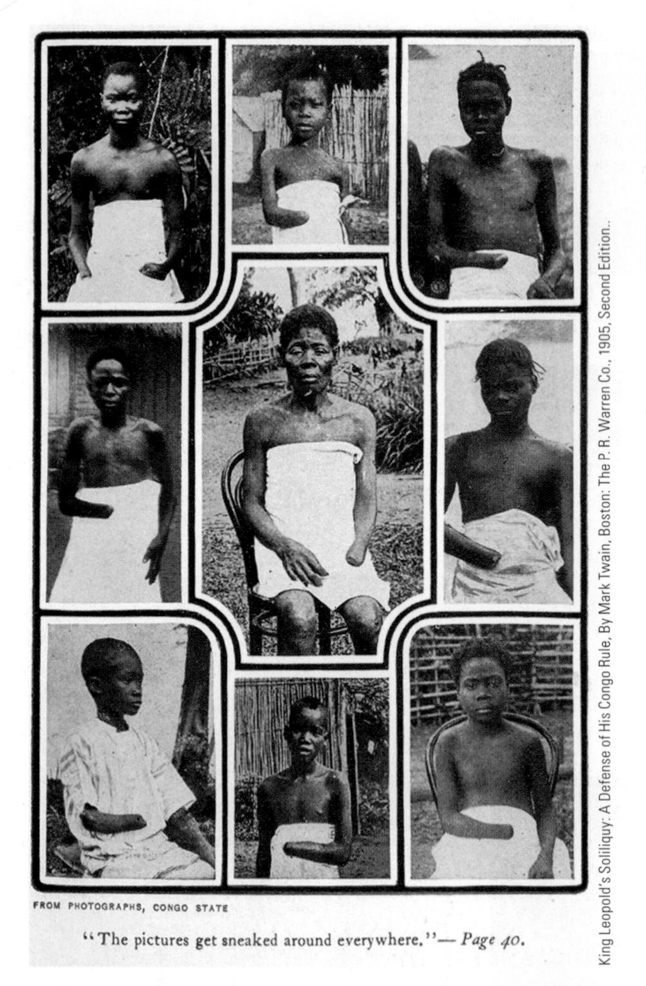

FROM PHOTOGRAPHS, CONGO STATE
"The pictures get sneaked around everywhere."— Page 40.
King Leopold's Soliloquy: A Defense of His Congo Rule. By Mark Twain, Boston: The P. R. Warren Co., 1905, Second Edition..

One of the most grisly policies of Leopold's rule was to sever a hand (often right) of Congolese who refused to gather rubber. Soldiers were under orders to turn in a hand for every bullet they fired, as proof that they were using bullets to kill people (and not animals or for target practice). Often soldiers would sever the hands of people to cover for other uses of bullets.

penalties—ranging in severity from a frown to imprisonment and even death—are, more often than not, applied (Merton 1957). A professor who misses class too many times will eventually be reported to a department chair or dean. An extreme example relates to the serious punishment meted out in the Congo when rubber gatherers and other forced laborers failed to perform roles as expected, no matter how absurd those expectations. To ensure that rubber gatherers met "expectations," family members were held hostage. Failing to gather enough rubber could mean death for a spouse, parents, or children. If a village refused to participate in forced labor, company troops shot everyone in sight so that nearby villagers were clear about expectations (Hochschild 1998, p. 165).

If we consider that people hold multiple statuses and that each status is enacted through its corresponding role set, we can identify at least two potential sources of stress and strain: role conflict and role strain.

Role Conflict **Role conflict** is a predicament in which the roles associated with two or more distinct statuses that a person holds conflict in some way. For example, people who occupy the statuses of college student and full-time employee often experience role conflict when professors expect students to attend class and keep up with coursework *and* employers expect employees to be available to work hours that leave little time for schoolwork. The student-employee must find ways to address this conflict between the two statuses, such as working fewer hours, quitting the job, skipping class, or studying less.

Role Strain **Role strain** is a predicament in which there are contradictory or conflicting role expectations associated with a single status. For example, doctors have an obligation to do no harm to their patients. At the same time, doctors have an obligation to pay office staff. Some physicians may feel pressured to recommend expensive and unnecessary medical treatments to generate sufficient revenue to stay in business.

Cultural Variations in Role Expectations

Role expectations associated with statuses vary across cultures. In the United States, patients expect those who occupy the status of doctor to use available technology such as X-rays, CAT scans, and MRIs to diagnose medical conditions; to prescribe drugs; and to recommend surgery. These expectations are shaped by a core cultural belief in the ability of science to solve problems. Western medicine uses all available tools of science to establish the cause of a disease, combat it for as long as possible, and ideally return the body to a healthy state. In view of this cultural orientation, it is not surprising that U.S. physicians rely heavily on technology to diagnose and cure disease. This reliance is reflected in the fact that the United States, with less than 5 percent of the world's population, consumes an estimated 52 percent of the world's pharmaceutical supply (World Health Organization 2004). This is equivalent to 3.7 billion prescriptions filled per year, or 12 prescriptions for every man, woman, and child (Kaiser Family Foundation 2009).

We can contrast the American expectations with those Congolese hold for the traditional African healer. The

role conflict A predicament in which the roles associated with two or more distinct statuses that a person holds conflict in some way.

role strain A predicament in which there are contradictory or conflicting role expectations associated with a single status.

social relationship between the African healer and patient is very different from that between the U.S. physician and patient. Like American physicians, traditional healers recognize and treat the organic and physical aspects of disease. But they also attach considerable importance to other factors: supernatural causes, social relationships (hostilities, stress, family strain), and psychological distress. This holistic perspective allows for a more personal relationship between healer and patient. Sociologist Ruth Kornfield observed Western-trained physicians working in the Congo's urban hospitals and found that success in treating patients was linked to the physicians' ability to tolerate and respect other models of illness and include them in a treatment plan. Among some ethnic groups in the Congo, when a person becomes ill, the patient's kin form a therapy management group and make decisions about treatment. Because many people in the Congo believe that illnesses result from disturbances in social relationships, the cure must involve a "reorganization of the social relations of the sick person that [is] satisfactory for those involved" (Kornfield 1986, p. 369; also see Kaptchuk and Croucher 1986, pp. 106–108).

Given these culturally rooted beliefs regarding the cause and treatment of disease, it is not surprising to learn that Europeans and Africans clashed over how to handle a sleeping sickness epidemic in the late 1920s and early 1930s. Recall that sleeping sickness is contacted when humans are bitten by infected tsetse flies. The most notable symptom is uncontrollable sleeping such that a person may fall asleep in the midst of activity, including work. As you might imagine, Europeans were concerned about this disease because the African workforce was affected.

Europeans assumed that sleeping sickness and other African diseases could be controlled, even eliminated, through testing, technology, and medicines. With regard to sleeping sickness, those techniques and technologies involved spinal taps to screen for the condition and the drug tryparsamide to treat it. About 80 percent of patients were "cured," but side effects included partial or complete loss of vision (Rockefeller University Hospital 2011). Most Congolese were suspicious of the European campaign against sleeping sickness. They believed that Europeans were responsible for the dramatic increase in cases of sleeping sickness and other diseases. Specifically, the Congolese believed that the forced labor was exhausting them, making them vulnerable to disease. Moreover, it was the dramatic disruptions to their way of life that put them in environments where they came in contact with the tsetse flies. In sum, they believed that this sleeping sickness was man-made. For the Congolese, diagnosis required careful investigation to identify the ultimate cause behind physical symptoms, and treatment focused on reintegrating the sick person into the social fabric (Lyons 2002). In *The Colonial Disease*, Maryinez Lyons (2002) argues that it "is vital to understand the profound importance and depth of this Congolese belief regarding cause and treatment" (p. 183). This is something the U.S. military doctors working in the DRC realize as they are paired with Congolese counterparts (see Sociological Imagination).

Europeans' approach to diagnosing sleeping sickness involved taking a blood sample through a finger prick or taking spinal fluid, shown here, through a lumbar puncture (spinal tap). In Africa, sleeping sickness was considered the AIDS of the early twentieth century.

People line up for sleeping sickness screening. In light of the labor needs at the time, the Belgian government and other Western players in Congo ventures considered diseased and dead Africans as economic losses. Thus, sleeping sickness posed a grave threat in need of control. Sleeping sickness campaigns were institutionalized through a network of rural clinics, screenings and injection centers, and hospitals. Between 1923 and 1938, an estimated 26 million exams were performed in the Belgian Congo (Lyons 2002).

Groups

Like statuses and roles, groups are an important component of social structure. A **group** consists of two or more people interacting in largely predictable ways and who share

group Two or more people interacting in largely predictable ways and who share expectations about their purpose for being.

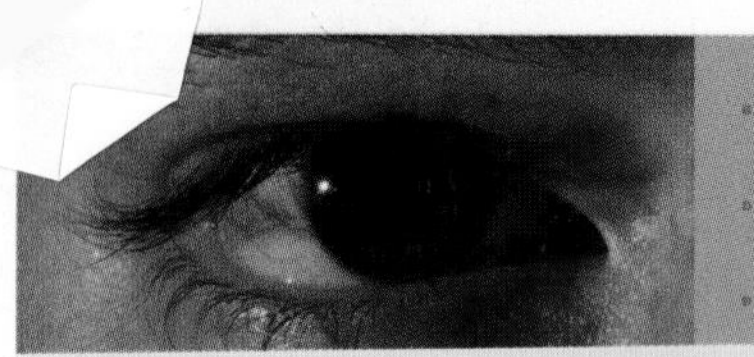

The Sociological Imagination Healing as a Cultural Collaboration

Sgt. James D. Sims

KINSHASA, Democratic Republic of Congo, September 9, 2010—Crowds gathered, some with pre-registered tickets in hand, others with just a hope of being seen by a healthcare professional in Kinshasa.

"I saw a crowd of people and asked what was going on," said Ousmane Kalotho Mutuala, a Kinshasa resident. "When they told me it was for medical care, I immediately went and got my friend who can barely see because his eyes are so bad and came back to try and get in."

The lines started forming hours before the humanitarian civic action site opened its doors for medical and dental care to the residents of Kinshasa. Residents that had tickets were registered in advance, ensuring they would be seen on a certain date. Even though some residents, like Mutuala, did not have tickets, medical providers saw them.

"Unfortunately there is a much bigger demand then what we have assets for," said Major Curt Kroh of Washburn, N.D., a physician assistant with the North Dakota National Guard's 814th Army Support Medical Company, which is based in Bismarck. "However, we stayed until we ran out of time and material."

Kroh is taking part in MEDFLAG 10, a joint medical exercise that allows U.S. military medical personnel and their Armed Forces of the Democratic Republic of the Congo (FARDC) counterparts to work side by side while providing humanitarian assistance to Kinshasa residents.

Approximately 25 FARDC and U.S. medical and dental personnel, and an additional 50 support staff, provided services. Over a four-day period, FARDC and U.S. medical personnel provided assistance to approximately 2,000 Congolese.

Patients were treated for various illnesses ranging from high blood pressure to malaria. The most common problem encountered was residents with eye problems, because they have never been examined, said Kroh. In addition to medical attention, dentists provided care ranging from basic oral hygiene to tooth extraction.

"The bulk of the medical care that was provided in the exam rooms was by FARDC doctors," said Kroh. "The FARDC doctors are very well involved in the treatment of the local population."

While all residents could not be seen and all problems could not be treated, residents were entered into the medical system and given referral letters for follow-up care.

Source: Staff Sergeant Kassidy Snyder (2010)

expectations about their purpose for being. Group members hold statuses and enact roles that relate to the group's purpose. Groups can be classified as primary or secondary. Primary groups are characterized by face-to-face contact and by strong emotional ties among members who feel an allegiance to one another. Examples of primary groups include the family, military units, cliques, and peer groups.

Secondary groups consist of two or more people who interact for a specific purpose. Thus, by definition, secondary group relationships are confined to a particular setting and specific tasks. Members relate to each other in terms of specific roles. People join secondary groups as a means to achieve some agreed-upon end, whether it be to cheer for a sports team, to achieve a status (college graduate), or to accomplish specific goals such as changing entrenched norms and fundamental assumptions. Secondary groups

secondary groups Two or more people who interact for a specific purpose. Secondary group relationships are confined to a particular setting and specific tasks.

can range from small to extremely large in size. They include a work unit, college classroom, parent–teacher associations, and churches. Larger secondary groups include fans gathering in a stadium, colleges, Wal-Mart employees, and activist organizations (see Working for Change: "Doing Good by Being Bad").

Within the DRC, large numbers of people are part of secondary groups, including the United Nations Peacekeeping Force believed to number 18,000, nongovernmental organizations such as Oxfam, World Vision, and Hope In Action seeking to address social problems resulting from the ongoing conflict. There are also corporations around the world whose products depend on minerals extracted from the Congo. A UN report listed a sample of 85 corporations who import DRC minerals through Rwanda alone. (Note that primary groups can form or be embedded within these secondary groups.)

Institutions

A fourth component of social structure is **institutions**, relatively stable and predictable social arrangements created and sustained by people that have emerged over time with the purpose of coordinating human activities to meet some need, such as food, shelter, or clothing. Institutions consist of statuses, roles, and groups. Examples of institutions include government, education, medicine, and sports (Martin 2004). Institutions have a number of important characteristics:

1. Institutions have a history. Institutions have standardized ways of doing things that have come to be viewed as custom and tradition. That is, most people accept these ways without question. In the United States, the military is an institution with a global reach. The United States has a military presence in more than 100 countries, including the DRC. The military is stationed in places where the United States has strategic and national security interests. In the DRC, the government as an institution has operated "as a system of organized theft," beginning with the government of Leopold II to Kabila, the present leader's government. All in power have used their office to exploit the country's natural resources and people for personal gain (Hochschild 2009). Likewise, outside governments and militarized group powers have always sought to control the DRC resources. In other words, exploitation of the DRC has been institutionalized.
2. Institutions continuously change. Over time, ways of doing things become outdated and are replaced by new ways. Change can be planned, orderly, forced, and chaotic. Change can come from within or from outside the institution. We know, for example, that the institution of medicine is always changing to accommodate new technologies, new research, evolving demographics, and so on. Likewise, the DRC has changed over time, but those changes have been largely fueled by consumer, corporate, and military interests in the country's resources. King Leopold exploited rubber and ivory when he controlled the Congo Free States. Today the country is exploited for those items and much more including coltan, copper, cobalt, diamonds, gold, and cassiterite; forests and wildlife, timber, coffee, ivory; livestock, tobacco, tea, palm oil, and land.
3. Institutions allocate scarce and valued resources in unequal ways. They also allocate privileged and disadvantaged status. These inequalities can be reflected in individuals' salaries, benefits, degree of autonomy, and

Staff Sgt. Kassidy Snyder

Sgt. James D. Sims

The U.S. military's reach extends to the DRC. Here Congolese citizens sit under a tent waiting to see U.S. and Congolese medical staff as part of a humanitarian and civic assistance outreach program. The second photo shows the commander of U.S. Army Africa, being greeted by the Armed Forces of the Democratic Republic of the Congo (FARDC) military police and music battalion.

institutions Relatively stable and predictable social arrangements created and sustained by people that have emerged over time with the purpose of coordinating human activities to meet some need, such as food, shelter, or clothing. Institutions consist of statuses, roles, and groups.

way, because the status they occupy requires them to do so. Coaches work to hide any doubts about whether they think their team can win an upcoming game. They engage in impression management because that is what coaches are expected to do. At other times, people are unaware that they are engaged in impression management because they are simply behaving in ways they regard as natural. Women engage in impression management when they put on makeup, dye their hair, or shave their legs, even if they never question the reason for doing so, which is to present themselves as something they are not. Men engage in impression management when they hide their emotions in stressful situations so that no one questions their masculinity.

Goffman (1959) judges the success of impression management by whether an audience "plays along with the performance." If the audience plays along, the actor has successfully projected a desired definition of the situation or has at least cultivated an understanding among the audience that they will pretend to play along. According to Goffman, there are times that an audience pretends to play along simply because that actor is "a representative of something" considered important to the society; that is, the audience gives deference, not "because of what they personally think of that actor but in spite of it" (p. 210). Most students, for example, manage to stay awake in class or at least give the appearance that they are paying attention, even when they think the teacher is not very good. They do so, not out of respect for the teacher they view as incompetent, but out of respect for the position that the teacher occupies.

Goffman (1959) recognized that there is a dark side to impression management that occurs when people manipulate their audience in deliberately deceitful and hurtful ways. Such was the case when King Leopold presented his interests in the Congo as purely philanthropic. He established the International African Association as a front for his profit making and exploitative ventures. Leopold's stated aim was to bring a humanizing influence to the Congo—an effort that included establishing a chain of medic posts and scientific centers across the region and abolishing Afro-Arab slave trade. In reality, Leopold pushed out Arab slave traders so he could control the labor and resources. Through Henry Morton Stanley, Leopold signed treaties with 450 African chiefs, who were unfamiliar with the written word and the concept of treaties that read like the following: "In return for one piece of cloth per month, give up . . . all their territories . . . such that all roads and waterways running through this country, the right of collecting tolls on the same, and all mining, fishing, and forest rights are to be the absolute right of the said Association" (Hochschild 1998, p. 72).

While impression management has a dark side, it can also be constructive. If people said whatever they wanted and behaved entirely as they pleased, social order would break down. Goffman argues that social order depends on people at least conveying the impression that they are living up to the societal expectations, standards, and agreements. According to Goffman, people in most social interactions weigh the costs of losing their audience against the costs of losing their integrity. If keeping the audience seems more important, impression management is deemed necessary. If being completely honest and upfront seems more important, we may take the risk of losing our audience.

The tension between revealing and concealing information comes into play when people test positive for HIV. If they disclose the test results, they risk discrimination, including loss of their jobs, insurance coverage, friends, and family (Markel 2003). This risk explains why many HIV-infected people fail to disclose their HIV status, even to sexual partners or when giving blood. In a study of 203 HIV-infected patients treated at two East Coast hospitals, Michael Stein (1998) and his colleagues found that more than half the group claimed to be sexually active. Of these 129 patients, 40 percent had not disclosed their HIV-positive status to partners with whom they had had sex in the past six months. In addition, only 42 percent reported that they always used a condom. In another study, almost one-third of a group of 304 HIV-positive blood donors indicated that they had donated blood because their colleagues had pressured them to do so.

Front and Back Stage

Goffman used a variety of concepts to elaborate on the process by which impressions are managed—including the idea of front and back stage. Just as the theater has a front stage and a back stage, so too does everyday life. The **front stage** is the area visible to the audience, where people feel compelled to present themselves in expected ways. Thus, when people step into an established social role such as a teacher in relation to students or a doctor in relation to patients, they step onto a front stage such as a classroom or an examining room (Goffman 1959). The **back stage** is the area out of the audience's sight, where individuals let their guard down and do things that would be inappropriate or unexpected in a front-stage setting. Because back-stage behavior frequently contradicts front-stage behavior, we take great care to conceal it from the audience. In the back stage, a "person can relax, drop his front, forgo his speaking lines, and step out of character" (Goffman 1959, p. 112).

The division between front stage and back stage can be found in nearly every social setting. Often that division

front stage The area visible to the audience, where people feel compelled to present themselves in expected ways.

back stage The area out of the audience's sight, where individuals let their guard down and do things that would be inappropriate or unexpected in a front-stage setting.

is separated by a door or sign signaling that only certain people such as employees can enter the back stage without permission or knocking. Goffman uses a restaurant as an example of a social setting that has clear boundaries between the back stage and front stage. Restaurant employees do things in the kitchen, pantry, and break room (back stage) that they would never do in the dining areas (front stage), such as eating from customers' plates, dropping food on the floor and putting it back on a plate, and yelling at one another. Once they enter the dining area, however, such behavior stops. Of course, a restaurant is only one example of the many settings in which the concepts of front stage and back stage apply.

In relation to the AIDS crisis, we can identify many settings that have a front stage and a back stage, including hospitals, doctors' offices, and blood banks. A U.S. government General Accountability Office (GAO 1997) study revealed that front stage–back stage dynamics affected the way blood donors answered screening questions. The study found that 20 percent of blood donors claimed they would have answered those questions differently if they had been in a more private setting. In other words, one in five donors did not give honest answers to screening questions because they were on the "front stage"—that is, others were within hearing distance and the donors feared disclosing answers that might be judged harshly. Such questions included the following: "Are you in good general health?" "Are you a male who has had sex with another male even once since 1977?" "Have you ever taken street drugs by needle, even once?" "Since 1977, have you ever exchanged sex for drugs or money?" "Have you had sexual contact with anyone who was born in or lived in Cameroon, Central African Republic, Chad, Democratic Republic of the Congo, Equatorial Guinea, Gabon, Niger, or Nigeria since 1977?"

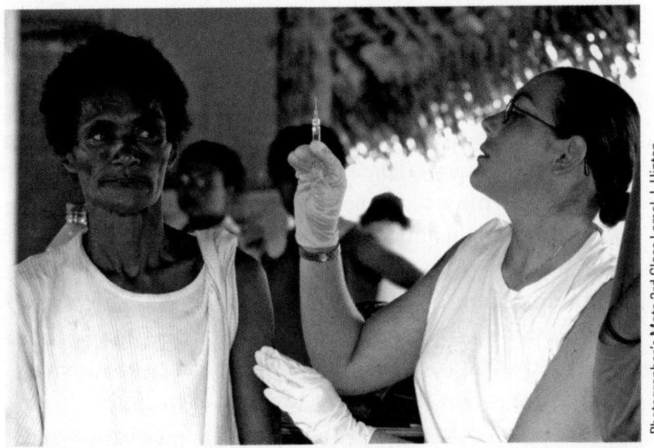

Photographer's Mate 3rd Class Lamel J. Hinton

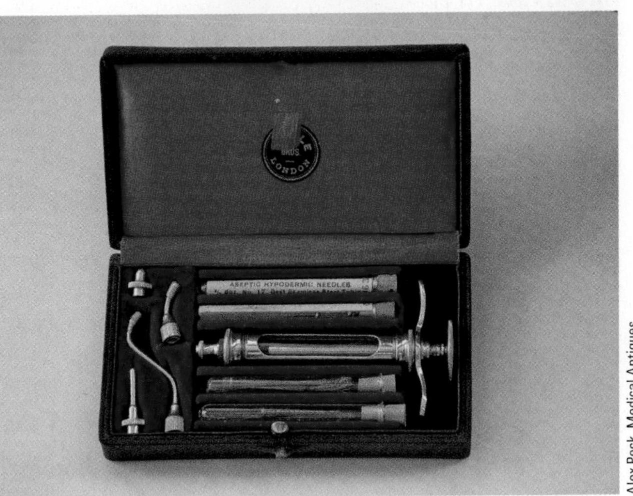

Alex Peck, Medical Antiques

Today when we draw blood and give injections, we use single, disposable syringes and needles. But disposable syringes did not become the norm until the mid-1970s. The case pictured shows hypodermic needles and syringes from the early twentieth century. Recall that tens of millions of African people were tested and treated for sleeping sickness, making it is highly unlikely that adequate sterilization of syringes and needles occurred between each use.

Attribution Theory

CORE CONCEPT 5 People assign causes to their own and others' behaviors. That is, they propose explanations for their own and other's behaviors, successes, and failures and then they respond accordingly.

Social life is complex. People need a great deal of historical, cultural, and biographical information if they are to truly understand the causes of even the most routine behaviors. Unfortunately, it is nearly impossible for people to have this information at hand every time they seek to explain a behavior. Yet, despite this limitation, most people attempt to determine a cause, even if they rarely stop to examine critically the accuracy of their explanations. As most of us know very well, ill-defined, incorrect, and inaccurate perceptions of cause do not keep people from forming opinions and taking action.

Attribution theory relies on the assumption that people make sense of their own and others' behavior by assigning a cause. In doing so, they often focus on one of two potential causes: dispositional (internal) or situational (external). **Dispositional causes** are forces over which individuals are supposed to have control—including personal qualities or traits, such as motivation level, mood, and effort. **Situational causes** are forces outside an individual's control—such as the weather, bad luck, and

dispositional causes Forces over which individuals are supposed to have control—including personal qualities or traits, such as motivation level, mood, and effort.

situational causes Forces outside an individual's immediate control—such as weather, chance, and others' incompetence.

What is HIV/AIDS?

According to the Centers for Disease Control (CDC, 2011), HIV (human immunodeficiency virus) is the virus that causes AIDS. This virus is passed from one person to another through exchanging infected blood or other body fluids, including semen, vaginal fluid, breast milk, cerebrospinal fluid surrounding the brain and spinal cord, synovial fluid surrounding bone joints, and amniotic fluid surrounding a fetus.

AIDS stands for acquired immunodeficiency syndrome.

Acquired means that the disease is not hereditary but develops after birth from contact with a disease-causing agent (in this case, HIV).

Immunodeficiency means that the disease is characterized by a weakening of the immune system.

Syndrome refers to a group of symptoms that collectively indicate or characterize a disease. In the case of AIDS, symptoms can include the development of certain infections or cancers, as well as a decrease in the number of certain cells in a person's immune system.

A physician makes a diagnosis of AIDS by using specific clinical or laboratory standards. The symptoms of AIDS are not unique to that condition. An HIV-infected person receives a diagnosis of AIDS after developing one of the 25 CDC-defined AIDS indicator illnesses. These illnesses include invasive cervical cancer, recurrent pneumonia, Kaposi's sarcoma, toxoplasmosis of the brain, and wasting syndrome due to HIV. An HIV-positive person who has not experienced one of the 25 indicator illnesses can receive an AIDS diagnosis based on tests showing that the number of CD4+ T cells per cubic millimeter of blood has dropped below 200.

Keep in mind that a diagnosis of AIDS depends on how the medical community defines the condition. Before 1993, many indicator illnesses were unique to the gay population with AIDS. Under the revised definition, HIV-positive women with cervical cancer are officially diagnosed as having AIDS. In 1993, the official definition of AIDS was revised to include HIV-related gynecological disorders, such as cervical cancer and HIV-related pulmonary tuberculosis or recurrent pneumonia as conditions that indicated AIDS. Before this definition change, physicians did not advise women with cervical cancer to be tested for HIV and did not treat HIV-positive women with this condition as if they had AIDS (Barr 1990; Stolberg 1996). The revised definition added 40,000 people to the total number diagnosed with AIDS.

others' incompetence. Usually people stress situational factors to explain their own failures and dispositional factors to explain their own successes (for example, "I failed the exam because the teacher can't explain the subject" versus "I passed the exam because I studied hard"). With regard to other people's failures or shortcomings, people tend to stress dispositional factors ("She doesn't care about school"; "He didn't try"). With regard to others' successes, people tend to emphasize situational factors ("She passed the test because it was easy"). Right or wrong, the attributions people make affect how they respond.

Many people stress dispositional traits to explain the cause of HIV/AIDS, arguing that certain groups of people behaved in careless, irresponsible, bizarre, or immoral ways. When I ask my students—most of whom were born after 1981 (the year AIDS was recognized as a disease)—about the origin of AIDS, they repeat these popular beliefs. With regard to Africa, students point to the evidence that the virus jumped from chimpanzees to humans—specifically from chimps to African hunters who were bitten or scratched by chimps. With regard to homosexuals, students point to high-risk and promiscuous behavior.

When evaluating the chimp-hunter and homosexual hypotheses, consider that chimps had been hunted for thousands of years. Why did the leap from monkey to human occur in 1930? We must consider situational factors that facilitated this leap. Think about the thousands of people fled into the forests and came in contact with chimps, the testing and inoculation campaigns with unsterilized needles, and a population that was weak and sick from forced labor. The point is that it took these factors to eventually push the virus onto the global scene 50 years later in 1981, the year it was first noticed (Moore 2004).

With regard to homosexual origins, we must ask the following: Are the categories male homosexual/bisexual and heterosexual mutually exclusive? If yes, that would mean that heterosexual sex is always penis-vaginal? One of the most comprehensive studies on sexual behavior in the United States found that between 13.2 and 37.5 percent of heterosexual respondents said they had engaged in anal sex with another sex partner at some time. Between 55.6 and 83.6 percent of the hetereosexuals surveyed indicated that they had had oral sex with another sex partner in the past year (Michael et al. 1994). An Urban Institute study financed by the U.S. government surveyed a representative sample of 1,297 males between the ages of 15 and 19 and found that two-thirds had had experience "with noncoital behaviors like oral sex, anal intercourse, or masturbation by a female" (Lewin 2000, p. 18A). The point is that we could probably learn more about how HIV spreads if we classified HIV cases by type of sex act (oral, anal, vaginal)

No Borders No Boundaries Imports/Exports of Blood and Blood Products

U.S. corporations supply about 60 percent of the world's plasma and blood products. For at least four years after HIV/AIDS was identified, U.S. blood bank officials continued to publicly affirm their faith in the safety of the country's blood supply, insisting that screening donors was unnecessary. Yet, these officials never revealed to the public the many shortcomings in production methods that could jeopardize the safety of blood. (By *shortcomings*, we mean deficiencies in medical knowledge and the level of technology rather than simply negligence.) Blood bank officials later argued that they practiced this concealment to prevent a worldwide panic. (In Goffman's terminology, blood bank officials did not want to "lose their audience.") Such a panic would have brought chaos to the medical system, which depends on blood products. This delay in implementing screening ultimately exposed many people to infection, especially hemophiliacs. Hemophiliacs were at higher risk of contracting HIV because their plasma lacks Factor VIII, a blood product that aids in clotting, or because their plasma contains an excess of anticlotting material (U.S. Centers for Disease Control 2010). In fact, we now know that 50 percent of hemophiliacs became infected with HIV from contaminated Factor VIII treatments before the first case of AIDS appeared in this group (*Frontline* 1993).

To truly understand the global AIDS epidemic, we need to consider the role of exported blood products in transmitting HIV. The amount of blood products exported from the United States to other selected countries as a percentage of the receiving country's total need is graphed below. These rates apply to today but also to 1981, the year that HIV was "discovered" and scientists first learned that it was in the blood supply. Notice that Japan imports 98 percent of its blood products from the United States—about 46 million units of concentrated blood products and 3.14 million liters of blood plasma (Yasuda 1994).

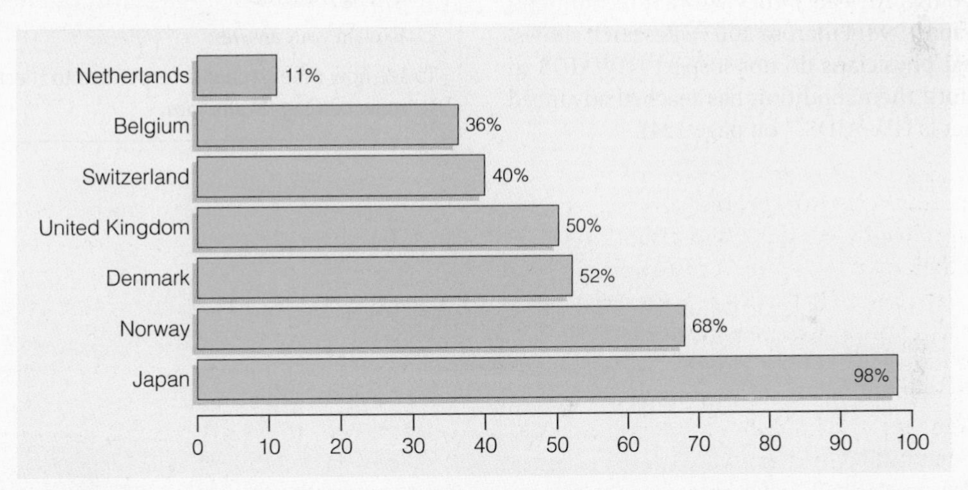

FIGURE 5.6 Amount of Blood Products Imported from the United States as a Percentage of Total Need

Sources: U.S. International Trade Administration (2010); International Federation of Pharmaceutical Manufacturers Associations (1981).

rather than identifying the categories of people having sex (homosexual/bisexual or heterosexual).

Attributing cause to dispositional factors—personal sexual practices, preferences, and appetite—works to reduce uncertainty about the source and spread of the disease. The rules are clear: If we can isolate "those people" who deserve or earned the disease, we are safe (Grover 1987; Sontag 1989). Such logic supports the naming of a scapegoat. A **scapegoat** is a person or group blamed for conditions that (a) cannot be controlled, (b) threaten a community's sense of well-being, or (c) shake the foundations of an important institution. Often the scapegoat belongs to a group that is vulnerable, hated, powerless, or viewed as different. The public identification of scapegoats gives the appearance that something is being done to protect the so-called general public; at the same time, it diverts public attention from those who have the power to assign labels. In the United States, the early identification

scapegoat A person or group blamed for conditions that (a) cannot be controlled, (b) threaten a community's sense of well-being, or (c) shake the foundations of an important institution.

of AIDS as the "gay plague" diverted attention from blood banks and the risks associated with medical treatments involving blood and blood products. Blood bank officials could maintain that the supply was safe as long as homosexuals abstained from giving blood.

From a sociological perspective, dispositional explanations that blame a group are simplistic and potentially destructive, not only to the group but also to the search for solutions. When a specific group and its behaviors emerge as targets, the solution is framed in terms of controlling that group. In the meantime, the problem can spread to members of other groups who believe that they are not at risk because they do not share the "problematic" attribute. This kind of misguided thinking about risk applies even to physicians and medical researchers. Attributions about who "should" have AIDS affect the way in which AIDS is defined and diagnosed. The definition and diagnosis, in turn, influence the statistics about who has AIDS. Such attributions also affect the content of the physician–patient interaction. In other words, if physicians do not believe a patient could be HIV-positive because he or she does not possess well-publicized attributes associated with the risk (homosexual, young, African), they will not recommend testing (Henderson 1998; Villarosa 2003). Research shows, for example, that physicians do not suspect HIV/AIDS in older patients until their condition has reached advanced stages (see "What Is HIV/AIDS?" on page 124).

In evaluating arguments about the origins of HIV/AIDS, we must recognize that we simply do not know how many people are infected with HIV in the United States or worldwide. To obtain information on who is actually infected, every country in the world would have to administer blood tests to a random sample of its population. Unfortunately (but perhaps not surprisingly), people resist being tested. A planned random sampling of the U.S. population sponsored by the CDC was abandoned after 31 percent of the people in the pilot study refused to participate, despite assurances of confidentiality (Johnson and Murray 1988).

Online Poll

I have changed the way I think about the emergence of HIV/AIDS since reading this account:

- ○ Strongly Agree
- ○ Agree
- ○ No change
- ○ Disagree
- ○ Strong Disagree
- ○ Explain your answer:

To see how other students responded to these questions, go to **www.cengagebrain.com.**

AIDSCOM
Education is not enough.

Courtesy of the National Library of Medicine

Summary of CORE CONCEPTS

When sociologists study interaction, they seek to understand the social forces that bring people together. And once people come together, sociologists seek to identify the factors that shape the interactions. This chapter applies sociological concepts and theories on social interaction to think about how a particular disease spreads through social interaction. That disease is HIV/AIDS. We consider how the disease emerged in Africa in 1930, to eventually affect people worldwide.

CORE CONCEPT 1 When sociologists study interactions, they seek to understand the larger social forces that bring people together in interaction and that shape the content and direction of that interaction.

Social interaction occurs when two or more people communicate and respond through language, gestures, and other symbols to affect one another's behavior and thinking. In the process, the parties involved define, interpret, and attach meaning to the encounter. When sociologists study social interaction, they seek to understand and explain the larger social forces that (1) bring people together in interaction, and (2) shape the content and course of interaction. With regard to HIV/AIDS, sociologists ask questions like: What forces increase the likelihood that HIV-infected and noninfected people will interact? What forces increase the likelihood that HIV-infected and noninfected people interact?

CORE CONCEPT 2 The division of labor is an important social force that draws people into interaction with one another and shapes their relationships.

Division of labor refers to work that is broken down into specialized tasks, each performed by a different set of workers specifically trained to do that task. The workers do not have to live near one another; they often live in different parts of a country or different parts of the world. Not only are the tasks geographically dispersed, but the parts and materials needed to manufacture products also come from many locations around the world. Division of labor is significant to understanding the forces that "pushed" African and European people together.

As the division of labor becomes more complex, solidarity or the social ties that bind people to one another shifts from mechanical to organic. Mechanical solidarity characterizes a social order based on a common conscience, or uniform thinking and behavior. Organic solidarity is a term applied to a social order based on interdependence and cooperation among people performing a wide range of diverse and specialized tasks. Durkheim maintained that societies and people become more vulnerable as the division of labor becomes more complex and more specialized. He was particularly concerned with the kinds of events that break down individuals' ability to connect with others in meaningful ways through their labor. Durkheim's theory is useful for placing HIV/AIDS in a broader context. For one, the European need for resources to support industrialization explains how Europeans and Africans became interdependent. Second, the forced division of labor and the means by which it was achieved are key to understanding how African people were made vulnerable to contracting the disease.

CORE CONCEPT 3 When analyzing any social interaction, or any human activity, sociologists locate the social structure in which it is embedded.

Social structures encompass at least four interrelated components: statuses, roles, groups, and institutions. The components help us to see how a social structure of exploitation emerged in the DRC and has been sustained over time. Taken together, we see that the exploitation of the DRC's people and resources by those inside and outside of the country has become institutionalized. That exploitation is, of course, key to understanding why HIV/AIDS rates are so high in the DRC.

CORE CONCEPT 4 Social interaction can be viewed as if it were a theater, people as if they were actors, and roles as if they were performances before an audience.

Social roles can be equated with the dramatic roles played by actors. In social situations, as on a stage, people engage in impression management, that is the management of the setting, their dress, their words, and their gestures to correspond to the impression they are trying to make or to an image they are trying to project. People alter their behavior depending on whether they are on the front stage (area visible to the audience, where people take care to create and maintain expected images and behavior) or back stage (area out of the audience's sight, where individuals let their guard down and do things that would be inappropriate or unexpected on the front stage). The concepts of impression management, front stage, and back stage help us to understand how Leopold engaged in impression management to disguise his exploitation of the then Congo Free State. These concepts also inform our understanding of why people give false answers to screening questions about sexual activities.

CORE CONCEPT 5 People assign causes to their own and others' behaviors. That is, they propose explanations for their own and other's behaviors, successes, and failures and then they respond accordingly.

In doing so, they often focus on dispositional (personal) or situational (external) causes. Dispositional causes are forces over which individuals are supposed to have control—including personal qualities or traits, such as motivation level, mood, and effort. Situational causes are forces outside an individual's control, such as the weather, bad luck, and others' incompetence. With regard to HIV/AIDS, it is clear that we have a better understanding of the disease origins and spread if we focus on situational factors that made African people vulnerable to the disease rather than searching for so-called personal decisions and behaviors as explanations. Also, the attributions we make about who has AIDS/HIV affects how we screen for and treat the condition.

Resources on the Internet

Login to CengageBrain.com to access the resources your instructor requires. For this book, you can access:

Access an integrated eBook, chapter-specific interactive learning tools, including flash cards, quizzes, videos, and more in your Sociology CourseMate.

Take a pretest for this chapter and receive a personalized study plan based on your results that will identify the topics you need to review and direct you to online resources to help you master those topics. Then take a posttest to help you determine the concepts you have mastered and what you will need to work on.

CourseReader

CourseReader for Sociology is an online reader providing access to readings, and audio and video selections to accompany your course materials.

Visit **www.cengagebrain.com** to access your account and purchase materials.

Key Terms

achieved statuses 114
ascribed statuses 114
back stage 122
dispositional causes 123
division of labor 106
dramaturgical model 121
front stage 122
group 117
impression management 121
institutions 119
master status 114
mechanical solidarity 109
organic solidarity 110
role 114
role conflict 116
role performance 115
role set 115
role strain 116
scapegoat 125
secondary groups 118
situational causes 123
social interaction 106
social status 114
social structure 114
solidarity 109
status set 114

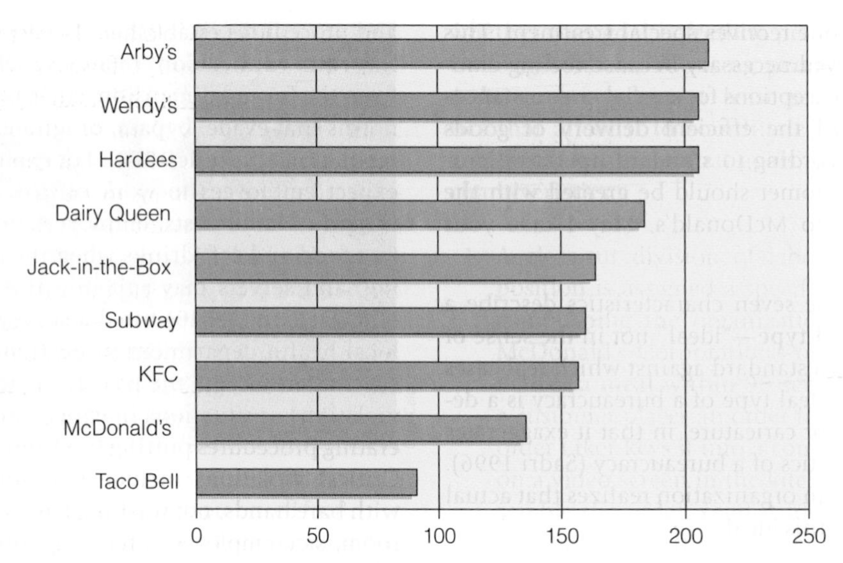

FIGURE 6.2 Number of Critical Violations at Nine Largest Fast Food Service Chains, 2009–2010

Source: Data from NBC Dateline (2010)

day, it can also have unintended, destructive consequences for workers, the public, and the environment. Rationalization applies to the dynamics underlying the never-ending quest to achieve a profit.

The growth and dominance of formal organizations goes hand in hand with a process known as rationalization. Weber defined **rationalization** as a process in which thought and action rooted in emotion (for example, love, hatred, revenge, or joy), superstition, respect for mysterious forces, or tradition is replaced by instrumental-rational thought and action. Through instrumental-rational thought and action, people strive to find the most efficient way to achieve a valued goal (Freund 1968).

One way to show how thought and action guided by emotion, tradition, superstition, or respect for mysterious forces differ from instrumental-rational thought and action is to compare two assumptions about farm animals. One assumption, portrayed by Matthew Scully (2003), presents animals (all animals, for that matter) as possessing complex emotions, including love and sorrow. Scully argues that humans have a moral obligation to treat animals with kindness and empathy. Cruelty toward animals is wrong, and when that cruelty "expands and mutates to the point where we no longer recognize the animals in a factory farm as living creatures capable of feeling pain and fear, . . . we debase ourselves" (Angier 2002, p. 9). Clearly, Scully believes that treatment of animals should be driven by emotion—kindness and empathy.

A second and contrasting assumption about farm animals and their purpose corresponds to the concept of instrumental-rational action. Here, how animals are treated cannot be separated from the desired goal of turning a profit. So factory farms, also known as concentrated animal feeding operations (CAFOs), raise thousands of cows and ten of thousands of chicken in tight quarters where they are fattened up for slaughter as quickly as possible. The living space is so tight that animals may not have room to lie down or turn around (Walsh 2009). Obviously, the more chickens, pigs, or cows a factory farm can house and the faster it can raise them, the more meat, eggs, and milk it can produce and sell. Egg and chicken suppliers raise chickens in crowded conditions that give each chicken an average of 49 square inches or 7 by 7 inches of space. Because chickens live in such close quarters, factory farm workers clip the wings and trim the beaks to prevent hens from injuring one another. Egg suppliers regularly practice "forced mating" by depriving hens of food and water for as long as two weeks (*Food Institute Report* 2000; Yablen 2000). Apparently this practice increases egg production.

Weber made several important qualifications regarding instrumental-rational thought and action. First, he used the term *rationalization* to refer to the way daily life is organized socially to accommodate large numbers of people, but not necessarily to the way individuals actually think (Freund 1968). So for example, most individuals believe that home-cooked meals are more nutritious than fast

rationalization A process in which thought and action rooted in custom, emotion, or respect for mysterious forces is replaced by instrumental-rational thought and action.

food, but they can't find the time to prepare food at home, nor can they resist the efficiency of fast food service:

> In a highly mobile society in which people are rushing, usually by car, from one spot to another, the efficiency of a fast food meal, perhaps without leaving one's car while passing by the drive-through window, often proves impossible to resist. The fast food model offers us, or at least appears to offer us, an efficient method for satisfying many of our needs. (Ritzer 1993, p. 9)

Second, rationalization does not assume better understanding or greater knowledge. In fact, Weber argues that the so-called primitive peoples can negotiate the environment in which they live without help from strangers and know how to acquire the food they consume (Freund 1968). People who live in an instrumental-rational environment typically know little about their surroundings (nature, technology, the economy). Consumers buy any number of products in the grocery store without knowing how they were made or what they consist of. People are not troubled by such ignorance; rather, they are content to let corporations set food choices.

Finally, when people identify a desired goal and decide on the means (actions) to achieve it, they seldom consider or dismiss less profitable or slower ways to achieve it. The problem may be as seemingly simple as growing potatoes that can be processed into French fries or potato chips that look and taste the same. In creating such a potato, people fail to consider that the demand for uniformity limits the varietal range of potatoes to those that are high-yielding, high in dry matter, low in sugar content, long and oval in shape, and uniform in color and flavor. With the help of science, such potatoes can be produced, but they require heavy doses of chemicals and threaten the longer-term viability of domestic potato production to meet commercial processing requirements (International Potato Center 1998).

The McDonaldization of Society

CORE CONCEPT 3 One organizational trend guided by instrumental-rational action is the McDonaldization of society, a process in which the principles governing fast food restaurants come to dominate other sectors of society.

Of course, McDonald's and other fast food service organizations are not the only formal organizations that strive to deliver a product or service by the most efficient means possible. Sociologist George Ritzer (1993) describes a larger organizational trend guided by instrumental-rational action: the "McDonaldization" of society.

Ritzer sees the **McDonaldization of society** as "the process by which the principles of the fast food restaurant are coming to dominate more and more sectors of American society as well as the rest of the world" (p. 1). Those principles are (1) efficiency, (2) quantification and calculation, (3) predictability, and (4) control. **Efficiency** is an organization's claim to offer the "best" products and services, which allow consumers to move quickly from one state of being to another (for example, from hungry to full, from fat to thin, from uneducated to educated, or from sleeplessness to sleep). **Quantification and calculation** are numerical indicators that enable customers to evaluate a product or service easily (for example, get delivery within 30 minutes, lose ten pounds in 10 days, earn a college degree in 24 months, limit menstrual periods to four times a year, or obtain eyeglasses in an hour). **Predictability** is the expectation that a service or product will be the same no matter where or when it is purchased. With regard to food products, this kind of predictability requires that they be genetically and chemically modified. If consumers expect cheese, for example, to have melted and to taste the same each time they eat it, it cannot be a naturally made cheese (Barrionuevo 2007). **Control** involves guiding, monitoring, and regulating the production and delivery of a service or product (for example, by assigning a limited task to each worker, by filling soft drinks from dispensers that shut off automatically, or having customers stand in roped off lines). Control is often achieved by replacing employees with technological innovations such as robots or computer-activated voices.

Ritzer's model allows for amazing organizational feats: each year, McDonald's handles 22 billion customer visits worldwide. Keep in mind that McDonald's does not produce any of the food items it sells—it thaws, assembles, fries, and warms food processed by other organizations. For example, Simplot, Lamb Weston, and McCain Food control 80 percent of the frozen French fry market in the United States. Simplot supplies McDonald's (ProPotato 2011, Schlosser 2002).

McDonaldization of society "The process by which the principles of the fast food restaurant are coming to dominate more and more sectors of American society as well as the rest of the world" (Ritzer 1993, p. 1).

efficiency An organization's claim of offering the "best" products and services, which allow consumers to move quickly from one state of being to another (for example, from hungry to full, from fat to thin, or from uneducated to educated).

quantification and calculation Numerical indicators that enable customers to evaluate a product or service easily.

predictability The expectation that a service or product will be the same no matter where or when it is purchased.

control The guiding or regulating, by planning out in detail, the production or delivery of a service or product.

No Borders No Boundaries Countries without a McDonald's

The first McDonald's restaurant opened in 1955, in Des Plaines, Illinois. Since then, operations have spread to more than 100 other countries. Franchise holders are McDonald's employees and independent businesses. That is, they operate their own restaurants but must adhere to strict operating guidelines (Waters 1998). One way the corporation builds its global identity across units is by requiring all franchise holders to attend a two-week course offered in 22 languages on quality control and management at one of its four Hamburger Universities or other 22 regional training centers (McDonald's Corporation 2011a).

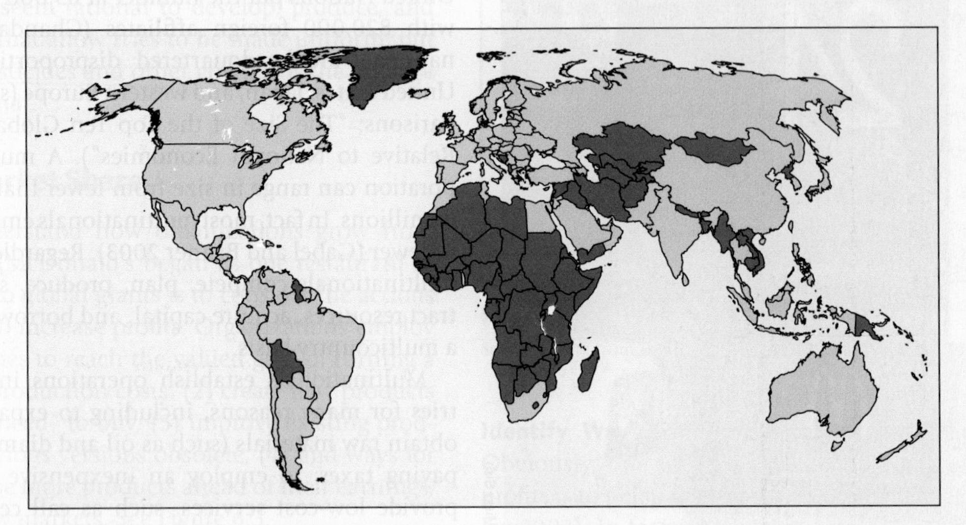

FIGURE 6.4 Countries in the World Without a McDonald's Franchise
The countries shaded in blue are countries that do not have a McDonald's franchise. What do you think that says about the country?
Source: Wikipedia (2009)

itself on foreign countries. Rather, he argues, governments around the world actively recruit the company.

In reality, we cannot make a simple evaluation that would apply to all multinationals. Obviously, on some level, multinational corporations "do spread goods, capital, and technology around the globe. They do contribute to a rise in overall economic activity. They do employ hundreds of thousands of workers around the world, often paying more than the prevailing wage" (Barnet and Müller 1974, p. 151). Critics, however, argue that, if anything, multinationals aggravate these problems, because the pursuit of profits creates social inequalities and ecological imbalances.

One can also argue that multinationals are not responsible for inequality or for other problems such as obesity. As one U.S. federal court judge stated: "a person knows or should know that eating copious orders of supersized McDonald's products is unhealthy, or may result in weight gain, it is not the place of the law to protect them from their own excesses. . . . Nobody forced them to eat at McDonald's" (Weiser 2003). McDonald's former CEO Jack Greenburg (2001) elaborated on the issue of healthy foods by arguing that "we're selling meat and potatoes and bread and milk and Coca-Cola and lettuce and everything else you can buy in a grocery store. What you choose to eat is a personal issue. Every nutritionist I've talked to says a

As measured by revenue, Wal-Mart is the world's largest global corporation, at $408 billion. This amount makes Wal-Mart the world's 28th largest economy in the world, as only 27 countries have a GNP larger than $408 billion (U.S. Central Intelligence Agency 2011).

Global Comparisons The Size of the Top Ten Global Corporations Relative to National Economies

One reason that the largest multinational organizations have great influence on the societies in which they operate is related to their size. Taken together, the combined annual revenue of the ten largest global corporations is about $2.34 trillion. Only five countries in the world—the United States, China, Japan, India, and Germany—possess a gross national product that exceeds that amount. The annual revenue of the world's largest corporation, Wal-Mart, exceeds $408 billion. Only 27 countries have a gross national product that exceeds that amount: the United States, China, Japan, India, Germany, the United Kingdom, Russia, France, Brazil, Italy, Mexico, South Korea, Spain, Canada, Indonesia, Turkey, Australia, Iran, Taiwan, Poland, the Netherlands, Saudi Arabia, Argentina, Thailand, South Africa, Egypt, and Pakistan.

The World's Largest Global Corporations, 2010

Corporation	Revenues (in $ millions)	Profits (in $ millions)	Headquarters
Wal-Mart Stores	408,214	14,335	Arkansas
Royal Dutch Shell	285,129	12,518	Netherlands
Exxon Mobil	284,650	19,280	Irving, Texas
BP	246,138	16,578	London, United Kingdom
Toyota Motor	204,106	2,256	Aichi, Japan
Japan Post Holdings	202,196	4,849	Tokyo, Japan
Sinopec	187,518	5,756	Beijing, China
State Grid	184,496	–343	Beijing, China
AXA	175,257	5,012	Paris, France
China National Petroleum	165,496	10,272	Beijing, China

Source: Data from *Fortune* 2010b.

balanced diet is the key to health. You can get a balanced diet at McDonald's. It's a question of how you use McDonald's. Nobody's mad at the grocery store because you can buy potato chips and pastries there. Nobody wants a full diet of that either."

Chris Caldeira

McDonald's menu includes such items as real fruit maple oatmeal, 1% low-fat milk, Fruit 'N Yogurt Parfait, apples, and a variety of salads. Still, its most popular items remain fries, Big Mac, Quarter Pounder, Chicken McNuggets, and Egg McMuffins.

To complicate matters, corporations claim that they merely respond to consumer tastes. For example, virtually all of the major fast food companies have introduced low-fat foods on their menus, and most have proven unpopular with consumers.

Nevertheless, many people question whether corporations should have the right to ignore the larger long-term effects of their products and business practices on people and the environment, even as they respond to consumer demand. The most profitable product for a corporation may prove costly for a society due to **externality costs**—hidden costs of using, making, or disposing of a product that are not figured into the price of the product or paid for by the producer. Yet, someone must eventually pay these costs (Lepkowski 1985). Such costs include those for cleaning up the environment and for medical treatment of injured workers, consumers, or other groups (see "The Obesity Epidemic in the United States" and Table 6.1).

externality costs Hidden costs of using, making, or disposing of a product that are not figured into the price of the product or paid for by the producer.

Culturally valued goal: One child per couple
Culturally valued means: One-child limit
Sources of structural strain: Preference for boys

Response	Goal Population control	Means One-child policy	Examples
Conformists	+	+	• Couples with no preference • "Ideologically sound" couples
Innovators	+	–	• Couples who abort female or unhealthy babies • Couples who abandon babies
Ritualists	–	+	• Couples who follow the rules but reject policies
Retreatists	–	–	• Couples who reject the idea of population control and hide the "extra" babies
Rebels	+/–	+/–	• Couples who replace official goals and means with new goals and means

+ Acceptance/achieve valued goals or means
– Rejection/fail to achieve valued goals or means

FIGURE 7.5 Merton's Typology Applied to China's One-Child Policy.

a child, accept the sex of the child (that is, do not abort female fetuses or kill a daughter), report the birth and sex of the child to appropriate agencies, and practice birth control to avoid conceiving other children.

We can apply Merton's typology of responses to structural strain to describe the reactions of Chinese couples to this one-child policy (Figure 7.5). Couples most likely to be conformists are those whose first child is a healthy son. Conformists would also include couples who have no preference as to the sex of their child and couples who are firmly committed to upholding the laws related to birth control because they see them as critical to China's quality of life (Bolido 1993, p. 6). Most people in urban China can be classified as conformists.

Innovators accept the culturally valued goal of one child per couple but reject the package of legitimate means to obtain this goal. Upon learning that she is expecting a child, a woman may undergo an ultrasound exam to learn the sex of the fetus. If it is female, the couple may decide to abort the fetus. Alternatively, upon the birth of a girl baby, the parents may kill her or have a midwife kill her. Such practices are blamed for the so-called "missing girls" problem. In response, many rural governments in China have launched "respect girls" campaigns backed by special incentives, such as free tuition to poor families with one or more girls (Yardley 2005).

A second category of innovators includes couples who abandon a baby considered unhealthy, a female child, or a second child born in violation of family planning regulations. These children fill the country's 67 state-run orphanages (Tyler 1996). (See No Borders, No Boundaries: "Immigrant Orphans Adopted by U.S. Citizens.")

Ritualists reject the one-child-per-couple goal of population control, but they adhere to the rules. They do not agree with the government policies, but they are afraid they will be punished if they do not follow them (Remez 1991).

Retreatists reject the one-child goal as well as the legitimate means open to them. This category of deviants include couples who continue having children until a son is born, adopt out girls, and hide the births of girls and second babies from party officials. In fact, Stirling Scruggs (1992) of the United Nations Population Fund argues that many of the so-called missing girls have just never been officially registered as having been born. Because of the high number of unreported births, some experts claim that it is impossible to know the actual size of China's population (Rosenthal 2000). In a sense, these retreatist parents are in society but, because of their secret, "not of it." So they can get a true picture of population size, the Chinese government has indicated that these undocumented children will be eligible for household registration cards—children must be registered to gain admission to schools and to find jobs as adults (China Daily 2010).

Rebels reject the culturally valued goal of population control as well as the legitimate means of achieving it; instead, they introduce new goals and new means. One

No Borders No Boundaries Immigrant Orphans Adopted by U.S. Citizens

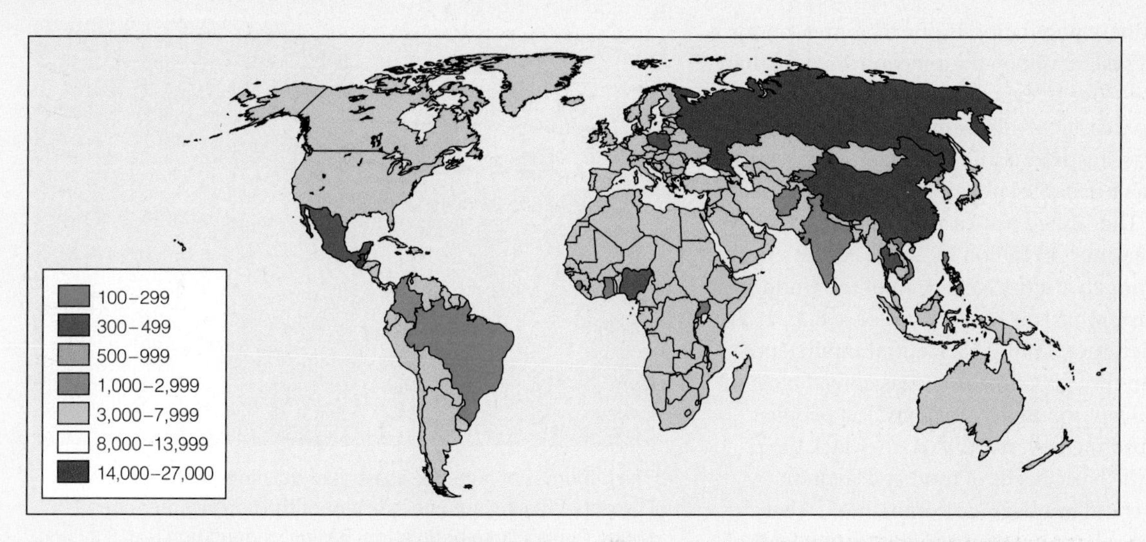

Country	# of Adoptions to U.S. 2005–2009
China	26760
Guatemala	17523
Russia	14079
Ethiopia	6428
South Korea	6084
Ukraine	3000
Kazakhstan	2565
Vietnam	2220
India	1658
Colombia	1484
Haiti	1366
Philippines	1349

Country	# of Adoptions to U.S. 2005–2009
Liberia	1104
Taiwan	984
Mexico	424
Nigeria	386
Poland	351
Thailand	308
Ghana	200
Brazil	187
Kyrgyzstan	132
Nepal	128
Uganda	123
Jamaica	116

FIGURE 7.6 Adoptions In U.S. Involving International Children by Country, 2005–2009
Study the list of countries from which Americans adopt international children. Notice that China leads the list. Why do you think Americans adopt children from abroad?
Source: Data from U.S. Department of State (2011)

could argue that this response applies to couples who belong to any of China's 55 ethnic minority populations, which make up 8.5 percent (114 million) of the total population (U.S. Central Intelligence Agency 2011). Most of these ethnic groups are exempt from the one-child policy. The government permits couples from such a group to have—depending on the group—two, three, or even four children. For these groups, the culturally valued goal is to increase the size of the minority populations, and the means of achieving that goal is exemption from the one-child policy. This option has prompted many couples to claim ethnic minority status (Tien et al. 1992). (See "The One-Child Policy in the Context of China's Population.")

Differential Association and Opportunities

CORE CONCEPT 8 Criminal behavior is learned; thus, criminals constitute a special type of conformist in that they conform to the norms of the group with which they associate.

Coined by sociologists Edwin H. Sutherland and Donald R. Cressey (1978), differential association explains how deviant behavior, especially juvenile delinquency, is learned. This theory states that exposure to criminal patterns and isolation from anticriminal influences put

The One-Child Policy in the Context of China's Population

The size of the Chinese population is one major reason for the country's rigid system of population control. More than 1.3 billion Chinese—one of every five people alive in the world—live in a space roughly the same size as the United States. After subtracting deserts and uninhabitable mountain ranges, China's habitable land area is about half that of the United States. The United Nations estimate that China's population will peak at 1.45 billion in 2030 (People's Daily Online 2010). Although almost 20 percent of the world's population lives in China, this country has only 6.3 percent of the world's agricultural land (U.S. Central Intelligence Agency 2011). Even though China manages to feed most of its population well, approximately 21.5 million people live in a state of absolute poverty; that is, they lack the resources to satisfy the basic needs of food and shelter. "Most Americans could not begin to comprehend what 21.5 million Chinese have to endure and what that level of deprivation is all about" (Piazza 1996, p. 47).

Another 150 million live in a state of substantial deprivation; that is, they live on less than $1.25 per day (China Daily 2010). It is believed that 240 million people from rural villages have migrated to the cities in search of work. "It is the largest migration in human history" (Gifford 2007, p. xvii, People's Daily Online 2010).

On average, 16.4 million births and 9.4 million deaths occur every year in China. If these patterns persist, the total population could increase by another 70 million people over the next 10 years. To complicate the situation, the population has expanded rapidly in the past 60 years, growing from approximately 500 million in 1949 to 1.33 billion in 2011 (Bureau of the Census 2011). Such rapid growth strains China's ability to house, clothe, educate, employ, and feed its people and has overshadowed its industrial and agricultural advancements. When one considers that approximately 1.3 billion people live under approximately 63 independent governments (not including island states) in the European Union and South America, it is easy to appreciate why the Chinese government is concerned (U.S. Central Intelligence Agency 2011). The one-child policy is associated (real or imagined) with a number of important changes in China, including the following: (1) the 4-2-1 problem when a child is cared for (and spoiled by) four grandparents and two parents, for whom the child must later care; (2) the Little Emperor/Empress syndrome, a situation in which only children who have been indulged become spoiled, rebellious, and self-centered; (3) a rise in childhood obesity and tooth decay as parents overfeed children; and (4) a rise in dog ownership as households acquire a dog to serve as a companion to only children and to people whose children have grown up (China Daily 2010, Wanli 2010).

Boni Li

The hundreds of high-rise apartment buildings in the Chaoyang District of Beijing are one indicator of the housing pressures facing China's large population. Beijing's population is 15.8 million.

people at risk of turning criminal. These criminal contacts take place within **deviant subcultures,** groups that are part of the larger society but whose members share norms and values favoring violation of that larger society's laws. People learn criminal behavior from closely interacting with those who engage in and approve of law-breaking activities. It is important to keep in mind, however, that contact with deviant subcultures does not by itself make criminals. Rather, there is some unspecified tipping point of exposure in which criminal influences offset exposure to law-abiding influences (Sutherland and Cressey 1978). If we accept the premise that behavior defined as criminal is learned, then criminals constitute a special type of conformist in that they are simply following the norms of the subculture with which they associate.

The theory of differential association does not explain how people make initial contact with a deviant subculture or the exact mechanisms by which people learn criminal behavior, except that individuals learn deviant subculture's rules the same way any behavior is learned. Sociologist

deviant subcultures Groups that are part of the larger society but whose members share norms and values favoring violation of that larger society's laws.

Terry Williams (1989) offers one example of how teenagers can make contact with a deviant subculture. He studied a group of teenagers, some as young as 14, who sold cocaine in the Washington Heights section of New York City. These youths were recruited by major drug suppliers because, as minors, they could not be sent to prison if caught. Williams argues that the teenagers were susceptible to recruitment for two reasons: They saw little chance of finding high-paying jobs, and they perceived drug dealing as a way to earn money to escape their circumstances. Williams's findings suggest that once teenagers become involved in drug networks, they learn the skills to perform their jobs the same way everyone learns to do a job. Indeed, success in an illegal pursuit is measured in much the same way that success is measured in mainstream jobs: pleasing the boss, meeting goals, and getting along with associates.

Williams's research suggests that criminal behavior is not simply the result of differential association with criminal ways. There are other factors at work, including **illegitimate opportunity structures**—social settings and arrangements that offer people the opportunity to commit particular types of crime. In the case of Williams's study, drug suppliers recruited 14-year-olds because minors would not go to prison. The larger society offers minors an opportunity structure that allows them to engage in criminal activity without risking the full penalties of the law. Opportunities to commit crimes are shaped by the environments in which people live or work (Merton 1957). That is, for someone to embezzle money, another person has to have entrusted a would-be embezzler with a large sum of money; the act of entrusting money has to occur before the embezzler can set money aside for some unintended purpose.

Illegitimate opportunity structures figure into the type of crimes people commit. Working as a pharmacist offers the opportunity to steal prescription drugs to sell or to feed a personal or friend's addiction. For addicts without money, the only option available to them is to hold up a pharmacy. White-collar and corporate criminals also benefit from differential opportunity. **White-collar crime** consists of "crimes committed by persons of respectability and high social status in the course of their occupations" (Sutherland and Cressey 1978, p. 44). **Corporate crime** is committed by a corporation through the way that it does business as it competes with other companies for market share and profits. In the case of white-collar crime, offenders are part of the system: They occupy positions in the organization that permit them to carry out illegal activities discreetly. In the case of corporate crime, everyone in the organization contributes to illegal activities simply by doing their jobs. Both white-collar and corporate crimes are often aimed at impersonal—and often vaguely defined—entities such as the tax system, the environment, competitors, and so on. Neither kind of crime is aimed at a specific person who can report the offense to the police (National Council for Crime Prevention in Sweden 1985).

White-collar and corporate crimes—such as the manufacturing and marketing of unsafe products, unlawful disposal of hazardous waste, tax evasion, and money laundering—are usually handled not by the police but by regulatory agencies (such as the Environmental Protection Agency, the Federal Bureau of Investigation, and the Food and Drug Administration), which have minimal staff to monitor compliance. Escaping punishment is easier for white-collar and corporate criminals than for other criminals.

The concept of an illegitimate opportunity structure challenges the belief that the uneducated and people in certain minority groups are more prone to criminal behavior than are those in other groups. In fact, crime exists in all social strata, but the type of crime, the extent to which the laws are enforced, access to legal aid, and the power to sidestep laws vary by social strata (Chambliss 1974). In the United States, police efforts are largely directed at controlling crimes against individual life and property (crimes such as drug offenses, robbery, assault, homicide, and rape) rather than patrolling office buildings looking for white-collar and corporate criminals. Less than one percent of all people sentenced to U.S. federal prisons seem to fall under the category of white-collar criminals, whereas 51 percent are classified as drug offenders (Federal Bureau of Prisons 2011).

In China, notions of differential association are the basis of the philosophy underlying rehabilitation: A deviant individual becomes deviant because of "bad"—education or associations with "bad" influences. To correct these influences, society must reeducate deviant individuals politically, inspire them to support the Communist Party, and teach them a love of labor. These principles guided Communist Party leaders in their handling of the Chinese students who participated in a series of pro-democracy demonstrations in Tiananmen Square in 1989. The government eventually suppressed these demonstrations violently, accusing the students of "counterrevolutionary revolt" and "bourgeois liberalization." Although these terms have never been clearly defined, they suggest wanton expressions of individual freedom, which threatened the country's stability

illegitimate opportunity structures Social settings and arrangements that offer people the opportunity to commit particular types of crime.

white-collar crimes Crimes committed by those with high status, respectable positions as they carry out the duties and responsibilities of their occupation.

corporate crimes Crimes committed by a corporation through the way that it does business as it competes with other companies for market share and profits.

Theories of Deviance: Summary

Theory	Central Question	Answer
Functionalist	How does deviance contribute to order and stability?	Deviance—especially the ritual of identifying and exposing wrongdoing, determining a punishment, and carrying it out—is an emotional experience that binds together members of groups and establishes a sense of community.
Labeling theory	What is deviance?	Deviance depends on whether people notice it and, if they do notice it, on whether they label it as such and subsequently apply sanctions/punishment.
Obedience to authority	How do rule makers get people to accept their definition of deviance and to act on those definitions?	Behavior that is unthinkable in an individual acting on his or her own may be executed without hesitation when authority figures command such behavior.
Constructionist	How do specific groups, activities, conditions, or artifacts come to be defined as problems?	Claims makers with the ability to define something as a problem, to have those claims heard, and to shape public responses play important roles in defining deviance and responses to it.
Structural strain	Under what conditions do people engage in behavior defined as deviant?	People engage in behavior defined as deviant when valued goals have no limits or clear boundaries, when legitimate means do not guarantee valued goals, and when the number of legitimate opportunities is in short supply.
Differential association opportunity	How is behavior defined as deviant learned? What role does opportunity to commit crimes play?	People learn deviant behavior through close associations and interactions with those who engage in and approve of criminal behavior.

and unity. Such expressions had to be curbed. In the government's view, students who demonstrated led a sheltered life. As a result, they ignored larger collective interests and concerns and failed to grasp the complexities of government reforms (Kwong 1988, pp. 983–984).

The government undertook a number of measures to persuade the students to learn more about the complexities of life and resist subversive ideas. These measures included sending them "to rural areas to teach them to endure hardship, work hard and appreciate the daily difficulties faced by China's mostly rural population" (Kristof 1989, p. Y1). The rationale was that proper ideological commitment could be instilled through association with the masses and through manual labor. Other measures included limiting the number of students entering the humanities and social sciences. In the year following the Tiananmen Square incident, almost no students were admitted to study academic subjects that government officials considered "ideologically suspect," such as history, political science, sociology, and international studies (Goldman 1989).

Online Poll

Can you think of a time when you obeyed the "orders" given by someone in a position of authority even though you did not approve of what that person was asking you to do?

○ Yes

○ No

To see how other students responded to these questions, go to **www.cengagebrain.com.**

Summary of CORE CONCEPTS

In this chapter, we focused on changing conception of deviance in the People's Republic of China. We used the sociological concepts and theories related to deviance to understand how something deviant at one time and place is not deviant in another. When sociologists study deviance, they do not focus on deviant individuals per se but on the context in which deviance occurs and on the reactions and judgments of others, including the larger society, the group whose norms are broken, and rule makers and enforcers.

CORE CONCEPT 1 The only characteristic common to all forms of deviance is that some social audience challenges or condemns a behavior or an appearance because it departs from established norms.

Deviance is any behavior or physical appearance that is socially challenged or condemned because it departs from the norms and expectations of a group. Conformity comprises behaviors and appearances that follow and maintain the standards of a group. All groups employ mechanisms of social control—methods used to teach, persuade, or force their members, and even nonmembers, to comply with and not deviate from norms and expectations.

CORE CONCEPT 2 Ideally, conformity is voluntary. When socialization fails to produce conformity, other mechanisms of social control—sanctions, censorship, or surveillance—may be used to convey and enforce norms.

Because socialization begins as soon as a person enters the world, one has little opportunity to avoid exposure to the culture's folkways and mores. That exposure may take place in schools. For example, Chinese preschoolers are taught to give constructive critiques of their classmates' work and to learn from critiques of their own and others' work. Chinese preschoolers are also taught to downplay interpersonal conflicts and play cooperatively with other children. In contrast, American preschoolers are taught to express themselves and expect praise. They are also taught to seek an adult to mediate conflicts with peers and decide who is right and wrong. In addition, American preschoolers are encouraged to express in words their feeling about conflict.

When conformity cannot be achieved voluntarily, other mechanisms of social control may be used to convey and enforce norms. These mechanisms include sanctions, censorship, and surveillance. Sanctions can be positive (reward oriented) or negative (punishment oriented), informal (spontaneous and unofficial), or formal (official, backed by force of law).

CORE CONCEPT 3 It is impossible for a society to exist without deviance. Always and everywhere there will be some behaviors or appearances that offend collective sentiments.

Durkheim argued that although deviance does not take the same form everywhere, it is present in all societies. He defined *deviance* as acts that offend collective norms and expectations. Durkheim believed that deviance is normal as long as it is not excessive, and that no society can exist without deviance. First, the ritual of identifying and exposing wrongdoing, determining a punishment, and carrying it out is an emotional experience that binds together the members of a group and establishes a sense of community. Second, deviance helps bring about necessary change.

CORE CONCEPT 4 Labeling theorists maintain that an act is deviant when people notice it and then take action to label it as a violation and apply appropriate sanctions.

Labeling theorists maintain that violating a rule does not automatically make a person deviant. That is, a rule breaker is not deviant (in the strict sense of the word) unless someone *notices* the violation and decides to take corrective action. Labeling theorists suggest that for every rule a social group creates, four categories of people exist: conformists, pure deviants, secret deviants, and the falsely accused. Labeling theorists pay particular attention to power and authority to evade and create laws to punish crime.

CORE CONCEPT 5 The firm commands of a person holding a position of authority over a person hearing those commands can elicit obedient responses.

In *Obedience to Authority*, Stanley Milgram sought to explain how large numbers of people follow orders to carry out atrocities. He learned that this level of obedience can be founded simply on the firm command of a person with a social status that gives him or her even minimal authority over a subject.

CORE CONCEPT 6 In an effort to understand how deviance is defined, sociologists take a constructionist approach to study the role of claims maker and claims-making activities.

Claims makers are people who articulate and promote claims and who tend to gain in some way if the targeted audience accepts their claims as true. Claims-making activities draw attention to a claim and involve a range of activities, from filing lawsuits to holding signs. The success of a claims-making campaign depends on a number of factors such as access to the media. When constructionists study the claims-making process, they ask who makes the claims, whose claims are heard, how audiences respond, and so on.

CORE CONCEPT 7 Deviant behavior is a response to structural strain, a situation in which a disconnect exists between culturally valued goals and legitimate means for achieving those goals.

The theory of structural strain takes three factors into account: (1) culturally valued goals defined as legitimate for all members of society, (2) norms that specify the legitimate means of achieving those goals, and (3) the actual number of legitimate opportunities available to people to achieve the goals. Structural strain occurs when the valued goals have unclear limits, people are unsure whether the legitimate means will allow them to achieve the goals, or legitimate opportunities for meeting the goals remain closed to a significant portion of the population.

Merton believed that people respond in identifiable ways to structural strain and that their response involves some combination of acceptance and rejection of the valued goals and means. He identified the following five responses—conformity, innovation (such as taking illegal performance-enhancing drugs to achieve success in sports), ritualism, retreatism, and rebellion.

CORE CONCEPT 8 Criminal behavior is learned; thus, criminals constitute a special type of conformist in that they conform to the norms of the group with which they associate.

The theory of differential association explains how deviant behavior, especially juvenile delinquency, is learned. This theory states that it is exposure to criminal patterns and isolation from anticriminal influences that put people at risk of turning criminal. If we accept the premise that behavior defined as criminal is learned, then criminals constitute a special type of conformist. Other factors are at work beyond simply learning criminal behavior. One factor involved illegitimate opportunity structures—the social settings, and arrangements that offer people the opportunity to commit particular types of crime including white-collar and corporate crimes. In China, notions of differential association are the basis of the philosophy underlying rehabilitation: A deviant individual becomes deviant because of "bad"—education or associations with "bad" influences. To correct these influences, society must reeducate deviant individuals.

Resources on the Internet

Login to CengageBrain.com to access the resources your instructor requires. For this book, you can access:

Sociology CourseMate

Access an integrated eBook, chapter-specific interactive learning tools, including flash cards, quizzes, videos, and more in your Sociology CourseMate.

Take a pretest for this chapter and receive a personalized study plan based on your results that will identify the topics you need to review and direct you to online resources to help you master those topics. Then take a posttest to help you determine the concepts you have mastered and what you will need to work on.

CourseReader

CourseReader for Sociology is an online reader providing access to readings, and audio and video selections to accompany your course materials.

Visit **www.cengagebrain.com** to access your account and purchase materials.

Key Terms

SOCIAL STRATIFICATION

8 With Emphasis on THE WORLD'S RICHEST AND POOREST

This chapter focuses on the world's richest and poorest people. Among other things the chapter describes how wealth, income, and other valued resources are unequally distributed among the 7 billion people living on the planet. The inequalities are especially dramatic when we compare the lives of those living in the 22 richest and 50 poorest countries. What do you think it would be like to have grown up in the United States—one of the world's richest countries—and, after graduating from college, volunteered to go to Mauritania, which is among the world's poorest countries? That is what Scott McLaren, a student in one of my sociology classes, did. Scott is holding the child wearing the baseball cap.

Scott McLaren

Why Focus On THE WORLD'S RICHEST AND POOREST?

When sociologists study social stratification, they focus on the connection between social location and their life chances. A person's social location is a product of the categories humans have created. Those categories relate to nationality (an American, a Mauritanian, a Mexican), race (black, white, Asian), gender (male, female, transgender), age (under five, over 65), sexual orientation, social class, occupation, education, and other characteristics. A person's social location affects the chances he or she will survive the first year of life, will live beyond age 75, will play soccer or football, become literate, speak more than one language, and so on. One way to directly experience how social location affects life chances is to do what a student from one of my sociology classes did: Scott McLaren, a person who is considered white, male, American, a college graduate, 20-something, and from a solid middle-class community in the Midwest with a median household income of $62,000, joined the Peace Corps and was assigned to the African country Mauritania. He lived in villages where people survived on the equivalent of $1.00 per day.

On September 27, 2008, I received a letter from Scott. He wrote:

> I want to take a few minutes and share with you some thoughts regarding my first few months in Mauritania. Before I begin, I would like to say how appreciative I am of the fate that my last semester held within it the basics of sociology. That discipline has given me yet another set of eyes to view this experience. I have found that culture shock is but a mild symptom of travel and service in the Peace Corps. What is the most painful symptom is what I call self-death awareness—the self, being my understanding of who I am as revealed and shaped through my home culture and interactions with others in my society of origin. This awareness has hit me so very hard this week and I believe that it is at the epicenter of the repeated tidal waves of emotion within me. . . . I am now the barely-able foreigner who has to ask where to pee. I depend completely on others to sustain my needs and it is humiliating. The self-death shock has been debilitating this week and limited me to self-loathing and pity for my situation. . . . Self-death is painful; I am now beginning to deconstruct the person I had worked to be known as. I will be reconstructed into a person, not of my own making or choosing. My new self will be molded to the needs of survival.

Sociologists seek to understand the patterns of inequality that exist in the world. In that sense, they are interested in which countries the richest and poorest people tend to live and why some countries have larger percentages of poor and wealthy than others. Likewise, sociologists are interested in the patterns of inequality that exist within countries. They are obligated to ask the kinds of questions Scott faced head-on each day of his service in the Peace Corps:

- How do we explain the extremes of wealth and poverty in the world?
- Why should a very small percentage of the world's population enjoy an inordinate share of the income, wealth, and other valued resources, whereas so many others struggle to survive?
- Can we assume capitalism and globalization will correct these dramatic inequalities, or should we rethink the way wealth and other valued resources are distributed?

Note that Scott's experiences represent one person's experience in the Peace Corps. Peace Corp volunteers are assigned to many locations around the world, from Albania to Zambia. Scott's experiences do offer lived accounts of ways in which inequality is lived and experienced. It is also important to note that Scott's sensation of "self-death" can be a phase in an adjustment process. Eventually, Scott moved beyond that stage and came to enjoy his two-year assignment.

Online Poll

Because you could not afford it, do you do without something that could help you be successful in college, such as a textbook, computer, a printer, or Internet connection?

○ Yes

○ No

To see how other students responded to these questions, go to **www.cengagebrain.com.**

The Extremes of Poverty and Wealth in the World

It is not easy to define a condition of poverty except to say that it is a situation in which people have great difficulty meeting basic needs for food, shelter, and clothing. Poverty can be thought of in absolute or relative terms. **Absolute poverty** is a situation in which people lack the resources to satisfy the basic needs no person should be without. Absolute poverty is often expressed as a state of being that falls below a certain threshold or a minimum. In this regard, the United Nations has set the absolute poverty threshold in developing countries at the equivalent of US$1.00 per day. The World Bank (2009), on the other hand, believes that threshold should be set at US$1.25 per day. According to the United Nations (UN) threshold, 1.1 billion people live in a state of absolute poverty. Based on the World Bank threshold, that number is 1.4 billion people.

There are other criteria by which to determine the number of people living in absolute poverty. For example, if we use the lack of access to a toilet as a measure of absolute poverty, one-third of the world's people—2.6 billion—do not have a decent place to go to the bathroom. Some use plastic bags and then throw them into ditches and streets; others squat in fields, yards, and streams (Dugger 2006). Scott McLaren described the absolute poverty he observed in Mauritania in this way: Absolute poverty is

- eating the same meal every day;
- living in the same quarters as goats and cows;
- not having access to clean drinking water or enough water to bathe;
- sleeping on a plastic mat between your body and the sand; and
- no longer believing that one's life can be improved.

Relative poverty is measured not by some objective standard, but rather by comparing a person's situation against that of others who are more advantaged in some way. When thinking of poverty in relative terms, one thinks not just about an inability to meet basic needs, but about a *relative* lack of access to goods and services that people living in a particular time and place have come to expect as necessities. Today, in the United States, examples of such goods and services that people have come to "need" might be an automobile, a smart phone, satellite television service, and the Internet.

Extreme wealth, on the other hand, is the most excessive form of wealth, in which a very small proportion of people in the world have money, material possessions, and other assets in such abundance that a small fraction of it (if spent appropriately) could provide adequate food, safe water, sanitation, and basic health care for the 1 billion poorest people on the planet. About how many people in the world are excessively wealthy? The World Institute for Development Economics Research of the United Nations University (2006) released a groundbreaking study

Chris Caldeira

In an information society, owning a basic cell phone with no smart features may put someone at a disadvantage. The person with a basic flip phone may experience a kind of poverty relative to those who can afford smart phones.

absolute poverty A situation in which people lack the resources to satisfy the basic needs no person should be without.

relative poverty A situation measured not by some objective standard, but rather by comparing against that of others who are more advantaged in some way.

extreme wealth The most excessive form of wealth, in which a very small proportion of people in the world have money, material possessions, and other assets (minus liabilities) in such abundance that a small fraction of it (if spent appropriately) could provide adequate food, safe water, sanitation, and basic health care for the 1 billion poorest people on the planet.

TABLE 8.1 The Distribution of Global Household Wealth

Researchers at the United Nations University have estimated how wealth is distributed among the adult population of the world. Of course, a significant percentage of the adult population have children or will have children who share in their wealth or poverty. We know that the world's poorest peoples have the greatest number of children. The table shows various income categories and the number of adults in each category. To be classified as extremely wealthy, a person must have a minimum wealth (assets minus liabilities) of $1 billion. Worldwide, only about 800 adults are categorized as "extremely wealthy." To be among the richest 1 percent, a person must have between $500,000 and $1 million (excluding the value of their home) in wealth. The richest 1 percent holds 40 percent of the world's wealth. Note that the poorest 50 percent of the world's adult population (1.9 billion people) possess less than 1 percent of the world's wealth.

Category	Estimated Number of Adults in Category	Minimum Wealth (Assets — Liabilities)*	Percent of All Household Wealth Held	Amount of Wealth Held
Extremely wealthy	1,200	$1 billion	Not known	Not known
Ultra rich	85,400	$30 million	Not known	Not known
Richest 0.025%	11.2 million	$1 million	Not known	Not known
Richest 1%	37 million	$500,000	40%	$49.6 trillion
Richest 10%	370 million	$61,000	85%	$105 trillion
Middle 40%	1.4 billion	$2,200	14%	$17 trillion
Poorest 50%	1.9 billion	<$2,200	<1%	$1.2 trillion

* Does not include the value of the house in which the person resides.
Source: Data from United Nations University 2006; *Forbes*. 2011. "The World's Billionaires." http://www.forbes.com/wealth/billionaires

estimating the world's total household wealth and describing how it was distributed. The study found that 1 percent of the world's adult population (or 37 million people) own more than 40 percent of the global household wealth (estimated to be $124 trillion). Contrast that with the poorest 50 percent of the world's adult population, who own less than 1 percent of global wealth (see Table 8.1).

CORE CONCEPT 1 When sociologists study systems of social stratification, they seek to understand how people are ranked on a scale of social worth and how that ranking affects life chances.

Social stratification is the systematic process of ranking people on a scale of social worth such that the ranking affects life chances in unequal ways. Sociologists define **life chances** as the probability that an individual's life will follow a certain path and will turn out a certain way. Life chances apply to virtually every aspect of life—the chances that someone will survive the first year of life after birth, complete high school and go on to college, see a dentist twice a year, work while going to school, travel abroad, be an airline pilot, play T-ball, major in elementary education, own 50 or more pairs of shoes, or live a long life. **Social inequality** describes a situation in which these valued resources and desired outcomes (that is, a college education, long life) are distributed in such a way that people have unequal amounts and/or access to them.

Lance Cpl. Chris Kutlesa

Lady Gaga is among the richest 85,400 people in the world. She earned $62 million between June 2009 and June 2010 (Telegraph 2011.)

social stratification The systematic process of ranking people on a scale of social worth such that the ranking affects life chances in unequal ways.

life chances The probability that an individual's life will follow a certain path and will turn out a certain way.

social inequality A situation in which these valued resources and desired outcomes (that is, a college education, long life) are distributed in such a way that people have unequal amounts and/or access to them.

Every society in the world categorizes and ranks its people. Almost any criterion can be used (and, at one time or another, probably has been used) to categorize people and assign them a status, whether it be based on hair color and texture, eye color, physical attractiveness, weight, height, occupation, sexual preference, age, grade point average, test scores, or many others. People's status in society can be ascribed or achieved.

Ascribed versus Achieved Statuses

Ascribed statuses are social positions assigned on the basis of attributes people possess through no fault of their own—those attributes are acquired at birth (such as skin shade, sex, or hair color), develop over time (such as height, weight, baldness, wrinkles, or reproductive capacity), or are possessed through no effort or fault of their own (such as the country into which one is born and religious affiliation if inherited from family). **Achieved statuses** are attained through some combination of personal choice, effort, and ability. In other words, people must act in some way to acquire an achieved status. Achieved statuses include a person's wealth, income, occupation, and educational attainment.

Ascribed and achieved statuses may seem clearly distinguishable, but that is not the case. One can always think of cases where people take extreme measures to achieve a status typically thought of as ascribed; they undergo a sex change operation, lighten their skin, or hire a plastic surgeon to look a different age. Likewise, one's ascribed statuses affect one's opportunities to achieve wealth, a college education, certain occupations, or high income. We can also raise questions as to what statuses children are able to achieve independent of their parents.

The various achieved and ascribed statuses hold **social prestige**, a level of respect or admiration for a status apart from any person who happens to occupy it. There is social prestige associated with a person's occupation, level of education, income, race, sex, age, and so on. The social prestige accompanying each of these characteristics can complicate the overall experience of prestige. As a case in point, the occupation of professional golfer—in the abstract—is a prestigious one. But that prestige can be complicated by the prestige assigned to race, sex, and age. Social prestige is further complicated by **esteem**, the reputation that someone occupying an ascribed or

Tim Hipps

Tiger Woods achieved the status of professional golfer. There is a level of prestige associated with that status that applies to any professional golfer. Woods's prestige as a golfer, however, is complicated by his other achieved and ascribed statuses. For example, Woods has achieved the statuses of a divorced father of two. His ascribed statuses include being male, classified as black in the U.S. and over 35 years of age. Taken together, the various statuses he holds enhance or detract from the prestige that comes with being a professional golfer. The esteem with which Woods was held, however, declined when his actions off the course led to a highly publicized divorce. As one measure of this decline, his 2010 earnings were $74.2 million, about $48.5 million less than in 2009, and his lowest earnings over the past decade (Myers 2010).

ascribed statuses Social positions assigned on the basis of attributes people possess through no fault of their own—those attributes are acquired at birth (such as skin shade, sex, or hair color), develop over time (such as height, weight, baldness, wrinkles, or reproductive capacity), or are possessed through no effort or fault of their own (such as the country into which one is born and religious affiliation "inherited" from parents).

achieved statuses Attained through some combination of personal choice, effort, and ability.

social prestige A level of respect or admiration for a status apart from any person who happens to occupy it.

esteem The reputation that someone occupying an ascribed or achieved status has earned from people who know and observe the person.

Scott McLaren

Scott McLaren

The top photo shows Scott on the day he graduated from college. The other shows him wearing a bubu. In the United States, there is considerable prestige associated with being a college graduate. In Mauritania, the quality of fabric or, in some cases, the color of a bubu corresponds to the position someone holds in the community and, by extension, signals some level of prestige. The average Mauritanian wears a white or light blue bubu (particularly in Moor culture) but will wear other colors to distinguish themselves during social events where people's position in the community is on display.

achieved status has earned from people who know and observe the person.

In a related vein, sociologists are also interested in the opportunities and constraints associated with ascribed and achieved statuses (white skin versus brown skin, blonde hair versus dark hair, single versus married, professional athlete versus high school teacher). The compensation guidelines for the September 11, 2001, attacks on the United States illustrate. The actual awards ranged from $250,000 (least valued life) to $7.1 million (most valued life). The "least valued" included single, childless adults age 65 and older with an annual income of $10,000. Under the guidelines, their relatives were awarded a one-time payment of $300,000. The "most valued" included married adults age 30 and younger with two children and an annual income of $225,000. Their relatives were awarded a one-time payment of $3,805,087. If the annual income had been $101,000 (rather than $225,000), the award would have been significantly less—$694,588 (Chen 2004; September 11 Victim Compensation Fund 2001).

Life Chances across and within Countries

Broadly speaking, what does it mean for a baby to be born in the United States as opposed to, say, Mauritania? For one thing, the country into which one is born has an incalculable effect on a person's aspirations about what they might accomplish in life. Scott McLaren has observed that Mauritanian boys, especially those who live in rural villages, "talk about herding large flocks of goats or owning lots of cattle. Cattle here are a sign of wealth and privilege. Many also talk about going to Nouakchott (the capital city) to open boutiques or become cab drivers. Very few really even mention going to university." By contrast, many American children no matter their situation in life are taught to believe they can be anything they want, however unrealistic that sentiment. And if they do not accomplish their dreams, it is largely their own fault.

We have no control over which of the world's 243-odd countries we are born in; in that sense, one's nationality is an ascribed status. Still, that country has important effects on life chances. The chances that a baby will survive the first five years of life depend largely on the country where he or she is born. Babies born in Sweden and Japan have some of the best chances of surviving their first five years, as fewer than 3 of every 1,000 babies born there die before reaching the age of 5. Babies born in Angola and Afghanistan have some of the worst chances of surviving that first year of life; some 175.9 babies die within the first five years of life (see Table 8.2).

One startling example of how life chances vary by country relates to consumption patterns. Table 8.3 shows countries with the highest and lowest known consumption on selected items. Note, on the whole, that people in

TABLE 8.2 Selected Life Chances in the Country with the Lowest and the Highest Infant Mortality Rates

In the African country of Angola, 17.6 percent of babies—175.9 of every 1,000 born—die before reaching age 1. In Sweden, 0.0027 percent—2.7 of every 1,000 babies—die before reaching age 1. In Angola, the average woman has 6 children, compared to the average Swedish woman, who has 1.1 children. Largely because of the high mortality rate among children, the life expectancy is 38.8 years in Angola—some 43 years shorter than the life expectancy in Sweden.

Life Chance	Angola	Sweden
Under-1 mortality rate per 1,000	175.9	2.7
Fertility rate (average number of babies born to woman over lifetime)	6.0	1.1
Life expectancy at birth	38.8	82
Percentage of population living in poverty	40.5	essentially 0

Sources: Data from CIA 2011, UNICEF 2009

TABLE 8.3 Selected Types of Consumption: Countries Believed to Have Highest and Lowest Rates of Per Capita Consumption

Type of Consumption	Highest Known Consumption Per Capita	Lowest Known Consumption Per Capita
Per Capita Calories Consumed per Day	United States-3,754 calories per person	Democratic Republic of the Congo-1,606 calories per person
Per Capita Liters of Bottled Water Consumed per Year	Italy-155 liters per person	South Africa-2.4 liters per person
Personal Motor Vehicles per 1,000 People	United States-765 motor vehicles per 1,000 people	Afghanistan Less than 1 vehicle per 1,000
Oil Consumption Barrel per Day per 1,000 People	Singapore-189.9 barrels per 1,000 people	Chad-31.7 barrels per 1,000 people
Televisions per 10,000 People	United States-7,400 televisions per 10,000	Eritrea-2 televisions per 10,000

Source: Data from Nationmaster.com (Retrieved 2011).
Note: Statistics represent the most recent year data is available.

the United States have the highest access to a high-calorie diet, cars, and television. Those who live in Democratic Republic of the Congo consume the fewest calories on average. People in Afghanistan have almost no access to motor vehicles.

Ranking countries according to specific kinds of life chances and consumption captures only part of the story. That is because life chances vary within countries as well. In the United States, on average, 8 babies per 1,000 live births die before reaching age 5. But the chances of survival vary by racial and ethnic classification (see Table 8.4).

One way of gauging the inequalities that exist within individual countries is to compare the incomes of their richest and poorest residents. From a global perspective, the African country of Sierra Leone, a former British colony, has the greatest inequality between the richest and poorest 20 percent of its population, with the richest segment receiving 58 times more income than the poorest. Azerbaijan, a largely socialist country that was once part of the Soviet Union, has the greatest equality, with the richest 20 percent receiving 2.5 times the income of the poorest 20 percent (World Bank 2011).

Within the United States, the average 2007 after-tax income of the richest 20 percent living in the United States is $198,300, or 11.2 times that of the poorest 20 percent, who average $17,700. That is, for every $1,000 of taxed

caste system Any form of stratification in which people are categorized and ranked by characteristics over which they have no control and that they usually cannot change.

class system A system of social stratification in which people are ranked on the basis of achieved characteristics, such as merit, talent, ability, or past performance.

TABLE 8.4 Infant Mortality Rates in U.S. by Race/Ethnicity

In the United States, an average of 6.8 babies per 1,000 live births die before reaching age 1. But the chances of survival vary by race and ethnic classification, with 13.7 babies classified as black and 9.2 babies classified as Native American dying in the first year of life for every 1,000 live births.

Race/Ethnicity	Under Age 1 Infant Mortality (per 1,000 live births)
American Indian/Alaska Native	9.2
Asian/Pacific Islander	4.8
Black/African American	13.7
White	5.6
Hispanic/Latino	5.9
Average, all groups	6.8

Sources: Data from U.S. Central Intelligence Agency 2011; Centers for Disease Control 2010.

income earned by the poorest one-fifth, the top one-fifth earns $11,203. When we compare the after-tax income of the top 1 percent with that of the bottom 20 percent, the inequality is even greater. That 1 percent's after-tax income is $1.3 million, or 73.4 times greater than the bottom 20 percent. To put it another way, for every $1,000 earned by the bottom 1 percent, the top 1 percent earn $74,559 (see Table 8.5).

To this point, we have shown that valued resources are distributed unequally both across countries and within countries. Next we will focus on the processes by which those resources are unevenly distributed.

Caste and Class Systems

Real-world stratification systems fall somewhere on a continuum between two extremes: a **caste system** (or "closed" system)—in which people's ascribed statuses (over which they have no control) figure most prominently in determining their life chances—and a **class system** (or "open" system)—in which merit, talent, ability, or past performance figure most prominently in determining their life chances.

> **CORE CONCEPT 2 Caste and class systems of stratification are opposite, extreme points on a continuum. The two systems differ in the ease of social mobility, the relative importance of achieved and ascribed statuses, and the extent to which those considered unequal are segregated.**

Most sociologists use the term *caste* to refer to any form of stratification in which people are categorized and ranked by ascribed characteristics over which they have no control. Whenever people are ranked and rewarded in this way, they are part of a caste system of stratification. A direct connection exists between caste rank and the opportunities to exercise power, acquire wealth, and secure valued opportunities. People in lower castes are labeled as innately inferior in intelligence, morality, ambition, and many other traits. Conversely, people in higher castes consider themselves to be superior in such traits. Moreover, caste distinctions are treated as if they are absolute; that is, the categories into which people are classified are viewed as unalterable and clear-cut. Finally, heavy restrictions constrain interactions among people in higher and lower castes. For example, marriage between people of different castes is forbidden. An excerpt from Scott McLaren's journal helps to clarify how

TABLE 8.5 Average After-Tax Income in the United States by Income Category, 1979 versus 2007 (in 2007 dollars)

The table compares after-tax income in the United States across six income categories at two points in time—1979 and 2007. Perhaps most striking is that the greatest financial gains within this time period have been made by those in the top fifth, including the top 1 percent. The financial gain made by those in the top 1 percent is 281 percent; for those in the top fifth, it is 95.0 percent.

Income Category	1979 Average After-Tax Income	2007 Average After-Tax Income	Percent Change	Dollar Change
Lowest Fifth	$15,300	$17,700	15.6%	$2,400
Second Fifth	$31,000	$38,000	22.6%	$7,000
Middle Fifth	$44,400	$55,300	25.4%	$11,200
Fourth Fifth	$57,700	$77,700	34.7%	$20,000
Top Fifth	$101,700	$198,300	95.0%	$96,600
Top 1 Percent	$346,600	$1,319,700	281%	$972,400

Source: Data from Congressional Budget Office 2010.

Lance Cpl. Jonathan G. Wright

Depending on where baby is born and to whom a baby is born, its chances of surviving to the first year of life varies dramatically. Almost 18 percent of babies born in Angola die before reaching age 1. In Sweden .0027 percent die.

caste systems in which gender is central operate in the rural Mauritanian village to which he has been assigned:

> There are very few girls in my village who can go to school. Our nearest school, which just opened this past year, is the first chance all of the kids in my area will have at getting an education. I'm praying that 2% literacy rate rises very soon. The girls are hard to get a handle on. Because of the cultural restrictions on men and women interacting, I can't really get close enough to any of them to really get to know them. The sisters in my family don't really talk much about what they'd like to do; they just talk about what they are doing in their daily chores.

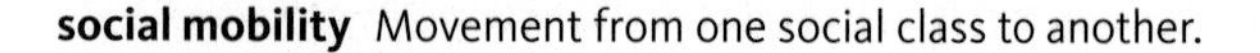

social mobility Movement from one social class to another.

Scott McLaren

In Mauritania, girls who live in rural areas are married off at relatively young ages, in some cases as early as 15. Over the past few decades, girls have begun to choose their partners, but the parents still hold the final say. Theoretically, girls can move away from home if they choose, but their support network becomes very limited, and the challenges of finding ways to support themselves are often very great.

Theoretically, in class systems, "people rise and fall on the strength of their abilities" (Yeutter 1992, p. A13). Although class systems contain inequality, that inequality is not related to ascribed statuses such someone's sex, skin shade, hair texture, and age. In theory, the inequality that does exist results from differences in talent, ability, and past performance. Thus, in class systems, people assume that they can achieve a desired level of education, income, and standard of living through personal effort; that they can raise their class position during their own lifetime; and that their children's class position can differ from (and ideally be higher than) their own.

Social Mobility

Within a class system, movement from one social class to another is termed **social mobility** (see Chart 8.1). When thinking about mobility in the United States, we often think about the possibility that children's economic status will exceed that of their parents. Many Americans believe that they live in a country in which it is possible to move from rags to riches. But how many people born poor become wealthy? Although data to answer this question are hard to come by, one study conducted by economist Paul Hertz (2006) followed 4,004 children into adulthood to calculate the odds of changing or maintaining economic status. The researcher averaged the income of each of the 4,004

CHART 8.1 Types of Social Mobility

Many kinds of social mobility exist in class systems of social stratification. This chart offers a list of the various types along with specific examples.

Type of Social Mobility	Definition	Example
1. Horizontal	A change in social situation that does not involve a change in social status	A waitress moves into a customer service position for an insurance agency
2. Vertical	A change in a person's social situation that involves a gain or loss of social status	See examples for the four types of vertical mobility that follow.
2A. Upward	A change in a person's social situation that involves a gain in social status	A medical student moves up in rank and becomes a physician
2B. Downward	A change in a person's social situation that involves a loss of social status	A wage earner loses a job, goes on unemployment, or otherwise moves down in rank
2C. Intragenerational	A change in social situation that involves a loss or gain of social status over the course of an individual's lifetime	A laid-off factory worker takes a job with a lower salary/a bank teller is promoted to a bank manager
2D. Intergenerational	A change in social situation that involves a loss or gain of social status relative to a previous generation	A son or daughter goes into an occupation that is higher or lower in rank and prestige than a parent's occupation

children's households for the years 1967 through 1971, and compared it to their household income when they became 40-year-old adults (averaged over four years, 1996 through 2000). The study found that the chances of moving from the lowest-income category to the highest or from the highest to the lowest are actually quite low (see Table 8.6).

Hertz also found that children from high-income households (top fifth of household income) receive more education and are healthier as adults, factors that

TABLE 8.6 Chances of Mobility by Income Category of Household in Which People Are Born

The table shows the relationship between a person's household income as a child (parents' quintile) and their household income as an adult (adult children's quintile). The highlighted numbers along the diagonal represent the percentage of adult children who remained in the same income grouping as their parents. To interpret the data in this table, ask the following kinds of questions: What are the chances that someone born in the lowest household income category (lowest fifth) will achieve a household income that is among the top 5 percent? The answer is 1.1 percent. What are the chances that someone born in the top 5 percent will retain that income status into adulthood? The answer is 21.7 percent. What are the chances that a person born in the top 5 percent income category will be among the lowest fifth as an adult? The answer is 2.9 percent.

Childhood Home (4-Year Average Household Income, 1967–1971)	Household Income as an Adult (4-Year Average Household Income, 1997–2001)					
	Lowest Fifth	Second Fifth	Third Fifth	Fourth Fifth	Top Fifth	Top 5 Percent
Lowest Fifth	**41.5**	24	15.5	13.2	5.9	1.1
Second Fifth	22.6	**25.8**	23.1	18.5	10	1.5
Middle Fifth	18.7	25.8	**24.1**	19.6	16.9	1.8
Fourth Fifth	11.1	19	20.7	**25.1**	24	5.6
Top Fifth	6.1	11.1	17.2	23.7	**41.9**	14.2
Top 5 percent	2.9	9	15.5	21.5	51.1	**21.7**

Source: Data from Hertz 2006.

surely contributed to them maintaining their economic status. In addition, Hertz found that children classified as white have advantages over children classified as black with regard to financial mobility. Specifically, 63 percent of black children who grow up in lowest-income households retain that status as adults, compared to the 32.3 percent of poor whites. Black children from the lowest-income households have a 3.6 percent chance of moving into the highest 20 percent income category as adults, compared to a 14.2 percent for poor white children.

Conceptualizing Inequality

Sociologists draw upon the theoretical traditions of functionalist, conflict, and symbolic interaction to think about inequality—why it exists, who benefits from inequality, and how it is enacted in interaction with others.

Functionalist View of Social Inequality

CORE CONCEPT 3 Functionalists maintain that poverty exists because it contributes to overall order and stability in society and that inequality is the mechanism by which societies attract the most qualified people to the most functionally important occupations.

Sociologists Kingsley Davis and Wilbert Moore (1945) wrote what is now considered the classic functionalist argument about why social inequality exists. The authors argue that social inequality—the unequal distribution of social rewards—is the device by which societies ensure that the best-qualified people fill the most functionally important occupations. So, in the United States, garbage collectors earn an average of $40,000 per year and surgeons earn an average of $227,000 per year (U.S. Bureau of Labor Statistics 2011a). From a functionalist point of view, the $187,000 difference in average salary represents the greater functional importance of a physician relative to a garbage collector.

How do Davis and Moore define functional importance? They offer two indicators:

1. The degree to which the occupation is functionally unique (that is, few other people can perform the same function adequately)
2. The degree to which other occupations depend on the one in question

Davis and Moore argue that, because people need little training and talent to be a garbage collector, the position is not functionally unique and thus does not need to be highly rewarded. However, with regard to the second indicator—the degree to which other occupations depend on garbage collectors—we must acknowledge that virtually every occupation and area of life depends on garbage

Lisa Southwick

From a functionalist point of view, sanitation workers are essential to the smooth operation of society, and yet they need not be rewarded highly because their job requires little training and skill.

collectors to maintain sanitary environments. Regardless, Davis and Moore argue that society still need not offer extra incentives to attract applicants for garbage collector because so many can do this job with little training. On the other hand, society does have to offer extra incentives to attract the most talented people to occupations such as physician that require long and arduous training and a high level of skills. Davis and Moore concede that the stratification system's ability to attract the most talented and qualified people is weakened when:

1. capable people are overlooked or denied access to the needed training;
2. elite groups control the avenues of training (by limiting admissions);
3. parents' influence and wealth (rather than the ability of their offspring) determine the status that their children attain.

Davis and Moore maintain, however, that society eventually adjusts to such shortcomings, as evidenced by the fact that medical schools, once dominated by white males, eventually began admitting people from groups previously denied access. Whenever there are shortages of applicants for occupations considered functionally important (such as teachers, nurses, and doctors), society adjusts by giving opportunities to those previously denied consideration. If such a step is not taken, the society as a whole will suffer and will be unable to meet its needs.

The Functions of Poverty

In another, now-classic essay, "The Functions of Poverty," written from a functionalist perspective, sociologist Herbert Gans (1972) asked, "Why does poverty exist?" He answered that poverty performs at least 15 functions, several of which are described here.

Fill Unskilled and Dangerous Occupations First, the poor have little choice except to take on the unskilled, dangerous, temporary, dead-end, undignified, menial work of society at low pay. In the United States, for example, there are 1.3 million residents in 17,000 certified nursing homes across the United States. Nursing home staff must lift patients, some of whom weigh more than they do, "help them in and out of baths, make beds, and take residents to and from the toilet" (Charney 2010). They have the highest rate of occupational-related injuries, especially to the back. They earn an average salary of $15,000 per year (U.S. Department of Health and Human Resources 2011).

Provide Low-Cost Labor for Many Industries The U.S. and many other economies depend on cheap labor from around the world and within its borders. Obviously, the lower the wage, the lower the associated labor costs to the hospitals, hotels, restaurants, factories, and farms that draw from the pool of laborers forced or "willing" to work at minimum wage or below. According to Pew Hispanic Center (2009) estimates, at least 12 million undocumented adults and children are living in the United States. Of these 12 million, 8.3 million are employed. Twenty-five percent of all farm workers, 17 percent of all construction workers, and 12 percent of all food preparation workers are believed to be undocumented. That is a "whole lot of cheap labor. Without it, fruits and vegetables would rot in the fields. Toddlers in Manhattan would be without nannies. Towels at hotels in states like Florida, Texas, and California would go unlaundered . . . bedpans and lunch trays at nursing homes would go uncollected" (Murphy 2004).

Serve the Affluent Affluent people contract out and pay low wages for many time-consuming activities, such as housecleaning, yard work, and child care. On a global scale, millions of poor women work outside their home countries as maids in middle- and upper-class homes. Consider that an estimated 460,000 Indonesian women from poor villages work as maids in Malaysia and Saudi Arabia. Many of these women must be trained in how to use toasters, vacuum cleaners, microwave ovens, refrigerators, and other appliances (Perlez 2004). Even the U.S. military depends on low-wage workers from the Philippines. Although only 51 Filipino troops served as part of the U.S.-led coalition in the Iraq War, more than 4,000 "serve food, clean toilets and form the backbone of the support staff for American forces. The military would be hard pressed to operate in Iraq without them" (Kirka 2004, p. 1A).

Volunteer for Drug Trial Tests The poor often volunteer for over-the-counter and prescription drug tests. Most new drugs, ranging from AIDS vaccines to allergy medicine, must eventually be tried on healthy human subjects to determine their potential side effects (such as rashes, headaches, vomiting, constipation, or drowsiness) and appropriate dosages. Money motivates people to volunteer as subjects for these clinical trials. Because payment is relatively low, however, the tests attract a disproportionate share of low-income, unemployed, or underemployed people (Morrow 1996).

Sustain Organizations and Employees Serving the Poor Many businesses, governmental agencies, and nonprofit organizations exist to serve poor people or to monitor their behavior, and, of course, the employees of these entities draw salaries for performing such work. The United States allocates about $1.6 billion toward food aid worldwide each year, but that money does not go directly to the poor; instead, it goes to corporations, agencies, and individuals who serve the poor. U.S. law mandates that American farmers must grow all government-donated food and that U.S.-flagged vessels must ship the food. Four agricultural corporations, five shipping companies, and seven food-aid charities receive a disproportionate share of food-aid monies. For example, $341 million of the $1.6 billion food-aid budget went toward paying packing, shipping, and storage costs. Critics argue that food should be grown in the countries where poor live. By one estimate, the United States is feeding 20 million fewer people a year because of costs lost to transportation (Dugger 2005b; Garber 2008).

Purchase Products That Would Otherwise Be Discarded Poor people use goods and services that would otherwise go unused and be discarded. Day-old bread, used cars, and secondhand clothes are purchased by or donated to the poor. In the realm of services, low-income communities purchase the labor of many professionals (such as teachers, doctors, and lawyers) who lack the competency to be hired in more affluent areas. Where do most of the estimated 307 million pieces of electronic equipment discarded each year such as cell phones, computer hard drives, and computer and TV monitors go? The Environmental Protection Agency (EPA) estimates that 50 to 80 percent is exported to countries like Ghana, China, Indonesia, and Pakistan (Frontline 2009). There, in some of the worst recycling facilities, people pick through e-waste, salvaging what they can by hand. Many are donated in the name of "bridging the digital divide" to poor countries. Each month, Nigeria alone receives 400,000 such computers, 75 percent of which are obsolete or nonfunctional and end up in landfills, where toxic materials pollute the surrounding environment, including groundwater (Dugger 2005c).

Gans (1972) outlines the functions of poverty to show how a part of society that everyone agrees is problematic

Scott McLaren

These boys live in the African country of Mauritania. Notice the little boy on the left is wearing a John Deere cap. Both are wearing clothes likely given away by Americans to a Goodwill store. The clothes come from what are called "dead toubab" stores (donated white people clothes). Given the racial and ethnic diversity in the United States, it is interesting that the clothes are seen as once belonging to white people.

and should be eliminated remains intact: It contributes to the supposed stability of the overall system. Based on this reasoning, the economic system as we know it would be strained seriously if we completely eliminated poverty; industries, consumers, and occupational groups that benefit from poverty would be forced to adjust.

A Conflict View of Social Inequality

CORE CONCEPT 4 Conflict theorists take issue with the premise that social inequality is the mechanism by which the most important positions in society are filled.

As you might imagine, conflict theorists challenge the fundamental assumption underlying the functionalist theory of stratification—that social inequality is a necessary device societies employ to ensure that the most functionally important occupations attract the best-qualified people. Sociologists Melvin M. Tumin (1953) and Richard L. Simpson (1956) point out that some positions command large salaries and bring other valued rewards even though their contributions to society are questionable. Consider the salaries of athletes. For the 2010–2011 season, the average NBA athlete earned $5.8 million, with the highest-paid athlete, Kobe Bryant, earning $24.8 million (Hooped Up 2010). We might argue that professional athletes deserve such enormous salaries because they generate income for owners, cities, advertisers, and media giants. The functionalist argument becomes less convincing, however, when we consider that U.S. school systems do not offer salaries high enough to attract the most qualified and gifted teachers. Consider that the median salary of a high school teacher is $52,200 (U.S. Bureau of Labor Statistics 2011b).

Tumin and Simpson maintain that the functionalist argument cannot explain why some workers receive a lower salary than other workers for doing the same job, just because they are of a different race, age, sex, or national origin. After all, the workers are performing the same job, so functional importance is not the issue. This question relates to issues of pay equity. For example, why do women working full-time as registered nurses in the United States earn a median weekly wage of $1,039, whereas their male counterparts earn $1,201? In fact, the U.S. Bureau of Labor Statistics (2010) data show only 8 of 300 occupational categories in which median weekly earnings for females exceed those of their male counterparts: counselors ($818 versus $808), occupational therapists ($1,094 versus $1,059), dental hygienists ($898 versus $897), food preparation and serving workers ($388 versus $369), nonfarm animal caregivers ($458 versus $455), bill and account collectors ($634 versus $612), file clerks ($583 versus $577), and stock clerks ($495 versus $482).

In addition to the issue of pay equity is the question of comparable worth: Should not women who work in predominantly female occupations (such as registered nurse, secretary, and day care worker) receive salaries comparable to those earned by men who work in predominantly male occupations that are judged to be of roughly comparable worth (such as those of a housepainter, a carpenter, and an automotive mechanic)? For example, assuming comparable worth, why should full-time workers at a child day care center (performing a traditionally female occupation) receive a median weekly salary of $398, whereas auto mechanics (performing a traditionally male occupation) earn $675 (U.S. Bureau of Labor Statistics 2010)?

Tumin and Simpson also ask how much inequality in salary is necessary to ensure that the best qualified people apply for what are considered the most important occupations. In the United States, the median base salary of the CEO of a large public (top 500) corporation is $9.6 million (Costello 2011). That salary is 190 times the median household income, which is $49,777 (U.S. Bureau of the Census 2009). If we compare CEO compensation with the wages of workers outside the United States, the inequality is even more dramatic.

Consider that the Wal-Mart CEO's total compensation package in fiscal year 2010 was $19 million, down from $28.2 million a year earlier (Reuters 2010). Notwithstanding the CEO's skills, the company's success can be partly attributed to the fact that it stocks its shelves with manufactured goods from the People's Republic of China. If the average Chinese factory worker earns about $3,600 per year, the Wal-Mart CEO earns about 5,277 times more than this worker. In contrast, the average full-time Wal-Mart hourly employee earns $10.11 per hour, or $21,000 per year. The CEO earns about 904 times more than the average clerk (Barboza 2010).

Conflict theorists ask if such high salaries are really necessary to make sure that someone takes the job of CEO over, say, the job of a factory worker. Probably not. Nevertheless, these high salaries are justified as necessary to recruit and retain the most able people to run a corporation in the context of a global economy. In *Wealth and Commonwealth*, William H. Gates Sr. (father of Bill Gates) and Chuck Collins (2003) critique one CEO who justified his "enormous compensation package" by stating, "I created about $37 billion in shareholder value." The CEO made no mention of the role the company's 180,000 employees played in creating shareholder value. Gates and Collins argue that "the problem with this individualistic way of assessing one's own contribution is that it is inaccurate and dishonest" (p. 113).

Finally, Tumin and Simpson argued that in societies characterized by a complex division of labor, it is very difficult to determine the functional importance of any occupation, because the accompanying specialization and interdependence make every position necessary to the smooth operation of society. Thus, to judge that physicians are functionally more important than garbage collectors fails to consider the historical importance of sanitation relative to medicine. Contrary to popular belief, advances in medical technology had little influence on death rates until the turn of the twentieth century—well after improvements in nutrition and sanitation had caused dramatic decreases in deaths due to infectious diseases.

A Symbolic Interactionist View of Social Inequality

CORE CONCEPT 5 Symbolic interactionists emphasize how social inequality is communicated and enacted in everyday encounters.

When symbolic interactionists study social inequality, they seek to understand the experience of social inequality; specifically, they seek to understand how social inequality is communicated and how that inequality shapes social interactions—interactions that involve a self-awareness of one's superior or inferior position relative to others. Social inequality is also conveyed through symbols that have come to be associated with inferior, superior, and equal statuses. In addition, there is a negotiation process by which the involved parties reinforce that inequality in the course of interaction or they ignore, challenge, or change it.

In the tradition of the symbolic interaction, journalist Barbara Ehrenreich (2001) studied inequality in everyday life as it is experienced by those working in jobs that paid $8.00 or less per hour. Ehrenreich, a "white woman with unaccented English" and a professional writer with a PhD in biology, decided to visit a world that many others—as many as 30 percent of the workforce—"inhabit full-time, often for most of their lives" (p. 6). Her aim was just to see if she could "match income to expenses, as the truly poor attempt to do everyday" (p. 6). In the process, Ehrenreich worked as a "waitress, a cleaning person, a nursing home aid, or a retail clerk" (p. 9). The only "lie" she told in presenting herself to others was that she had completed three years of college. Yet no supervisor or coworker ever indicated to Ehrenreich that they found her "special in some enviable way—more intelligent, for example, or clearly better educated than most" (p. 8). Ehrenreich reflected on her lack of specialness in this way: "Low-wage workers are no more homogeneous in personality or ability than people who write for a living, and no less funny or bright. Anyone in the educated classes who thinks otherwise ought to broaden their circle of friends" (p. 8).

Ehrenreich's on-the-job observations show the many ways inequality is enacted. Working as a retail clerk in ladies' wear, one of Ehrenreich's jobs is to put away the returns—"clothes that have been tried on and rejected . . . there are also the many items that have been scattered by customers, dropped on the floor, removed from the hangers and strewn over the racks, or secreted in locations far from their natural homes. Each of these items, too, must be returned to its precise place matched by color, pattern, price, and size" (p. 154).

Ehrenreich tells of a colleague who becomes "frantic about a painfully impacted wisdom tooth and keeps making calls from our houses (we are cleaning) to try and locate a source of free dental care" (p. 80). She tells of a colleague who would like to change jobs but the act of changing jobs mean "a week or possibly more without a paycheck" (p. 136); then there is the colleague making $7.00 per hour at K-Mart thinking about trying for a $9.00-per-hour job at a plastics factory. Ehrenreich also tells of "single mothers who live with their own mothers, or share apartments with a coworker or boyfriend" and of a woman "who owns her own home, but she sleeps on a living room sofa, while her four grown children and three grandchildren fill up the bedrooms" (p. 79).

Explaining Inequalities across Countries

Figure 8.1 in the Global Comparisons box shows the poorest and the richest countries in the world. How did the poorest countries become so poor? To answer this question, sociologists draw upon two views: modernization theory and dependency theory.

Global Comparisons The World's 22 Richest and 50 Poorest Economies

The countries highlighted in orange represent the world's 50 poorest economies. Notice that most are concentrated in Africa. The world's 22 richest countries are highlighted in blue. In the sections that follow, we will try to understand possible reasons for the differences in wealth. There are 846 million people living in the 50 poorest countries. The combined gross domestic product (GDP) of these 50 countries is $429 billion or $507 per person. There are 1.24 billion living in the richest 22 countries. The combined GDP is $39.5 trillion or $31,854 per person.

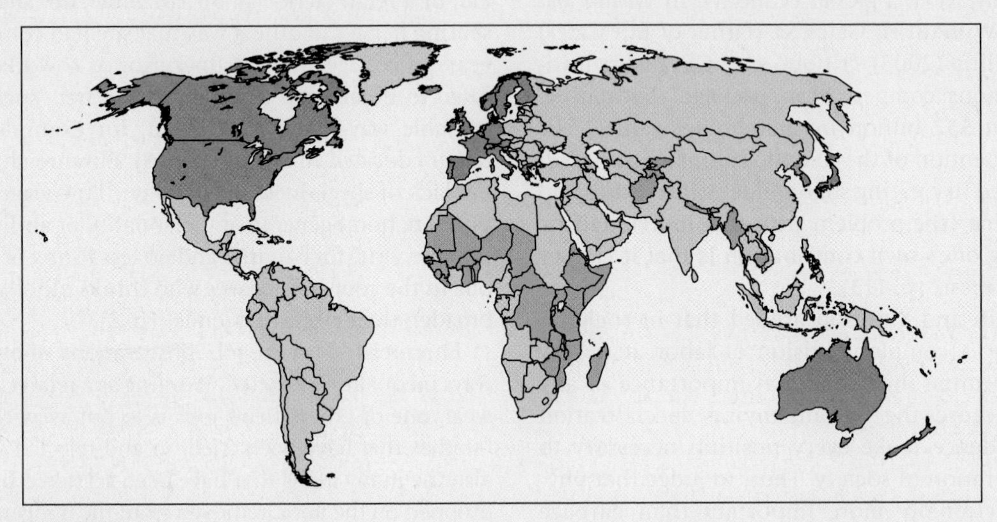

FIGURE 8.1
Source: Data from World Bank

Modernization Theory

CORE CONCEPT 6 Modernization theory holds that poor countries are poor because they have yet to develop into modern economies and that their failure to do so is largely the result of internal factors such as a country's resistance to free-market principles or to the absence of cultural values that drive material success.

Modernization is a process of economic, social, and cultural transformation in which a country "evolves" from preindustrial or underdeveloped status to a modern society in the image of the most developed countries (that is, western European and North American countries). A country is considered modern when it possesses the following eight characteristics (after each characteristic are comments from Scott McLaren about how the country to which he has been assigned—Mauritania—stands):

1. *A high proportion of the population lives in and around cities such that the society is urban-centered.* Scott notes that there has been a steady migration to the cities as infrastructure becomes more available. The number of new villages surrounding cities is growing. With the establishment of Nouakchott as the capital, rural people have migrated there looking to open boutiques or to find work so they can send money home.
2. *The energy to produce food, make goods, and provide services does not come primarily from physical exertion (human and animal muscle) but from inanimate sources of energy such as oil and electricity.* Scott notes that, in Mauritania, few areas have access to heavy machinery; however, much of the farming/gardening is done by hand. Government agencies sometimes provide large rice plantations with machines to help with the harvest.
3. *There is widespread access to goods and services, which are features of productive economies with high standards of living.* Scott observes that the capital city—Nouakchott—has some very wealthy families that can

modernization A process of economic, social, and cultural transformation in which a country "evolves" from preindustrial or underdeveloped status to a modern society in the image of the most developed countries.

afford many of the nuances we in the United States enjoy. The regional capitals have steady access to a broad range of products and services, although the quality or quantity of those services is far lower than those found in Nouakchott.

4. *People have a voice in economic and political affairs.* Scott has found that the people have very little voice. The country is dominated by a small elite, and although ideals of democracy have been planted, democracy does not exist in practice.
5. *Literacy is widespread and there is a scientific, rather than secular, orientation to solving problems.* Scott notes that although some government statistics report relatively high literacy, the reality in rural areas is that literacy is not widespread. There is a continually expanding school system that is striving to eventually include all people, but an infrastructure needs to develop.
6. *A system of mass media and communication is in place that offsets the influence of the family and local cultures.* Scott notes that the Mauritanian government controls the media. Little unapproved news is seen or heard. There is some access to outside news sources via radio, satellite TV, and the Internet.
7. *There are large-scale, impersonally administered organizations such as government, businesses, schools, and hospitals that reduce dependence on family for child care, education, and social security.* Scott believes that some government-funded organizations are set up to help, but their funding is limited and their reach is largely confined to the immediate areas around the capital.
8. *People feel a sense of loyalty to a country (a national identity), not to an extended family and/or tribe* (Naofusa 1999). Scott believes that, in Mauritania, loyalty to family is much stronger than to country. If Scott had to rank people's loyalties, people's sense of loyalty and connection to family is first, followed by a loyalty to culture, then ethnicity, and lastly country.

Modernization theorists seek to identify the conditions that launch underdeveloped countries on the path to modernization and to identify the stages through which those countries must pass to reach modernization. The road to modernization begins with a tradition-oriented way of life (stage 1) dominated by kinship-related obligations and loyalties that modernization theorists claim discourage change and personal mobility. Stage 1 is also characterized by a level of productivity limited by an inaccessibility to modern science, including its applications and frame of mind (Rostow 1960). W. W. Rostow (1960), an economic advisor to President John F. Kennedy and a proponent of modernization theory, described tradition-oriented cultures as possessing a long-run fatalism fueled by the "pervasive assumption that the range of possibilities open to one's grandchildren would be just about what it had been for one's grandparents.

Scott McLaren

This village school in Afghanistan enrolls more than 600 students. It has no modern conveniences, and these girls are allowed to attend on the condition that their father allows it. Modernization theorists would argue that a country will not modernize until the decision is out of the fathers' hands and becomes a government mandate.

According to Rostow's model of modernization, the next stages are the pre-takeoff (stage 2), the takeoff (stage 3), the drive to maturity (stage 4), and the age of high mass consumption (stage 5). Western countries can set the preconditions for takeoff by jump-starting modernization (stage 2) through foreign aid and investments that include technology transfers (for example, fertilizers, pesticides), birth control programs, loans, cultural exchange, and medical interventions (for example, inoculation programs). Ideally, these interventions "shock" the traditional society and hasten its undoing so that the country "takes off." Such interventions set into motion the ideas and sentiments by which a "modern alternative to the traditional society" evolves "out of the old culture" (Rostow 1960). The developing countries can hasten modernization through appropriate government reforms and policies. Eventually—perhaps as many as 60 years later—the developing country will reach a final state of modernization characterized by technological maturity and high mass consumption.

Tech. Sgt. Scott T. Sturkol

The U.S. government sent school buses to Afghanistan. How might this change the Afghan culture? Consider that before the school bus, schoolchildren walked to school with family members. What does it mean for family when children now spend time on a bus with just their peers?

According to Rostow, modernization involves a transformation of cultural beliefs and values away from those that supposedly support fatalism and collective orientation to those that support a work ethic, deferred gratification, future-orientation, ambition, and individualism (important attitudes and traits believed to be essential to the development of a free-market economy or capitalism). As the country modernizes, "the idea spreads, not merely that economic progress is possible, but that economic progress is a necessary condition for some other purpose . . . be it national dignity, private profit, the general welfare, or a better life for the children."

Dependency Theory

CORE CONCEPT 7 Dependency theory holds that, for the most part, poor countries are poor because they are products of a colonial past.

Dependency theorists challenge the basic tenet of modernization theory—that poor countries fail to modernize because they reject free-market principles and because they lack the cultural values that drive entrepreneurship. Rather, dependency theorists argue that poor countries are poor because they have been, and continue to be, exploited by the world's wealthiest governments and by the global and multinational corporations that are based in the wealthy countries. This exploitation began with colonialism.

Colonialism is a form of domination in which a foreign power uses superior military force to impose its political, economic, social, and cultural institutions on an indigenous population so it can control their resources, labor, and markets (Marger 1991). The age of European colonization began in 1492, with the voyage of Christopher Columbus.

By 1800, Europeans had learned of, conquered, and colonized much of North America, South America, Asia, and coastal parts of Africa, setting the tone of international relations for centuries to come (see Figure 8.2 in No Borders, No Boundaries). During this time, European colonists forced local populations to cultivate and harvest crops and to extract minerals and other raw materials for export to the colonists' home countries. When indigenous populations could not meet the colonists' labor needs, the colonists imported slaves from Africa or indentured workers from Asia and Europe. In fact, an estimated 11.7 million enslaved Africans survived their journey to the "New World" between the mid-fifteenth century and 1870 (Chaliand and Rageau 1995, Conrad 1996, Holloway 1996).

The scale of social and economic interdependence changed dramatically with the Industrial Revolution, which gained dramatic momentum in Britain around 1850, and then spread to other European countries and the United States. The Industrial Revolution even drew people from the most remote parts of the world into a process that produced unprecedented quantities of material goods, primarily for the benefit of the colonizing countries. Between 1880 and 1914, pursuit of and demand for raw materials and labor increased dramatically. This period, known as the Age of Imperialism, saw the most rapid colonial expansion in history. During this time, rival European powers (such as Britain, France, Germany, Belgium, Portugal, the Netherlands, and Italy) competed to secure colonies and influence in Asia, the Pacific, and especially Africa (see, again, No Borders, No Boundaries).

Consider as one measure of the extent of colonization that, during the twentieth century, 130 countries and territories gained political independence from their colonial masters. That process of gaining political independence is known as decolonization. **Decolonization** is a process of undoing colonialism such that the colonized country achieves independence from the so-called mother country. Decolonization can be a peaceful process by which the two parties negotiate the terms of independence, or it can be a violent disengagement that involves civil disobedience, insurrection, or armed struggle (war of independence). Once independence is achieved, civil war between rival factions often takes place as each seeks to secure the power relinquished by the colonizer. Some scholars argue that the Americas (which include the United States,

colonialism A form of domination in which a foreign power uses superior military force to impose its political, economic, social, and cultural institutions on an indigenous population so it can control their resources, labor, and markets.

decolonization A process of undoing colonialism such that the colonized country achieves independence from the so-called mother country.

This map of Africa shows the decade in which each country gained independence from its "mother" country. The colonizing country exploited the labor and resources of the colony. After independence, the ties between the two countries do not end, however.

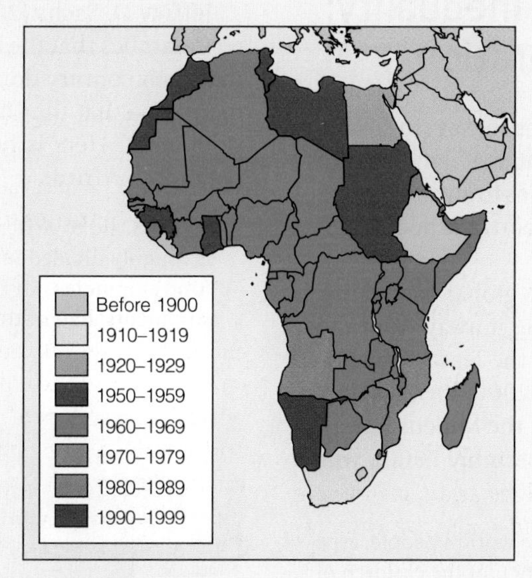

FIGURE 8.2 African Country by Decade of Independence

Source: Data from U.S. Central Intelligence Agency (2011)

Canada, and Central and South America) are technically still colonized lands because the indigenous peoples were not the ones to revolt and declare independence; rather, it was the colonists and/or their descendants who revolted and declared their independence. Those who took power simply continued colonizing and exploiting the land and resources belonging to indigenous peoples (for example, people now known as Native Americans) and others (that is, the enslaved and indentured peoples forced to immigrate) (Cook-Lynn 2008, Mihesuah 2008).

Gaining political independence does not mean, however, that a former colony no longer depends on its colonizing country. In *How Europe Underdeveloped Africa*, Walter Rodney argues that in the end, the African continent—90.4 percent of which was once controlled by colonial powers—has been "consigned to the role of producer of primary products for processing in the West" (Obadina 2000). Examples of primary products include products that are mined (gold) or extracted (oil) from the earth, as well as fish and agricultural products. Mauritania, a former colony of France, gained independence in the 1960s. The country has rich deposits of iron ore that account for 40 percent of that country's total exports, mostly to European Union countries and China. Its coastal waters are considered to be among the richest fishing areas in the world, but overexploitation by foreigners is threatening to deplete a major source of revenue (U.S. Central Intelligence Agency 2011).

This continuing economic dependence on former colonial powers is known as **neocolonialism**. In other words, neocolonialism is a new form of colonialism where more powerful foreign governments and foreign-owned businesses continue to exploit the resources and labor of the postcolonial peoples. Specifically, resources still flow from the former colonized countries to the wealthiest countries that once controlled them. Some critics would argue that the U.S. military presence in and around the continent of Africa is a form of neocolonialism.

In 2007, the U.S. Department of Defense put the continent of Africa under one military command, known as Africom. Prior to this, the command of the African continent was split between the European and Pacific Commands. The Pentagon's decision has faced resistance and criticism because "many African leaders questioned the formation of the command, calling it a U.S. grab for African resources—while others felt the command represented the

neocolonialism A new form of colonialism where more powerful foreign governments and foreign-owned businesses continue to exploit the resources and labor of the post-colonial peoples.

militarization of U.S. foreign policy." The official U.S position is that Africom "allows the U.S. military to help the Africans help themselves, provide security, and to support the far larger U.S. civilian agency programs on the continent" (Garacome 2008, U.S. African Command 2008).

A Response to Global Inequality: The Millennium Declaration

CORE CONCEPT 8 Structural responses to global inequality include transferring modest amounts of wealth from highest-income countries to lowest-income countries through foreign aid and fair-trade policies.

Clearly one obvious way to reduce global inequality is to redistribute wealth by transferring it away from those with the most wealth to those with the least. The United Nations has devised such a plan. In 2000, the 189 United Nations member countries endorsed the Millennium Declaration (United Nations General Assembly 2000), which states:

> As leaders we have a duty . . . to all the world's people, especially the most vulnerable and in particular the children of the world, to whom the future belongs. . . . We will spare no effort to free our fellow men, women, and children from the abject and dehumanizing conditions of extreme poverty, to which more than a billion are currently subjected. . . . We also undertake to address the special needs of the least developed countries including the small island developing states, the landlocked developing countries, and the countries of sub-Saharan Africa.

The Millennium Project set 18 targets and 60 measures of success to be reached by 2015, including these:

- Halve the proportion of people whose income is less than $1 a day.
- Halve the proportion of people who suffer from hunger.
- Reduce by three-quarters the maternal mortality ratio.

Clearly these are ambitious targets. Their success hinges on at least two major commitments from the world's 22 richest countries:

1. Increase current levels of foreign or development aid to 0.7 percent (seven-tenths of 1 percent) of their annual GNP in the Millennium Project.
2. Eliminate the subsidies, tariffs, and quotas that put the lowest income economies' products at a disadvantage in the global marketplace.

Increase Development Aid from Richest Countries

According to the UN the United States is the largest donor of foreign aid in absolute dollars ($26.3 billion). But that amount is deceiving when we consider that the population size of the United States is about 310 million. Per person spending on foreign aid translates into $85 per person. The European Union, by contrast, spends $145 per person. If we consider foreign aid as a percentage of gross national income (GNI), the United States is tied with Japan as the least generous (see Figure 8.3).

Jeffrey D. Sachs (2005), the UN Millennium Project director argues that the biggest myth most Americans hold is that their country donates a large amount of money. Sachs maintains that the United States spends very little on development. Here is how Sachs assesses U.S. financial assistance to Africa:

> The U.S. aid to Africa is $3 billion this year. That $3 billion is roughly divided into three parts: The first is emergency food shipments. Of the billion or so in emergency food shipments, half of that, roughly $500 million, is just for

Cpl. Thomas Childs

US NAVY

A review of U.S. State Department budget suggests that the bulk of the foreign aid assistance goes not toward development but toward other nondevelopment-related programs such as disaster and famine relief and military training apart from overall defense spending.

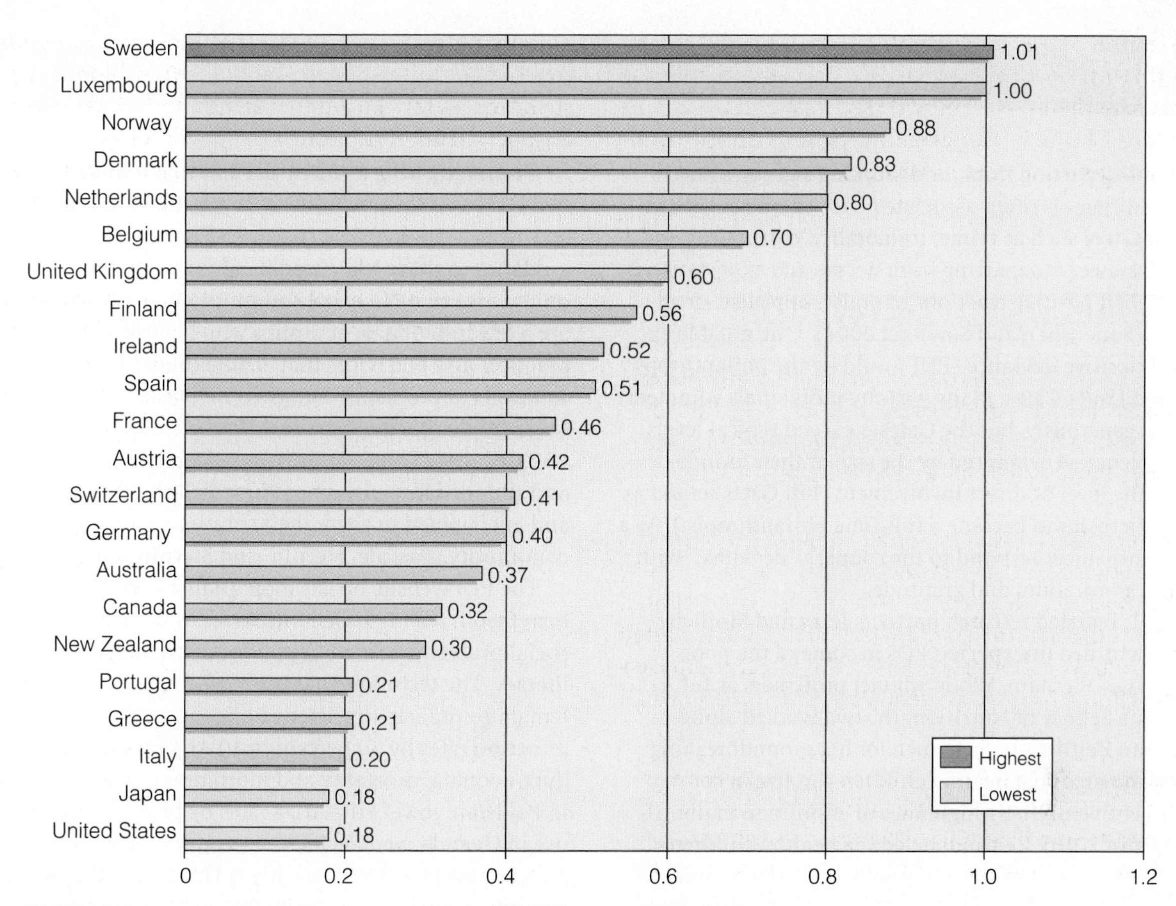

FIGURE 8.3 Foreign Assistance as a Percentage of Gross National Income, 2010
The bar lines show that the United States falls very short of meeting the 0.7 percent UN-recommended donation target. Given that the U.S. GDP is $14.62 trillion; 0.7 percent of that is $102 billion. Currently, the U.S. donates 0.18 percent, which is $26.3 billion, $75.5 billion short of the targeted amount.

Sources: Data from U.S. Department of State (2011), Central Intelligence Agency (2011), United Nations (2010)

transport costs. So the commodities are maybe half a billion dollars. That's not development assistance, that's emergency relief. The second billion is the AIDS program, now standing at about $1 billion. That, on the whole, is a good thing. I would call it a real program. It's providing commodities; it's providing relief. It started late and it's too small, but it's there. The third billion is everything else we do for child survival, maternal survival, family planning, roads, power, water and sanitation, malaria; everything is the third $1 billion. Most of that, approaching 80 percent, is actually for American consultant salaries. There's almost no delivery of commodities, for example. There's essentially zero financing to help a country build a school or build a clinic or dig a well.

Jeffrey Sachs's assessment contrasts starkly with the American public's belief that their government contributes about 20 percent of its $3.82 trillion federal budget (or $764 billion) in foreign aid (U.S. Office of Management and Budget 2011). Could the UN and Jeffrey Sachs be mistaken? According to U.S. Department of State data, they are not. A review of the State Department budget shows that $3.8 billion goes toward development assistance, and $8.5 billion goes toward global health and child survival. Other line items include contributions to military training ($5.4 billion), narcotics control ($2.1 billion), and to crisis intervention programs such as refugee assistance ($2.1 billion) and Pakistan Counterinsurgency Capability Fund ($1.2 billion) (U.S. Department of State 2011). It could also be that Americans count the $800 billion spent each year by the Department of Defense as foreign aid. Supporters of the U.S. level of giving argue that the United States gives assistance in other, less officially recognized forms, including private donations, wage remittances from immigrants working in the United States, and trade investments. Hundreds of creative and successful programs are aimed at reducing poverty (see Working for Change: "Positive Deviants as Change Agents").

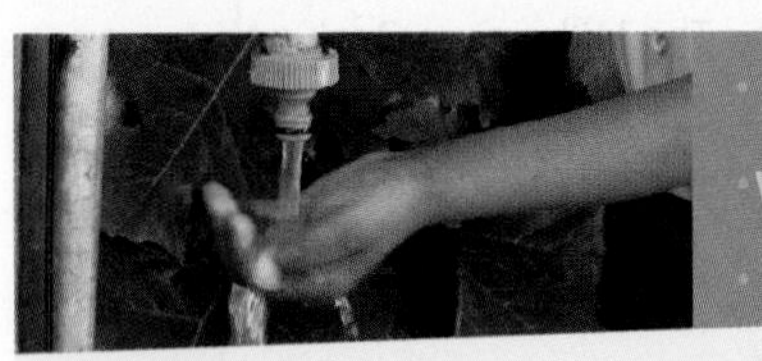

Working for Change Positive Deviants as Change Agents

Deviance—departing from, or disrupting a group's norms or expectations—is often associated with something that is bad or negative such as crime, immorality, corruption, and mischief. However departing from norms and expectations can also elicit positive reactions of praise, applause, or admiration (Sonenshen and Spreitzer 2004). One notable example of positive deviance (PD) would be the philanthropy of Bill and Linda Gates. Many wealthy individuals volunteer or donate generously, but the Gateses exceed typical levels of benevolence as evidenced by the size of their foundation and the level of direct involvement (Bill Gates retired as CEO of Microsoft to become a full-time philanthropist). As a consequence, most respond to the couple's "deviance" with applause, admiration, and gratitude.

In 1991, married research partners Jerry and Monique Sternin identified unexpected PDs in some of the poorest villages of Vietnam. While adjunct professors at Tufts University's School of Nutrition, the two worked alongside Marian Zeitlin who is known for her groundbreaking research showing that healthy children can live in communities with extremely high rates of childhood malnutrition (Sternin 2010). Zeitlin labeled the healthy children's caregivers positive deviants because, even though these caregivers were from impoverished households, they managed to accomplish something out of the ordinary. Zeitlin maintained that if she could identify the norms these caregivers were breaking, she would have the basis for a new health initiative (Sternin 2010).

After Jerry assumed the position of director of Save the Children, Vietnam, he and Monique moved to Vietnam, a country where 65 percent of children were malnourished, eager to test out Zeitlin's approach in four Hanoi villages. The couple collaborated with villagers to identify a handful of households with healthy-weight children. Eventually, the Zeitlins were able to determine what these families were doing. The mothers were feeding their children tiny shrimps and crabs found in rice paddies. Other families routinely discarded this food source because they assumed young children could not digest them (Sternin 2010). Because rice is Vietnam's primary agricultural crop, everyone in the villages had access to paddies so the challenge was to convince families with malnourished children to adopt the behavior of the positive deviants. Six months later, after several group sessions and a concerted effort to spread the word, many caregivers adopted the practice, resulting in a 74 percent reduction in number of malnourished children age 3 and younger (Positive Deviance Initiative 2011).

Upon returning to Tufts, the Sternins founded the Positive Deviance Initiative (PDI). Today, PDI has matured into an international, online network that provides resources and guidance to those interested in applying the model. PDI rests on the idea that "in every community or organization, there are a few individuals or groups who have found uncommon practices and behaviors that enable them to achieve better solutions to problems than their neighbors who face the same challenges and barriers" (Pascale, Sternin, and Sternin 2010, p. 206). Thus, PDI involves establishing a collaborative and supportive environment in which the PDs are identified and encouraged to share their solutions with others in their community (Pascale, Sternin, and Sternin 2010).

The PDI website boasts more than 41 countries that have benefited from PD-based initiatives to solve an array of social problems related to public health, nutrition, and illiteracy. The technique has been used to prevent 1,000 female genital circumcisions in Egypt, increase student retention rates by 50 percent in 10 Argentinean schools, reduce neonatal mortality and morbidity in Pashtun villages in Pakistan, lower HIV/AIDS rates by promoting condom use in Uganda, and decrease sex trafficking among young girls in East Java. Most recently, a U.S. hospital applied the concept in response to the 90,000 to 100,000 annual deaths in the United States caused by methicillin-resiliant Staphylococcus aureus (MRSA) (Gertner 2008). With support from the Veterans Administration (VA), the PDI model was applied. The end result was the discovery of the *Palmer method*, named for the hospital orderly identified as the positive deviant. Now universally taught in medical facilities across the country, the Palmer method involves taking gowns worn by the hospital staff who handle patients infected with MRSA, rolling them into a tight ball, and stuffing them inside their protective gloves, thereby reducing the chances that contaminated gowns lying around on floors and in baskets will facilitate the spread the infection (Gertner 2008). If PDI has not been employed to address MRSA outbreaks, Mr. Palmer's work habits would likely have been overlooked as simply eccentric. PDI has received international recognition and praise from reputable organizations such as USAID, UNICEF, NGOs, Peace Corp, and even the World Bank. The *New York Times* also recognized PDI in 2008, by highlighting it in its prestigious "Year in Ideas" magazine, an annual publication that pays tribute to the world's greatest social inventions and ideas (Gertner 2008).

Source: Written by Ashley Novogroski, Northern Kentucky University, Class of 2010.

End Subsidies, Tariffs, and Quotas

Even though the wealthiest countries have agreed to eliminate tariffs, subsidies, and quotas on products imported from the poorest countries, the wealthiest have resisted dismantling a system of trade structured to their advantage (Bradshear 2006). In particular, the United States, Japan, the European Union, and other high-income countries continue to subsidize agriculture and other sectors, such as steel so that producers in these countries are paid more than world market value for their products. Considerable attention has been given to agricultural subsidies, which give farmers in wealthy economies an estimated $376 billion in support (United Nations 2010). It is well documented that those subsidized are the large agricultural corporations, not small farmers. For example, Riceland Foods (the top recipient) received $554 million in subsidies between 1995 and 2009 (Environmental Working Group 2011).

In addition, the wealthiest countries apply tariffs and quotas to many imported items, thereby increasing their cost so those imports are higher or equal to the price of the domestic versions. Consider sugar. The European Union protects its sugar industry so that domestic producers earn double, sometimes triple, the world market price. Because of subsidies, 6 million tons of surplus sugar each year is dumped into the world market at artificially low prices. Of course, sugar is not the only protected commodity. The United States, although not a major exporter of sugar, applies tariffs and quotas on imported sugar cane and other sugars grown in Brazil, Vietnam, and elsewhere to protect its domestic sugar producers (Thurow and Winestock 2005). Subsidies, tariffs, and quotas are designed to keep products dumped onto the world market artificially low and prices on protected products artificially high.

Such policies affect not just workers in the poorest countries, but also workers in wealthy countries. For example, Brach and Kraft Foods closed their U.S.-based candy plants and outsourced more than 1,000 jobs to Argentina and Canada, where sugar can be purchased at lower, world-market prices (Kher 2002).

Ending subsidies, tariffs, and quotas may sound like a solution until one looks closely at real cases. For example, the United States imposed a 35 percent tariff on tires from China after a surge in tire imports from that country had the effect of lowing tire production in the United States, from 218.4 million to 160.3 million tires per year. The U.S. government argued that China was subsidizing domestic tire production by undervaluing its currency. Its low labor costs also gave China an unfair competitive advantage (Chan 2010). In this situation, one would hardly blame the United States for imposing tariffs. Despite these complications, as of 2010, 81 percent of the goods imported from the poorest countries by the richest countries now enter duty-free. The UN goal is 97 percent (United Nations 2010).

Progress Toward Reaching Millennium Goals (at Midpoint)

Recall that the UN Millennium goals were set in 2000, with the goal of achieving them by 2015. In 2008—the halfway point—the UN released a report highlighting the areas in which progress has been made and those in which greater efforts must be focused. Some areas of progress include the following:

- The Millennium Project is on track to reduce absolute poverty by half for the world as a whole.
- Deaths from measles have fallen from approximately 750,000 in 2000 to 250,000 in 2006, and an estimated 80 percent of children in developing countries receive measles vaccines.
- The use of ozone-depleting substances has been virtually eliminated.
- The private sector has increased access to some critical and essential drugs.
- Mobile phone technology has increased dramatically throughout the developing world. (UN 2009b, p. 4)

These successes notwithstanding, the UN (and, by extension, the world community) are failing to meet a number of goals that include the following:

- Reduce the proportion of people in sub-Saharan Africa living on less than $1 per day by 50 percent.
- Alleviate childhood undernourishment and its effects.
- Reduce maternal mortality.
- Improve sanitation of 2.5 billion people, or 50 percent of the developing world's population.
- Prevent carbon dioxide emissions from increasing. (United Nations 2009b, p. 4)

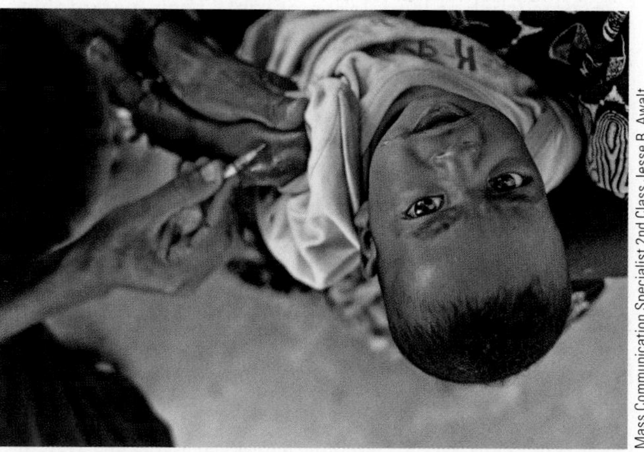

Mass Communication Specialist 2nd Class Jesse B. Awalt

The United Nations appears to be on track to meet its goal of vaccinating children against measles. It is not on track to meet a number of other goals, such as alleviating childhood malnutrition.

Criticism of the Millennium Declaration

Critics of the UN recommendation to increase aid and eliminate trade barriers say that is only a first step. For one, other factors make it difficult for poor economies to compete—one being that employers take advantage of desperate workers "willing" to work for less. As the factory worker's wages in China and elsewhere increased to about $150 per month, Vietnam became an attractive location as the minimum wage in its largest cities was $75 a month. In 2010, Intel moved semiconductor production from its facilities in China, Malaysia, and the Philippines to Ho Chi Minh City, Vietnam (Arnold 2010).

A second reason that increasing development and reducing trade barriers are not sufficient steps for reducing global inequality relates to **brain drain**, the emigration of the most educated and talented people from the poorest to richest countries. Brain drain encompasses those who are or who plan to be hospital managers, nurses, accountants, teachers, engineers, political reformers, and other skilled and educated segments. One estimate of brain drain shows that from 30 percent (Sri Lanka) to 84 percent (Haiti) of college-educated citizens from the poorest countries live abroad in high-income countries (Dugger 2005c).

The rich economies facilitate brain drain with immigration policies that give preference to educated, skilled foreigners. The British Medical Association has grown so concerned about the shortage of health care workers around the world and the migration of such workers from poor to rich countries that it has called for all countries to expand their capacity to train and retain their own physicians and nurses.

Analyzing Social Class

CORE CONCEPT 9 Social class is difficult to define. It depends on many factors, including people's relationship to the means of production, their sources of income (such as land, labor, or rent), their marketable abilities, their access to consumer goods and services, their status group, and their membership in political parties.

Sociologists use the term **class** to designate a person's overall economic and social status in a system of social stratification. They see class as an important factor in determining life chances. We begin with the writings of Karl Marx and Max Weber, who represent the "two most important traditions of class analysis in sociological theory" (Wright 2004, p. 1).

Karl Marx and Social Class

In *The Communist Manifesto*, published in 1848 and written with Friedrich Engels, Marx observed that the rise of factories and mechanization as a means of production created two fundamental classes: those who owned the means of production (the bourgeoisie) and the largely propertyless workers (proletariat) who must sell their labor to the bourgeoisie (Allen and Chung 2000). For Marx, then, the key variable in determining social class is source of income.

In *Das Kapital*, Marx names three classes, each of which is comprised of people whose revenue or income "flow from the same common sources" (p. 1,032). For wage laborers, the source is wages; for capitalists, the source is profit; for landowners, the source is ground rent. In the *Class Struggles in France 1884–1850*, Marx named another class, the **finance aristocracy**, who lived in obvious luxury among masses of starving, low-paid, and unemployed workers. The finance aristocracy includes bankers and stockholders seemingly detached from the world of "work." Here is how Marx described the finance aristocracy's source of income: "it is a source of income created from nothing—without labor and without creating a product or service to sell in exchange for wealth." The finance aristocracy speculates or employs people who know how to speculate for them. "But while speculation has this power of inventiveness, it is at the same time also a gamble

This 1883 political cartoon captures Karl Marx's vision of social class. It shows four of the wealthiest people who have ever lived—Cyrus Field, Jay Gould, Cornelius Vanderbilt, and Russell Sage. These men were wealthier in real dollars than those considered the world's richest today. The four are seated on bags of money atop a large raft, which is being held afloat by millions of workers.

Library of Congress Prints and Photographs Division Washington, D.C.

brain drain The emigration from a country of the most educated and most talented people.

class A person's overall economic and social status in a system of social stratification.

finance aristocracy Bankers and stockholders seemingly detached from the world of "work."

and a search for the 'easy life'; as such it is the art of getting rich without work." According to Marx, the financial aristocracy appropriates to themselves "public funds or private funds without giving anything equivalent in exchange; it is the cancer of production, the plague of society and of states" (Marx 1856, Bologna 2008, and Proudhon 1847).

Max Weber and Social Class

Karl Marx clearly states that social class is based on people's relationship to the means of production, a relationship that determines the sources of their income. For Max Weber, the basis extends beyond someone's relationship to the means of production. According to Weber (1947), people's class standing depends on their marketable abilities (work experience and qualifications), their access to consumer goods and services, their control over the means of production, and their ability to invest in property and other sources of income. According to Weber, people completely lacking in skills, property, or employment, or who depend on seasonal or sporadic employment, constitute the very bottom of the class system. They form the **negatively privileged property class**.

Individuals at the very top—the **positively privileged property class**—monopolize the purchase of the highest-priced consumer goods, have access to the most socially advantageous kinds of education, control the highest executive positions, own the means of production, and live on income from property and other investments. Weber viewed class as a series of rungs on a social ladder, with the top rung being the positively privileged property class and the bottom rung being the negatively privileged property class. Between the top and the bottom of this social-status ladder is a series of rungs. He argued that a "uniform class situation prevails only among the negatively privileged property class." We cannot speak of a uniform situation regarding the other classes, because people's class standing is complicated by their occupation, education, income, group affiliations, consumption patterns, and so on.

Weber states that class ranking is complicated by the status groups and political parties to which people belong. He defines a **status group** as an amorphous group of people held together by virtue of a lifestyle that has come to be "expected of all those who wish to belong to the circle" and by the level of social esteem accorded them (Weber 1948, p. 187). This shared lifestyle can encompass leisure activities, eating, time devoted to sleeping, occupation held, and friendships.

Weber's definition suggests that wealth, income, and occupation are not the only factors that determine an individual's status group. Simply consider that some people possess equivalent amounts of wealth yet hold very different statuses due to their upbringing and education. In addition to status group, people can also belong to **political parties**—organizations "oriented toward the planned acquisition of social power [and] toward influencing social action no matter what its content may be" (Weber 1982, p. 68). Parties are organized to represent status groups and their interests. The means to secure power can include violence, vote canvassing, bribery, donations, the force of speech, and fraud. Examples of political parties include the Tea Party, National Organization for Women (NOW), Promise Keepers, the United Auto Workers (UAW), the National Rifle Association (NRA), and American Association of Retired Persons (AARP).

Journalist 2nd Class Alexis R. Brown

Bodybuilders are a status group in that they have developed a lifestyle around maximizing the size and appearance of their muscles. They are held together by a shared way of living, including eating high-protein foods and sleeping at least eight hours of the day to help muscles recuperate and build efficiently after intense training.

Income and Wealth as Indicators of Social Class

Although social class is a complex concept, we know that wealth and income are key components. When studying wealth and income, sociologists ask some key questions:

- *What distinguishes one social class from another?* Once sociologists settle on the number of class categories relevant to their analysis, they work to identify objective criteria by which to classify people into those

negatively privileged property class Weber's category for people completely lacking in skills, property, or employment or who depend on seasonal or sporadic employment; they constitute the very bottom of the class system.

positively privileged property class Weber's category for the people at the very top of the class system.

status group Weber's term for an amorphous group of people held together both by virtue of a lifestyle that has come to be expected of "all those who wish to belong to the circle" (Weber 1948, p. 187).

political parties According to Weber, "organizations oriented toward the planned acquisition of social power [and] toward influencing social action no matter what its content may be."

TABLE 8.7 Three Hypothetical Distributions of Wealth

In the Norton and Ariely (2011) study, 5,522 respondents were presented with three hypothetical distributions of wealth and asked to choose under which they would like to live. In hypothetical distribution #1, the richest 20 percent of households have 36 percent of all wealth; the poorest 20 percent have 11 percent. In hypothetical distribution #2, the wealthiest 20 percent of households have 84 percent of all wealth, and the bottom 20 percent have 0.1 percent. In hypothetical distribution #3, the wealth is divided evenly among all five household groups. Very few respondents chose hypothetical distribution #2, and virtually no one chose the third hypothetical. Almost everyone in the study chose distribution #1.

	Hypothetical Distribution #1	Hypothetical Distribution #2	Hypothetical Distribution #3
Richest 20 percent	36%	84%	20%
2nd 20 percent	21%	11%	20%
3rd 20 percent	18%	4%	20%
4th 20 percent	15%	.2%	20%
Poorest 20 percent	11%	.1%	20%

Source: Data from Norton and Ariely 2011.

categories. Should amount of income or accumulated wealth be used to determine one's social class? **Income** refers to the money a person earns usually on an annual basis from salary or wages. **Wealth** refers to the combined value of a person's income and other material assets such as stocks, real estate, and savings minus debt. If using wealth to divide the 113.6 million U.S. households into five classes consisting of 22.7 million households each, the top one-fifth controls 88 percent of the wealth (the equivalent of $43.6 trillion) and the bottom 22.7 million households hold no wealth after debt is paid out (Di 2007, U.S. Bureau of the Census 2009).

- *How much income or wealth qualifies someone to be upper, middle, or lower class?* The Pew Research Center (2008) defines social class in this way: A person is considered middle-income class if he or she lives in a household with an annual income that falls within 75 percent to 150 percent of the median household income. A person with a household income above that range is upper-income class; a person whose household income is below that range is the low-income class (see Figure 8.4). In 2009, the median household income in the United States was $61,082. Seventy-five percent of $61,082 is $45,765; 150 percent of $61,082 is $91,623. So a middle-class household is one with an annual income between $45,765 and $91,623. Of course, this represents just one way to conceptualize social class—in the 2008 presidential elections, then-candidate Barack Obama defined the upper limits of middle-class status as $250,000 with no lower limit specified.

  Pew researchers found that the "greater the income, the higher the estimate of what it takes to be middle class." Those with household incomes between "$100,000 and $150,000 a year believe, on average, that it takes $80,000 to live a middle-class life." Conversely, those with household incomes of "less than $30,000 a year believe that it takes about $50,000 a year to be middle class" (Pew Research Center 2008, p. 15).
- *How do people believe wealth is and should be distributed? How does that compare with reality?* Researchers Michael Norton and Dan Ariely (2011) asked a sample of 5,522 U.S. respondents a series of questions about how wealth is and should be distributed. The respondents shared a social and demographic profile that represented that of the United States. Before answering questions, respondents were given a definition of wealth—"net worth or the total value of everything someone owns minus any debt that he or she owes. A person's net worth includes his or her bank account savings plus the value of other things such as property, stocks, bonds, art collections, etc., minus the value of things like loans or mortgages." Respondents were then shown three hypothetical distributions of wealth and asked to choose under which one they would like to live. Ninety percent of respondents indicated that they did not want to live in a country where the top 20 percent controlled 84 percent of the wealth. Most people chose to live in a country where the top fifth controlled 36 percent of the wealth.

income The money a person earns, usually on an annual basis through salary or wages.

wealth The combined value of a person's income *and* other material assets such as stocks, real estate, and savings minus debt.

The Disadvantaged in the United States

One in eight people, or 14.3 percent of the U.S. population (43.6 million people), is officially classified as living in poverty (see Figure 8.4). To determine who lives in poverty, the U.S. Bureau of the Census sets a dollar-value threshold that varies depending on household size and age (under 65 and 65 and over). If the total household annual income is less than the specified dollar value, then that household is considered as living in poverty. The poverty threshold for a four-person family consisting of one adult and three children is $21,832. The poverty threshold for a single person under age 65 is $11,161. How does the U.S. government determine the various thresholds? The formula was set in 1963, and is based on the estimated daily cost per person of a nutritionally adequate diet; that estimate is then multiplied by 3. The resulting number is the amount of money (threshold) a person needs each day to live outside of poverty. That daily cost is multiplied by 365 to determine the yearly amount. Those living in poverty can be broadly classified into one of three geographic groups: inner city, suburban, or rural poor.

Inner-City Poor

What happens to people when a factory closes? What happens to the jobs in surrounding communities that serviced the people who worked at that factory? Detroit automobile plants, built in 1907, could employ as many as 40,000 people, enough employees to sustain a department store, two schools, and a grocery store on the premises. The Detroit plants began to close in 1957, laying off 130,000 autoworkers by 1967 as car companies restructured their operations, relocating plants to the suburbs and automating production facilities (Sugrue 2007). What happened to those who worked in the stores and schools? Of course, a second wave of restructuring and layoffs began in the 1970s and continues though today, as Detroit automakers steadily moved operations to overseas locations and downsized their operations in the United States. Beginning in 2008, the global economic crisis pushed Ford, General Motors, and Chrysler to restructure even further in response to foreign competition and declining car sales. GM closed 16 of 47 operating plants, laid off 23,000 production and 10,000 white-collar workers, and shut down 50 percent of its 6,200 dealerships (Goldstein 2009). This restructuring and downsizing was on the mind of sociologist William Julius Wilson when he wrote *The Truly Disadvantaged in the United States.*

> **CORE CONCEPT 10** Economic or occupational restructuring can devastate people, leaving them without jobs and disrupting the networks of occupational contacts crucial to moving affected workers into and up job chains.

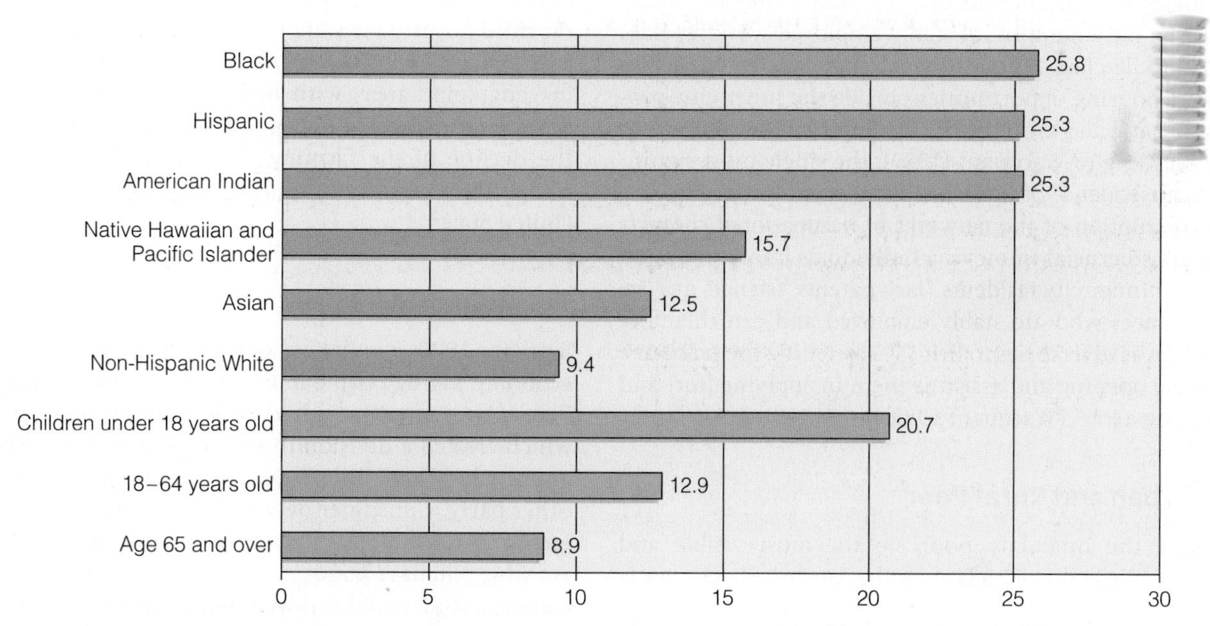

FIGURE 8.4 Percentage of Population in Poverty by Selected Characteristics, 2009
Poverty rates vary according to race, ethnicity, age, and sex. What percentage of non-Hispanic whites live in poverty? What percentage of Native Americans live in poverty? What percentage of children under 18 do?

Source: Data from U.S. Bureau of the Census (2010)

In *The Truly Disadvantaged* and other related studies, sociologist William Julius Wilson (1983, 1987, 1991, 1994) describes how structural changes in the U.S. economy beginning in the 1970s helped create what he termed the "ghetto poor," who are now known as the inner-city poor or **urban underclass**—diverse groups of families and individuals living in the inner city who are "outside the mainstream of the American occupational system and consequently represent the very bottom of the economic hierarchy" (Wilson 1983, p. 80). Those economic transformations include:

- The restructuring of the American economy from a manufacturing base to a service and information base.
- The rise of a labor surplus marked by the entry of women and the large "baby boom" segment of the population into the labor market.
- A massive exodus of jobs from the cities to the suburbs.
- The transfer of manufacturing jobs out of the United States.
- The transfer of customer service and information jobs out of the United States over the past decade (see Sociological Imagination).

Wilson (in collaboration with sociologist Loic J. D. Wacquant) studied Chicago to illustrate this point. (The same point applies to every large city in the United States.) In 1954, Chicago was at the height of its industrial power. Between 1954 and 1982, however, the number of manufacturing establishments within the city limits dropped from more than 10,000 to 5,000, and the number of jobs declined from 616,000 to 277,000. This reduction, along with the out-migration of stably employed working-class and middle-class families, which were fueled by access to new housing opportunities outside the inner city, profoundly affected the daily life of people left behind.

According to Wacquant (1989), the single most significant consequence of these historical economic events was the "disruption of the networks of occupational contacts that are so crucial in moving individuals into and up job chains." Inner-city residents "lack parents, friends, and acquaintances who are stably employed and can therefore function as diverse ties to firms . . . by telling them about a possible opening and assisting them in applying [for] and retaining a job" (Wacquant 1989, pp. 515–516).

Suburban and Rural Poor

Because the inner-city poor are the most visible and most publicized underclass in the United States, many Americans associate poverty with minority groups or urban areas. In fact, the suburban poor outnumber the urban poor by 1.2 million. Many of the suburban poor were pushed out of the city when factories and other businesses closed; they headed to the suburbs in search of jobs and low-cost housing (Jones 2006).

The rural poor in the United States are another population that needs attention. Demographers William P. O'Hare and Kenneth M. Johnson (2004) estimate that approximately 2.6 million rural children can be classified as underclass. In fact, 48 of the 50 counties with the highest child poverty rates are rural. Like their urban counterparts, members of the rural underclass are concentrated in geographic areas with high poverty rates. They, too, have felt the effects of economic restructuring, including the decline of the farming, mining, and timber industries and the transfer of routine manufacturing out of the United States.

Chris Caldeira

When factories shut down, the employees lose their jobs. Those who work for businesses that supported the factory workers lose their jobs as well.

The Indebted

Since the 1970s, credit has helped drive the U.S. and global economy, giving people money to spend that they did not have. Many people acquired unmanageable levels of debt, which created a division in society between the debt-free and the indebted. Simply put, *debt* is money owed to another party. Consumer debt is one way to fuel economic growth because credit puts money in the hands of consumers who purchase goods and services. Some of the most common sources of borrowed money are credit cards, payday loans, and other financing arrangements (two years same as cash, no payments for two years). Typically, these "short-term" credit sources are financed at higher interest rates than are mortgage, car, and student loans. Debt becomes a problem when borrowers cannot make payments

urban underclass The group of families and individuals in inner cities who live "outside the mainstream of the American occupational system and [who] consequently represent the very bottom of the economic hierarchy" (Wilson 1983, p. 80).

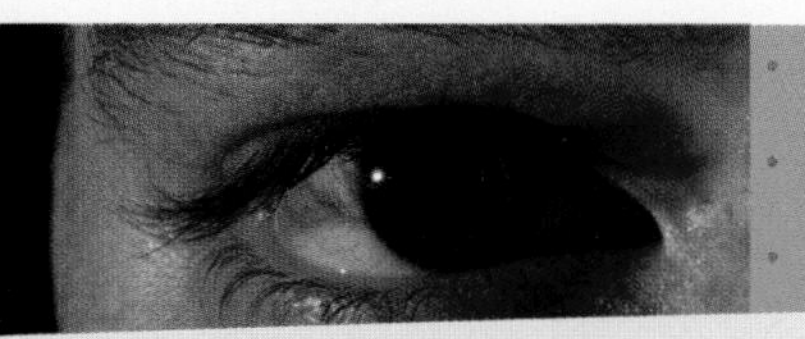

The Sociological Imagination The Personal Impact of Economic Restructuring

Between December 2007 and April 2009, there were 5.7 million (net payroll) jobs lost in the United States. My students, many of whom are older adults or 18- to 26-year-olds working more than 20 hours a week, shared the experience of losing their job at this time:

- About a month ago, I found out my job was being eliminated. It was one of the worst days of my life. I had worked for my employer, an investment company, for 14 years. It was my first real job and I worked my way up in the company to a position that I truly loved. I was really in a great place, and even thought I would retire from that company! I was loyal, dedicated, and gave 100 percent each day. So needless to say, I was heartbroken when I found out I was part of the "RIF" (reduction in force). Not to mention the fact that my husband and I are in the process of trying to adopt two children. The timing couldn't have been worse.
- Today, before coming to class, I was called into the office and told I was to be laid off. I work for one of the largest banks in the United States. Already I have experienced a wide range of reactions from pretending to be totally confident about my future to what I can only describe as losing my mind. I even yelled at my husband, saying that I wished he would be more upset and stop telling me everything would be fine. My layoff wasn't totally unexpected, but to be told by my manager that the work I did is "above and beyond" that of other employees but that my services just aren't needed anymore is overwhelming. I have laughed, cried, yelled, and sat in silence. It feels like I am living through my own funeral—people stare or offer condolences.
- My boyfriend works as a driver for a major package delivery corporation that has cut out all domestic deliveries and now only delivers international. The company has dropped from 150 full-time drivers to 11. My boyfriend is number 12, which means he is on call when other drivers are overloaded or when someone is off or on vacation. So far, he has been called in to work every day, but every day he worries about whether he will be needed. He is on call day or night, so the company might call at 6:00 a.m. or 5:00 p.m. His life completely revolves around his job. He is very cautious about spending money. He needs a new car but is afraid to take on the payment. I feel sorry for him, because he's in a terrible situation.
- During the start of the economic crisis, I was working for an acrylic manufacturer when orders really slowed down due to the declines in new home sales. Fortunately, a job opportunity came up with the post office and I took it. Now my work hours have been reduced at the post office due to the decrease in mail volume. Within the past six months, I went from working 50-plus hours per workweek down to about 40 hours. I feel fortunate that I haven't lost full-time status.

or do not have the money to make payments large enough to reduce the overall amount owed. Although debt temporarily frees borrowers from their financial constraints in the short term, it can severely constrain their life chances if it becomes unmanageable. Often, the borrowers least able to afford credit and to pay off credit card debt each month are subjected to the highest interest rates.

The Pew Research Center conducted a nationally representative sample of 2,000 adults to learn the extent of debt problems in the United States. The survey found that one in seven adults (14 percent) have experienced a debt problem at some point in their lives that was so severe they used a debt consolidation service or declared bankruptcy. That percentage varies by income, with 8 percent of those with household incomes $100,000 or greater experiencing a debt problem and 19 percent of adults earning $30,000 or less experiencing such a problem (see Figure 8.5).

Americans who are late making credit card payments pay an estimated $15 billion in penalty fees a year. One in every five credit card holders carries over debt each month and pays interest rates of 20 percent or more (Baker 2009). President Barack Obama (2009) argues that Americans have a responsibility not to use credit cards to live beyond their means, but he also points out that "We're lured in by ads and mailings that hook us with the promise of low rates while keeping the right to raise those rates at any time for any reason—even on old purchases."

Payday loans represent a lending practice that can trap its users in a cycle of debt. Payday loan companies offer credit in the form of cash advances to be repaid when borrowers receive their next paycheck (usually one to two weeks later). Typically, the interest charged is equivalent to an annual percentage rate exceeding 400 percent. Borrowers who fail to repay the loan in full at the designated time can renew the loan for an added fee (plus interest). Although the research on payday loans is limited, existing data suggest that a large fraction of payday loan customers roll over their principal multiple times. Political scientist Robert Mayer (2009) sought to create a profile of the person who uses payday loan services. Mayer thought of an ingenious way to gain insight into this industry, its clients, and lending practices. He realized that when debtors file

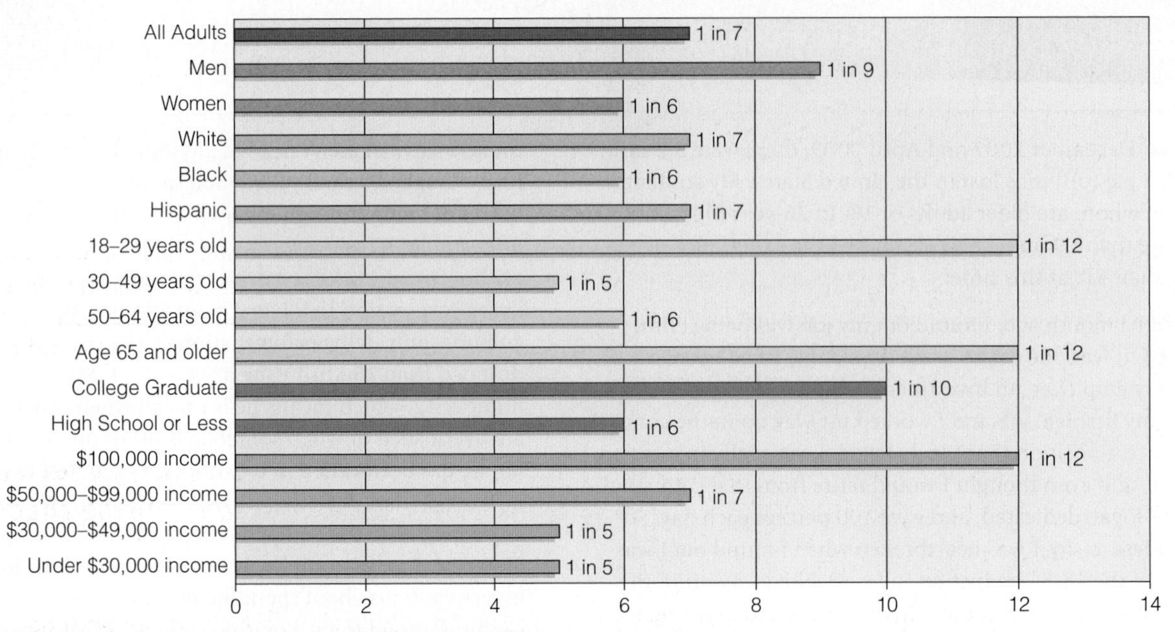

FIGURE 8.5 Chances of Ever Having a Debt Problem
The figure shows the chances of someone having a debt problem by selected population. Here, a debt problem was defined as debt so severe that a person has used a debt consolidation service or has filed for bankruptcy. Which category has the lowest chance of having a debt problem? Which category has the highest chance of having a debt problem?

Source: Data from PEW Research Center (2009)

for bankruptcy, they are required to list the names of creditors from whom they are seeking protection, including the amount of debt and the date the debt was incurred. Mayer examined a sample of 500 bankruptcy petitions filed by residents of one U.S. county—Milwaukee County—in 2004. It is worth noting that there are 66 licensed payday services in that county, the equivalent of one store for every 10,000 adults. Mayer found that 15.2 percent (n = 76) of petitions listed a debt owed to one payday loan company and 10.6 percent listed debts owed to two or more payday loan companies, with some petitions listing as many as nine loans. Overall, about 50 percent of the petitioners owed payday lenders more than the amount of their next paycheck. Mayer then created a profile of those filing for bankruptcy who list payday loans as a source of debt. The annual median income was $26,573, with an average pay day lenders debt of $928. Almost half were single mothers. Fifty three percent resided in a majority black neighborhood.

Online Poll

Look at Table 8.7. Under which hypothetical distribution of wealth would you prefer to live?

- Hypothetical 1
- Hypothetical 2
- Hypothetical 3

To see how other students responded to these questions, go to **www.cengagebrain.com.**

Summary of CORE CONCEPTS

This chapter focuses on the world's richest and poorest people. Among other things, we followed the journey of Scott McLaren who, after graduating from college, joined the Peace Corp and was assigned to Mauritania. We learned how wealth, income, and other valued resources are unequally distributed among the 7 billion people on the planet. We learned that there is a pattern to inequality and we looked at sociological theories that seek to explain the extremes of wealth and poverty in the world and the efforts to correct that inequality.

CORE CONCEPT 1 When sociologists study systems of social stratification, they seek to understand how people are ranked on a scale of social worth and how that ranking affects life chances.

Every society in the world stratifies its people according to ascribed and achieved statuses. The various achieved and ascribed statuses hold social prestige. Sociologists give special attention to stratification systems in which ascribed statuses shape people's life chances and access to achieved statuses. The country into which we are born has important effects on our life chances. For example, the chances that a baby will survive the first year of life depend largely on the country where it is born. There are also variations within countries.

CORE CONCEPT 2 Caste and class systems of stratification are opposite, extreme points on a continuum. The two systems differ in the ease of social mobility, the relative importance of achieved and ascribed statuses, and the extent to which those considered unequal are segregated.

Real-world stratification systems fall somewhere on a continuum between two extremes: a caste system (or "closed" system), in which people are ranked by ascribed statuses (over which they have no control), and a class system (or "open" system), in which people are ranked by merit, talent, ability, or past performance.

CORE CONCEPT 3 Functionalists maintain that poverty exists because it contributes to overall order and stability in society and that inequality is the mechanism by which societies attract the most qualified people to the most functionally important occupations.

Functionalists see the unequal distribution of social rewards as the mechanism by which societies ensure that the best-qualified people fill the most functionally important occupations. In addition, functionalists maintain that poverty performs many functions that contribute to overall social order and stability. As one example, the poor have no choice but to take the unskilled, dangerous, temporary, dead-end, undignified, menial work of society at low pay.

CORE CONCEPT 4 Conflict theorists take issue with the premise that social inequality is the mechanism by which the most important positions in society are filled.

From a conflict point of view, inequality is an ineffective and flawed mechanism for filling the most important positions if only because many occupations that command large salaries and valued rewards contribute little to the good of the overall society. The issues of comparative work and pay equity remind us that some occupational categories command greater rewards for no apparent reason. Finally, there are many cases where the rewards offered exceed the amount likely needed to attract the best qualified people into a valued occupational category such as professional athlete or CEO.

CORE CONCEPT 5 Symbolic interactionists emphasize how social inequality is communicated and enacted in everyday encounters.

Social inequality is conveyed through symbols that have come to be associated with inferior, superior, and equal statuses. Symbolic interactionists focus on the process whereby inequality is enacted in daily life and interactions. In the tradition of the symbolic interactionist, journalist Barbara Ehrenreich worked at many different kinds of low-paying jobs as a way of learning about the work, family, and social lives of those who work in jobs with little status, benefits, and pay ($8.00 or less per hour).

CORE CONCEPT 6 Modernization theory holds that poor countries are poor because they have yet to develop into modern economies and that their failure to do so is largely the result of internal factors such as a country's resistance to free-market principles or to the absence of cultural values that drive material success.

Modernization is a process of economic, social, and cultural transformation in which a country "evolves" from underdeveloped status to a modern society in the image of the most developed countries (that is, western European and North American countries). A country is considered modern when it possesses certain characteristics including widespread literacy, an urban-centered environment, mechanized farming, and a system of mass communication.

CORE CONCEPT 7 Dependency theory holds that, for the most part, poor countries are poor because they are products of a colonial past.

Dependency theorists challenge the basic tenet of modernization theory. They argue that poor countries are poor because they have been, and continue to be, exploited by the world's wealthiest governments and by the global and multinational corporations that are based in the wealthy countries. This exploitation began with colonialism and continues today.

CORE CONCEPT 8 Structural responses to global inequality include transferring modest amounts of wealth from highest-income countries to lowest-income countries through foreign aid and fair-trade policies.

The United Nations' plan to reduce global inequality hinges on the world's 22 richest countries (1) investing 0.7 percent (seven-tenths of 1 percent) of their annual gross national income in foreign aid, and (2) implementing a trading system that eliminates the subsidies, tariffs, and quotas that put the poorest economies at a disadvantage in the global marketplace. To this point, some of the richest countries have exceeded the amount of foreign aid, but most have failed to deliver. Agricultural subsidies are a major problem, but tariffs have been eliminated for most products produced in the world's poorest countries.

CORE CONCEPT 9 Social class is difficult to define. It depends on many factors, including people's relationship to the means of production, their sources of income (such as land, labor, or rent), their marketable abilities, their access to consumer goods and services, their status group, and their membership in political parties.

Marx alerts us that the key variable in determining social class is source of income. For wage laborers, the source is wages; for capitalists, the source is profit; for landowners, the source is ground rent. The finance aristocracy's source of income is created without labor and without creating a product or service. For Weber, the basis for a person's class also depends on marketable abilities (work experience and qualifications), access to consumer goods and services, control over the means of production, and ability to invest in property and other sources of income. Social class spans a range of categories, with the "negatively privileged" property class at the bottom and the "positively privileged" property class at the top. Class ranking is also affected by the status group and political parties to which people belong.

CORE CONCEPT 10 Economic or occupational restructuring can devastate people, leaving them without jobs and disrupting the networks of occupational contacts crucial to moving affected workers into and up job chains.

Major changes in a society's economic or occupational structure—such as factory closings or a decline in farming—can have life-devastating consequences for the affected groups. Such structural changes in the United States help explain the situation of the inner-city, suburban, and rural poor, whose networks of occupational contacts (so crucial to moving them into and up job chains) have collapsed. We can also think of lending practices in structural terms. Since the 1970s, credit has helped drive the U.S. and global economy, giving people money to spend that they did not have. Many people acquired unmanageable levels of debt, which created a division in society between the debt-free and the indebted.

Resources on the Internet

Login to CengageBrain.com to access the resources your instructor requires. For this book, you can access:

Sociology CourseMate

Access an integrated eBook, chapter-specific interactive learning tools, including flash cards, quizzes, videos, and more in your Sociology CourseMate.

Take a pretest for this chapter and receive a personalized study plan based on your results that will identify the topics you need to review and direct you to online resources to help you master those topics. Then take a posttest to help you determine the concepts you have mastered and what you will need to work on.

CourseReader

CourseReader for Sociology is an online reader providing access to readings, and audio and video selections to accompany your course materials.

Visit **www.cengagebrain.com** to access your account and purchase materials.

Key Terms

absolute poverty 182
achieved statuses 184
ascribed statuses 184
brain drain 202
caste system (or "closed" system) 186
class 202
class system (or "open" system) 186
colonialism 196
decolonization 196
esteem 184
extreme wealth 182
finance aristocracy 202
income 204
life chances 183
modernization 194
negatively privileged property class 203
neocolonialism 197
political parties 203
positively privileged property class 203
relative poverty 182
social inequality 183
social mobility 188
social prestige 184
social stratification 183
status group 203
urban underclass 206
wealth 204

9 RACE AND ETHNICITY

With Emphasis on BRAZIL

This is a photo of Brazil's soccer team. Study the physical features of the athletes. How many of the players appear to be black? How many appear to be white? You might be surprised to learn that few, if any, of these players would define themselves as black. In this chapter, we consider why this is the case. When sociologists study race, they study the process by which racial and ethnic categories are created and how people are assigned to them. They study how racial categories come to be, the meanings assigned to them, and their effects on life chances, race relations, and identity.

Why Focus On BRAZIL?

Like the United States, Brazil is considered a melting pot of culturally and racially diverse peoples. Brazil's ideas about race, however, are very different from those of the United States. Since 1600, the United States has worked hard to make everyone who has lived in and immigrated to the country fit into one of its official and ever-changing racial categories. Over 400 years, as many as 2,000 distinct groups of indigenous peoples were placed in the single category "Native American." The millions of immigrants from Europe eventually became "white." The peoples from all of Central and South America (except for Brazil) became "Hispanic." Those of African descent became "black." Those from the Far East, Southeast Asia, and the Indian subcontinent were lumped together into the category "Asian." The peoples of Hawaii and other Pacific Islands (such as American Samoa and Guam) were eventually lumped into the category of "Native Hawaiian and Pacific Islander." The guiding, yet erroneous, assumption was that everyone should fit neatly into one category. For most of its history, the United States was a country that discouraged sexual relationships and marriage between whites and nonwhites, but especially between whites and blacks. (Note: That does not mean that such relationships did not exist. In fact, they were quite common, especially during slavery.)

In contrast, the Brazilian government did not present race as categorical. From the beginning, the Portuguese colonizers were officially encouraged to "marry" the conquered indigenous and enslaved African peoples. As a result, the Brazilian idea of race held that Africans, native peoples, and Europeans had mixed to the point that race was no longer important. The interracial mixing, however, was driven by a broader purpose, and that purpose was to whiten its population. To achieve this goal, the Brazilian government officially recruited immigrants from Europe to whiten the gene pool. As a result of 500 years of policies supportive of whitening and racial mixing, most Brazilians do not see themselves as a particular race; rather, they see themselves on a continuum of color with black and white as endpoints (Telles 2004).

In 2000, the United States government, after hundreds of years of sorting people into clear-cut racial categories, now allows people to identify with more than one of 63 racial categories. Likewise, in 2001, after hundreds of years supporting multiracial identities, Brazilian public universities have instituted affirmative action policies that now require applicants to identify with one of two racial categories—white or black ("Negro").

Online Poll

In terms of race, how do you self-identify?

- ○ As one race, white
- ○ As one race, black
- ○ As one race, Native American
- ○ As one race, Asian
- ○ As one race, Hawaiian/Pacific Islander
- ○ As some other single race, please specify: ____________
- ○ As being biracial or multiracial, please specify: ________

For the most part, does your self-identity match the race others see you as?

- ○ Yes
- ○ No
- ○ Don't know

To see how other students responded to these questions, go to **www.cengagebrain.com.**

Race

CORE CONCEPT 1 Sociologists define race as human-created or constructed categories that assume great social importance.

Although on some level we can say that **race** has something to do with skin shade, hair texture, eye shape, and geographical origins of ancestors, it is so much more than that. When sociologists study race, they study its social importance—the meanings assigned to physical traits, the rules for placing people into racial categories, and the effect race has on opportunities in life. We know that race is human created if only because racial categories vary across time and place. This variation suggests that it is people who "determine what the categories will be, fill them up with human beings, and attach consequences to membership" (Cornell and Hartmann 2007, p. 26). And it is people who embrace, resist, and change them.

To grasp the idea that race is human constructed, think about President Barack Obama, who is considered the first black president of the United States. Do you think that odd considering he described his father as a Kenyan immigrant who was "black as pitch" and his Kansas-born mother as "white as milk" (Obama 2004)? For Obama to be considered black, we decide that certain ancestors (those from Africa, specifically Kenya) are much more important than others (the Kansas-born ones). To call him black, we also have to discount that his skin shade is much lighter than that of his biological father and closer to that of his mother. Finally, we have to emphasize the African or black historical experience as more important that the European or white one in shaping Obama as a person.

In Brazil, it is highly unlikely that Barack Obama would be considered a member of a distinct racial group known as black. Brazilians very likely think of Obama as a cộr (color) along a white–black continuum. For administrative purposes, the Brazilian government uses three broad categories (more like segments on a continuum)—branco (white), pardo (brown), and preto (black), with branco and preto considered ends on that continuum. The three categories apply to 99 percent of the country's population. Two other categories that apply to the remaining one percent are amarelo (yellow) and indigena (indigenous). According to Brazilian point of view, Obama would be brown and perhaps white.

race Human-constructed categories that assume great social importance. Those categories are typically based on observable physical traits (for example, skin shade, hair texture, and eye shape) and geographic origin believed to distinguish one race from another.

Sgt. 1st Class JR Williams, 10th Combat Aviation Brigade

In the United States, because of his physical appearance and African ancestry, President Barack Obama is seen as belonging to the race of just his father. In Brazil, it is unlikely that Obama would be considered black; rather, he would likely be considered brown or perhaps white.

The differences in the way people in Brazil and the United States view Obama force us to think about the nature of racial categories. Although it may seem natural to divide people into racial categories, upon close analysis, it is illogical. First, there are no sharp lines to mark the physical boundaries that distinguish one race from another. In fact, it is difficult to specify the exact point at which some hair texture or skin shade marks a person as white and another as black. That is because no clear line separates so-called black from white skin or tightly curled hair from wavy or straight hair. Although the Brazilian system recognizes a continuum of hair textures and skin shades, it does not specify a place on the continuum that marks the point at which the white (branco) segment gives way to the brown (pardo) and brown gives way to the black (preto) segment.

Second, millions of people in the world are products of sexual unions between people of different races. Obviously, the offspring of such unions cannot be one biological race. Even if we devised rules for classifying each child as a single race, the biological reality does not support this conclusion. Third, the diversity of people within any one racial category is so great that knowing someone's race or color tells us little about him or her. For example, in the United States, people expected to identify as Asian include those who have roots in very different places: Cambodia, India, Japan, Korea, Malaysia, Pakistan, Siberia, the Philippines, Thailand, Vietnam, and dozens of other Far Eastern or Southeast Asian countries. Similar diversity exists within populations labeled as "black or African American," "white," or "American Indian and Alaska Native." Likewise the diversity of people within white (branco), brown (pardo), and black (preto) categories is too great to draw any meaningful conclusions about the people within each segment of the continuum.

Jason Eric Dustin, Courtesy of Joan Ferrante

These four brothers who live in the United States are offspring of the same biological parents pictured. Yet they do not appear to belong to the same race. In fact, it is difficult to specify the exact point or line at which hair texture or skin shade marks one brother as white and another as black. Likewise, if the brothers lived in Brazil, no line would mark the place where white becomes brown and brown becomes black.

Missy Gish

According to racial formation theory, people do not question basic assumptions that they hold about race. In the United States, most people would likely label one brother white and the other black even though they both come from the same genetic pool. Most Americans do not question the logic of this practice.

Finally, racial categories are problematic because they vary across time and place. We must remember that racial categories are not static; they shift over time. We have seen that the United States and Brazil employ different systems to divide humanity—one uses a continuum of racial categories; the other uses a clear-cut category scheme. Racial categories do not just vary by place, however, they vary over time within the same society. Over the past 200+ years, the United States has used as few as two and as many as 63 racial categories.. In addition, it has changed the labels assigned to racial categories (for instance, from "Negro" to "black"). The Brazilian Census has counted and used four racial categories since 1950: white (branco), brown (pardo), black (preto), and amarelo (yellow). In 1991, it added the category indigena (indigenous) (Telles 2004, Bailey 2008).

Racial Formation Theory

Sociologists are interested in how something that makes no logical sense has come to assume such great importance. Sociologists Michael Omi and Howard Winant (1986) offer racial formation theory as a way to understand the significance of race. The two theorists argue that anyone who lives in the United States (or elsewhere for that matter) must learn to "see" its racial categories—that is, they must learn to see arbitrary physical traits, such as skin color and hair texture, as meaningful and significant (see The Sociological Imagination). Moreover, they must develop what the two sociologists call **racial common sense,** shared ideas about each race believed to be so obvious or natural they need not be questioned. For most people in the United States, it is "common sense" that people can be placed in a racial category even in the face of clear evidence that they are not (for example, Barack Obama). In the United States, people must learn to "see" people with any discernable physical traits suggestive of African descent as black.

In contrast, there is a different kind of common sense in Brazil. In that country, it is common sense to see race not in terms of clear-cut categories, but as a continuum. It is also common sense to see people who have any discernable physical traits suggestive of European descent as evidence that that person is not black (Telles 2004).

Racial common sense extends beyond "seeing" racial categories as natural ways to divide humanity. It also involves unquestioned assumptions held about those associated with a specific racial category. Most people are reluctant to discuss commonsense thinking about race for fear that it will open a minefield of misunderstanding (Norris 2002). Through her comedy, Margaret Cho (2002) offers insights about racial common sense as it relates to those classified as Asian American. Specifically, for many in the United States, it is common sense to believe that people classified as Asian were born elsewhere. As Cho puts it, "when people say 'where are you from?' it's a really loaded question because when I answer 'Oh, I'm from

racial common sense Shared ideas believed to be so obvious or natural about racial groups that they need not be questioned.

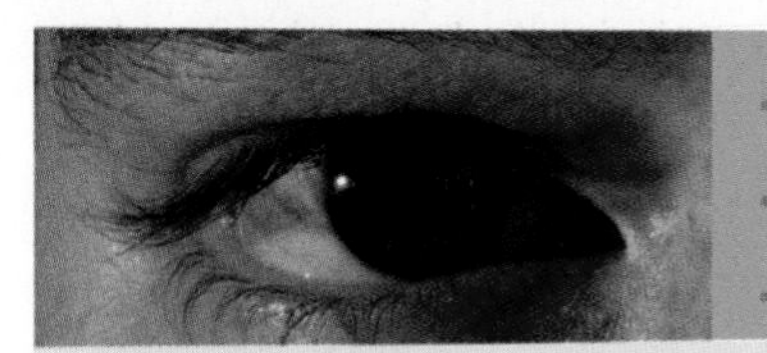

The Sociological Imagination Learning to See Race

It is important to keep in mind that racial identity is something that is learned; it is not automatic that one is aware of or knows his or her race. In fact, a friend once told me that, when her daughter was about 6 years old, she was obsessed for a time about remembering her race. She would come home and say, "Mom, Mom, I can't remember, am I African American or Native American?"

The following statements come from interviews my students conducted with someone who appeared a different race than them. It is worth noting that white students selected primarily blacks to interview, whereas black students selected other nonwhites. I did not have a situation where nonwhite students chose to interview whites. What do you make of that pattern?

- I asked R. from India if she could tell me the story of when she first realized what her race was. She explained, "I came to USA at age 12. They gave me options to choose, and to me Indian was like American Indian. So I asked my uncle who told me to go with Asian. I remembered this and stuck with it. There were many options and at the advice of my family, I go by Asian."
- G. who appears "black" remembers that, although he was surrounded by whites, he never really saw himself as a different race. He thought of himself as just one of the kids playing on the playgrounds. He became more aware of his race in early elementary school. He remembered his father getting defensive when he told him about class modules on race relations.
- It must have been at a family event when T. was three or four years of age. He remembers a family reunion at a park. Everyone was enjoying themselves and each other's company. At the other part, he noticed a white family doing the same. He said he wondered why there weren't any white people with his family or in his family.
- When J. was young he remembers meeting his grandmother who lived in Alabama. He was confused when he first met her because she had very dark hair but she was also white. Both of J's parents appear black, so he was confused as to why his grandmother was white. He never really discussed the concept of race with his grandmother, although he did ask her what race she was because he was so confused.
- As a young child, K. did not pay attention to race. The idea of race presented itself to him in his teenage years. He recalled dating girls who appeared white. In particular, he remembers a girl who was a dancer for the school. She told him that she was not allowed to date black guys and if he had been white, it would be completely different.

San Francisco,' they always come back with 'No, uhm, I mean where are you really from?'"

Omi and Winant (2002) maintain that racial common sense understanding of race persists even when we meet people who defy expectations and assumptions. When that happens, we simply assume that they are exceptions to the rule. So an Asian born in the United States is an exception to the rule, thereby proving the rule exists.

Sociologist Charles A. Gallagher (2009) puts it this way: "race exists because we say it exists" (p. 2). Once racial labels and categories are put in place, it became easy to reify them. **Reify** means to treat them as if they are real and meaningful and to forget that they are made up. When we reify categories, we act as if people *are* those categories. When people do things or appear in ways that don't fit their assigned racial category, we act as if something is wrong with them or that they are the exceptions to the rule rather than questioning the category scheme.

reify Treating labels and categories as if they are real and meaningful and to forget that they are made up.

Racial Categories in the United States and Brazil

CORE CONCEPT 2 Most societies, if not all, have created racial categories and rules for placing people into those categories. Keep in mind that, when subjected to scrutiny, category schemes rarely, if ever, make sense.

Category schemes involve dividing people into racial categories that are implicitly or explicitly ranked on a scale of social worth.

U.S. Racial Categories

Today, the U.S. government recognizes five official racial categories (OMB 2011) plus a sixth category, "other race"—the category of last resort for those who resist identifying, or cannot identify, with one of the five official categories:

- American Indian or Alaskan Native—a person having origins in any of the original peoples of North, Central, or South America and who maintains tribal affiliation or community attachment

- Asian—a person having origins in any of the original peoples of the Far East, Southeast Asia, or the Indian subcontinent
- Black or African American—a person having origins in any of the black racial groups of Africa
- Native Hawaiian or Other Pacific Islander—a person having origins in any of the original peoples of Hawaii, Guam, Samoa, or other Pacific Islands
- White—a person having origins in any of the original peoples of Europe, the Middle East, or North Africa

In critiquing this category scheme, notice that "Black or African American" is the only racial category that does not refer to original peoples. In fact, the words *original peoples* are replaced with *black racial groups of Africa*. If the words *original peoples* were included in the definition of black or African American, every person in the United States would have to claim this racial category. We know from archaeological evidence that all humans evolved from a common African ancestor.

As a further critique, ask if there is any group for which there appears to be no category. For example, where might people known as Hispanic/Latino and Arab fit? As we will learn, the U.S. government does not consider Hispanic to be a race; it considers it an ethnic group and maintains that Hispanics can be of any race. As for those of Arab ancestry, the U.S. government labels them white (see "Why Are People of Arab Ancestry Considered White?").

Until the 2000 census, the U.S. system of racial classification required that an individual identify with only one racial category. Although Americans have always acknowledged racial mixture unofficially by using (often derogatory) words like *mixed, mulatto, half-breed, mongrel,* and *biracial,* the government still insisted that everyone, including those with more than one racial background, identify with just one category. This practice changed with the 2000 census, when for the first time in its history, the United States allowed people to identify with two or more of its six official racial categories. This change represents a monumental shift in thinking about race. Still, when given a chance to identify with more than one race, very few people in the United States do so (see Table 9.1).

Because the U.S. government now allows people to identify with one or more race, there are six one-race categories and 57 permutations of those six. In other words, there are 15 two-race categories (for example, "Black-Asian"), 24 three-race categories (for example, "White-Black-Asian"), 10 four-race categories (for example, "White-Black-Native Indian/Alaska Native-Asian"), 7 five-race categories, and 1 six-race category. It is important to point out that, although the U.S. government has yet to decide what to call people who identify with more than one race, one thing is clear: the government has stated that it will not classify them as multiracial to date and it has never employed such a term in describing its population (OMB 2011).

TABLE 9.1 Number and Percentage of People in the United States By Race, 2009

What percentage of the U.S. population identifies as one race? What percentage identifies as two or more races? Are you surprised at how few identify with more than one race?

One race	299,501,383	97.60%
White	229,773,131	74.80%
Black	38,093,725	12.40%
American Indian and Alaska Native	2,457,552	0.80%
Asian	13,774,611	4.50%
Native Hawaiian and other Pacific Islander	454,001	0.10%
Some other race	14,948,363	4.90%
Two or more races	7,505,173	2.40%
White and Black	1,824,890	0.60%
White and American Indian/Alaska Native	1,776,923	0.60%
White and Asian	1,308,745	0.40%
Black and American Indian/Alaska Native	284,322	0.10%
Some other combination	852,996	0.30%

Source: Data from U.S. Bureau of the Census 2009.

Why Are People of Arab Ancestry Considered White?

We have noted that the United States defines people of Arab ancestry as white. Yet, in light of the September 11, 2001, attacks on the United States, one must question whether people of Arab ancestries are really viewed and treated as white. The U.S. Bureau of the Census explored the idea of creating a separate racial or ethnic category for this group because it recognized that many people of Arab descent do not think of themselves as "white," nor are they treated as such. In the end, the bureau decided *against* creating a special racial or ethnic category. But it did issue a special report on this population. In fact, the bureau made history on December 3, 2003, when it released a report on the Americans of Arab *ancestry*. This report was historic because it was the first time the agency published a brief on a subpopulation not officially designated as a racial minority (Arab Institute 2003).

In the brief, the bureau defined *Arab* as anyone reporting their ancestry on the 2000 census to be Algerian, Alhuceman, Arab, Bahraini, Bedouin, Berber, Egyptian, United Arab Emirates, Iraqi, Jordanian, Kuwaiti, Kurdish, Lebanese, Libyan, Middle Eastern, Moroccan, North African, Omani, Qatari, Palestinian, Rio de Oro, Saudi Arabian, Syrian, Tunisian, or Yemeni. The census bureau pointed out that this definition of Arab includes some peoples, such as Kurds and Berbers, who do *not* necessarily identify themselves as Arab, while excluding other peoples who do identify themselves as Arab such as Mauritanians, Somalis, and Sudanese (U.S. Bureau of the Census 2003).

Here we might pause and ask: Why does the U.S. government classify people of Middle Eastern and Arab ancestry as white? Although this question has no clear answer, some critics argue that the Middle East holds important symbolic value that whites (or at least whites who had the power to make their classification) hope to associate with their racial group. For example, the Middle East is the birthplace of Christianity and the site of many important biblical cities (such as Jerusalem, in Israel and the Palestinian territories, and Babylon, in Iraq). The Middle East boasts the Egyptian pyramids, considered to be among the wonders of the world. Finally, the Middle East has renowned geographic landmarks associated with the cradle of civilization, including ancient Mesopotamia (the land between the Tigris and Euphrates Rivers, in what is now Iraq).

Some highlights from this report follow:

- Approximately 1.2 million people reported being of Arab ancestry.

Joe Seer/Shutterstock.com

According to the census bureau's definition of Arab, a number of celebrities in the United States would be classified as such, including Steve Jobs (cofounder, Apple, Inc., who has a Syrian father) and Marlo Thomas (actress, whose father is of Lebanese ancestry). Pictured here is Paula Abdul, former *American Idol* judge whose father is of Syrian ancestry.

- About half of the Arab population was concentrated in five states: California, Florida, Michigan, New Jersey, and New York.
- Arabs represented 30 percent of the population in Dearborn, Michigan.
- People claiming Lebanese, Syrian, and Egyptian ancestries account for about three-fifths of the Arab American population.
- An estimated 75 percent of Arab Americans are Christian. Many immigrants from Middle Eastern and Arab countries belonged to religious minorities in their home countries (Hertz 2003).

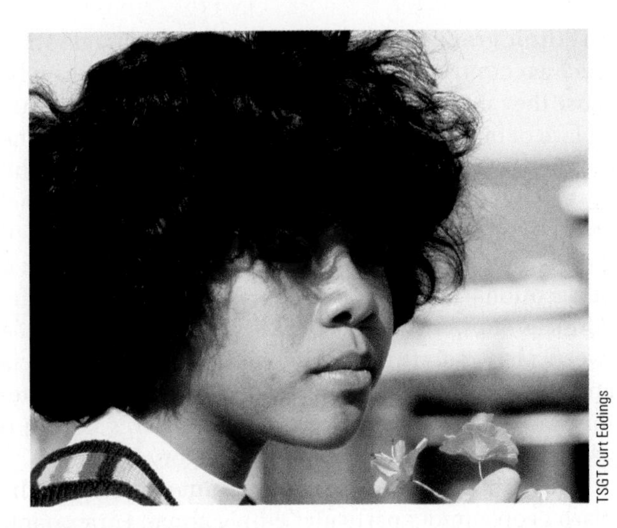

To which racial categories would you assign these two people? The man on the left is considered Laotian, so under the U.S. system he would be classified as Asian. The girl on the right is considered Korean and black so she could be assigned to Asian and black categories. Before 2000, what race would she be? How do you think each would be classified under the Brazilian system?

Brazilian Racial Categories

To understand how Brazilians see race, we must consider three category schemes: the official categories used by the Brazilian Census Bureau (the IBGE), used in popular language, and promoted by the black consciousness movement (Telles 2004) and Bailey (2009).

The IBGE employs five categories: white (branco), brown (pardo), black (preto), yellow (amarelo), and indigenous (indìgena). Only about one percent of the Brazilian population is considered Asian (yellow) or indigenous. The remaining 99 percent are considered white, brown, or black (see Table 9.2). It is important to recall that Brazilians present race as segments on a continuum.

TABLE 9.2 Population of Brazil by Color or Race, 2009

What percentage of Brazilian population identifies as brown? What percentage as black? as white? Where might browns fit under the U.S. system of racial classification?

Color or Race	Number	Percentage
Total	193,733,795	100.00
White	92,992,221	48.43
Black	13,251,391	6.84
Brown	84,855,402	43.8
Yellow	1,123,656	0.58
Indian	542,454	0.28
No declaration	135,613	0.07

Source: Data from Instituto Brasileiro de Geografia e Estatística 2010.

A second system of racial classification relates to the popular language used to refer to race or color. Telles (2004) found that when presented with an open-ended question asking people their race, Brazilian answered with 135 distinct terms, 45 of which were used only once or twice. Ninety-seven percent of respondents used one of six popular terms to describe their color: branco/white (42 percent), moreno/no clear race (32 percent), pardo/brown (7 percent), moreno claro/light of no clear race (6 percent), preto/black (5 percent), Negro (3 percent), and claro/light (3 percent). Telles was intrigued by the fact that more than one-third (38 percent) used the term *moreno* or *moreno claro*. *Moreno,* a term not used by the Brazilian census, means a "colored person" of ambiguous or no clear race. *Moreno claro* means a person of light color.

The third system of racial classification is a two-category scheme employed by those in Brazil's black consciousness movement—someone is either negro or blanco. That movement seeks to dismantle ideas of race as a continuum, to destigmatize "blackness," and to challenge the unspoken assumption that brown is superior to black. Black consciousness movement activists argue that ambiguous racial identities have discouraged browns and blacks from mobilizing to fight well-documented discrimination and prejudices. Thus, the movement encourages all people who see themselves as moreno and pardo to identify as Negro.

Ethnicity

CORE CONCEPT 3 Like race, when sociologists study ethnicity, they are interested in studying the processes by which people make ethnicity important (or not).

An **ethnic group** consists of people within a larger society (such as country) who possess a group consciousnesses because they share or believe they share a common ancestry, a place of birth, a history, a key experience or some other distinctive social traits they have defined as the "essence of their peoplehood" (Schermerhorn 1978, p. 12; Cornell and Hartman 2007). The thing shared may be a religion, a style of dress, a language, a shared experience of persecution—anything that sets them apart from other ethnic groups or the society at large. Distinguishing between race and ethnicity is complicated because racial and ethnic identities are intertwined. In the United States, for example, people who consider themselves Korean or Chinese are assigned to the racial category "Asian."

It is also difficult to specify the unique markers that place people into a particular ethnic group. For example, does having a Chinese first name and last name make someone an ethnic Chinese? Or is it the ability to speak Mandarin or Cantonese that does? What if the person is bilingual? Does that make him or her less ethnic Chinese? Because markers of ethnicity are imprecise, one way to determine someone's ethnicity is to simply ask, "What is your ethnicity?" Self-identification, however, is also problematic because people's sense of ethnicity can range in intensity from nonexistent to all-encompassing (Verkuyten 2005).

In claiming ethnicity, people may point to ancestors they have never met (for example, "My great-grandfather was from Portugal.") or to a distinct community from their past (for example, "Growing up, I lived in Puerto Rico."). For some, ethnicity is based on a sentimental connection that may manifest itself in rooting for a particular soccer team (that is, the Brazilian soccer team). For others, ethnicity is a complete lifestyle that involves being born in a particular place, speaking the language, dressing a particular way, and interacting primarily with others in that ethnic group.

Ethnic identity is also affected by **selective forgetting**, a process by which people don't know about, forget, dismiss, or fail to pass on a connection to one or more ethnicities. Forgetting an ethnicity is affected by larger societal forces. For example, in the United States, some races have more freedom than others in claiming an ethnic identity. People in the United States classified as white have a great deal of freedom to claim a European ethnic identity. But it is unusual for those who appear white to claim a non-European ethnicity, such as Kikuyu, one of Kenya's many ethnic groups. People in the United States classified black have even less choice; they are expected to identify as simply "black" or as of African descent, even when they know their specific African and non-African ancestors (Waters 1994). In fact, most Americans almost never think of nonwhites as having a European ethnicity. For example, most Americans would likely dismiss as irrelevant the connections President Barack Obama has to his Swiss, German, or Irish ancestors (Dacey 2010).

People's sense of ethnicity can shift through a process known as **ethnic renewal**. This occurs when someone discovers an ethnic identity, as when an adopted child learns about and identifies with newly found biological relatives or a person learns about and revives lost traditions. Ethnic renewal includes the process by which people take it upon themselves to find, learn about, and claim an ethnic heritage (see The Sociological Imagination: "Selective Forgetting").

ethnic group People within a larger society (such as a country) who possess a group consciousnesses because they share or believe they share a common ancestry, a place of birth, a history, a key experience, or some other distinctive social traits they have defined as the "essence of their peoplehood" (Schermerhorn 1978, p. 12; Cornell and Hartman 2007).

selective forgetting A process by which people forget, dismiss, or fail to pass on a connection to one or more ethnicities.

ethnic renewal This occurs when someone discovers an ethnic identity, as when an adopted child learns about and identifies with newly found biological relatives or a person learns about and revives lost traditions.

involuntary ethnicity When a government or other dominant group creates an umbrella ethnic category and assigns people from many different cultures and countries to it.

ethnicity People who share, believe they share, or are believed by others to share a national origin; a common ancestry; a place of birth; distinctive concrete social traits (such as religious practices, style of dress, body adornments, or language); or socially important physical characteristics (such as skin color, hair texture, or body structure).

EVARISTO SA/AFP/Getty Images

The Macuxi are one of the estimated 230 indigenous ethnic groups in Brazil. The group lives in northeastern Brazil. Their ancestral land was confiscated by farmers some 30 years ago, and the group has struggled for at least 30 years to reclaim it. A supreme court decision ruled in Macuxi's favor, but farmers have armed themselves and have refused to leave the land (Amnesty International 2008).

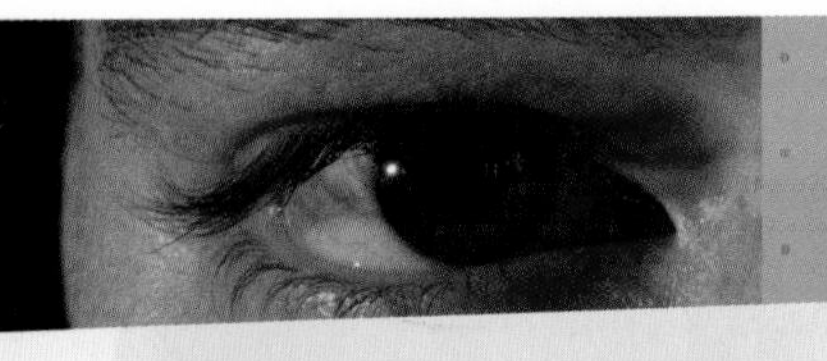

The Sociological Imagination Selective Forgetting

Chris Caldeira

Chris Caldeira

For most of his life, Don Caldeira who was born and raised in the United States, celebrated his Portuguese ethnicity. To instill a sense of ethnic pride in his children, he did things like hang the Portuguese flag in the family room and teach his children to recognize other people who were Portuguese by noticing the spelling of their last name—(for example, "the *-eira* is what makes it. . . . Names like Meredith Vieira or Bobby Ferreira are Portuguese!"). Don would also tell his children that "we are the only Caldeira listed in the phone book." He made that fact a point of pride as it suggested that his family was special. When his daughter Chris moved to Massachusetts, she looked up the name *Caldeira* in the phone book and saw hundreds listed. She tore the page out to show her dad. Apparently, at some point a large Portuguese community had immigrated to the Cape Cod area.

It was not until Don was in his 70s, after his sister gave him a copy of their mother's birth certificate, that he learned his mother was classified as "Portuguese native Creole." Like most Americans, the process known as selective forgetting shaped his ethnic identity. For whatever reason, his mother and sister "forgot" to tell him. If he lived in Brazil, it is unlikely that he would have thought of himself as simply Portuguese, even if his sister and mother "forgot" to pass on that fact on. In Brazil, everyone is assumed multiracial. Now, in the spirit of ethnic renewal, Don makes a point of telling his grown children they are Portuguese *and* Creole.

Involuntary Ethnicity

Then there is the phenomenon of **involuntary ethnicity**. In this situation, a government or dominant group creates an ethnic category for a specific group. Often the people assigned to that category come from many different cultures and countries. In the United States, the ethnic category Hispanic is illustrative. Of the thousands of ethnic groups that live in the United States, the U.S. Bureau of the Census *officially* recognize just two—(1) "Hispanic or Latino" and (2) "Not Hispanic or Latino." The term *Hispanic*, created in 1970, applies to people from, or with ancestors from, 21 Central and South American countries that were once former colonies of Spain. Each of these 21 countries consists of peoples with distinct histories, cultures, and languages. To complicate matters even further, the history of Central and South America intertwined with that of Asia, Europe, the Middle East, and Africa. Because of this interconnected past, these 21 countries are populated not by one ethnic group known as "Hispanics," native- and foreign-born people, immigrants, nonimmigrant residents, and people from every conceivable ancestry, not just Spanish (Toro 1995). To those classified as Hispanic, the label *Hispanic* is likely very confusing because it forces them to identify with conquistadors and settlers from Spain, who imposed their culture, language, and religion on indigenous people and on the African peoples they enslaved.

With regard to ethnicity, the Brazilian government seeks to identify only the ethnicity of the 1 million or so indigenous peoples who live within its borders. After establishing that a person is racially indigenous (for example, Do you consider yourself indigenous?), the Brazilian census asked: (1) "What is your ethnic group or people you belong to? (2) Do you speak an indigenous language in your housing unit? and (3) Do you speak Portuguese in your housing unit?" (IBGE 2010). In Brazil, an estimated 230 indigenous ethnic groups speak 180 languages, representing one percent of the total population (Osava 2010).

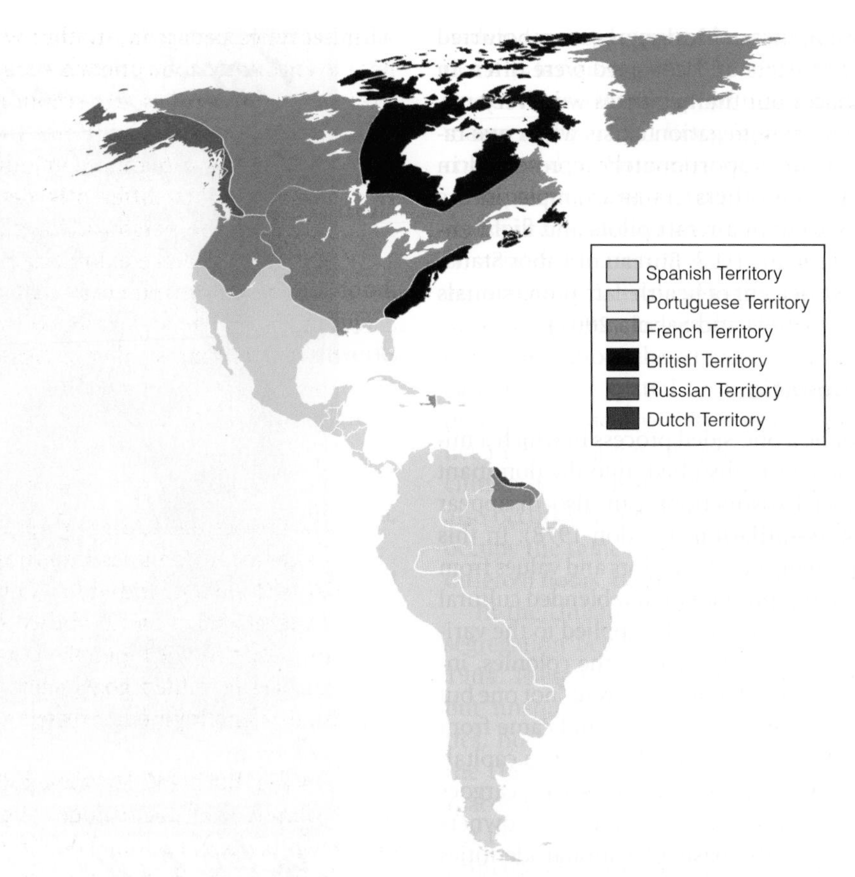

FIGURE 9.4 The map shows how North, Central, and South America were divided up by European and Russian powers in a process known as colonization. Colonization is a situation where a foreign power uses superior military force to impose its political, economic, social, and cultural institutions on an indigenous population to control their resources, labor, and markets (Marger 2012).
Source: Data from U.S. Central Intelligence Agency (2011)

differences in athletic ability, IQ scores, and taste in music. From a racist point of view, there is no other possible explanation for these inequalities.

Racism has probably always existed. Modern racism, however, emerged as a way to justify European exploitation of people and resources in Africa, Asia, and the Americas. Between 1492 and 1800, Europeans learned of, conquered, and colonized much of North America, South America, Asia, and Africa and set the tone for international relations for centuries to come. Among other things, this exploitation took the form of enslavement and colonialism (see Figure 9.4).

scientific racism The use of faulty science to support systems of racial rankings and theories of social and cultural progress that placed whites in the most advanced ranks and stage of human evolution.

Under colonialism, local populations were forced to cultivate and harvest crops and to extract minerals and other raw materials for export to the colonists' home countries. Racism helped justify this exploitation of nonwhite peoples and their resources by pointing to the so-called superiority of the white race. More precisely, the exploitation was justified by **scientific racism**, the misuse of science to create findings that supported systems of racial rankings and theories of social and cultural progress that placed whites in the most advanced ranks and stage of human evolution.

The eugenics movement, popular during the early decades of the twentieth century, included racist elements. Eugenics was presented as an applied science with the purpose of identifying ways to improve the genetic composition of populations. With regard to race, eugenicists named and ranked racial groups, placing the white race at the apex of the hierarchy and black at the bottom. In addition, some eugenists maintained that racial mixing created degenerates (Telles 2004).

In the United States, eugenic policies supported by scientific racism were behind laws created to protect white racial purity, including prohibiting sexual relationships between whites and nonwhites and banning interracial marriage. Eugenic thinking gave support to the Immigration Act of 1924 that sought to restrict inferior racial stocks from immigrating to the United States.

Eugenicists maintained that tropical climates contributed to mental and physical inferiority and, by extension, pointed to Brazil's population as an illustration of biological degeneracy dooming that country's fate to one of perpetual underdevelopment (Telles 2004). This assessment hit Brazil's political elite hard as many were from mixed-race ancestries. As a solution, Brazil's scientists, intellectuals, and political elite embraced whitening as a solution. While still embracing the doctrine of white superiority, they maintained that "black and mulatto inferiority could be overcome by miscegenation" and that "race mixture would eliminate the black population, eventually resulting in a white or mostly white Brazilian population" (Telles 2004, p. 28; see "Working for Change: Lester Ward").

Prejudice, Stereotyping, and Discrimination

> **CORE CONCEPT 9 Prejudice, stereotyping, and discrimination are tools that preserve the system that benefits advantaged race and ethnic groups.**

A **prejudice** is a rigid and usually unfavorable judgment about an out-group that does not change in the face of contradictory evidence. That judgment is applied to anyone who belongs to that out-group. Prejudices are based on **stereotypes**—simplistic generalizations about out-groups and applied to anyone who belongs to that out-group. Stereotypes interfere with a person's ability to recognize new or unexpected information that contradict that generalization. Stereotypes give the illusion that people are predictable and that one knows and has the right to construct images of "the other" (Crapanzano 1985).

Stereotypes are supported and reinforced in a number of ways. In **selective perception**, prejudiced people notice only those supposed facts that support their stereotypes. When subjected to close scrutiny, the facts prove unfounded. For example, many people use facts to support stereotypes that white men can't jump and are slow, based on the facts. After all, the facts show that only a small portion of professional basketball and track athletes are white. If one takes the time to look beyond the sport of basketball, it is easy to identify sports where the white athletes who dominate them possess extraordinary speed and leaping ability. Surprisingly, one such sport is weightlifting (see photos on p. 234).

Stereotypes persist in another way: When prejudiced people encounter someone who contradicts a stereotype, they see that person as an exception to the rule. In fact, such encounters merely reinforce the stereotype. Finally, stereotypes survive because prejudiced people evaluate the same behaviors differently depending on the race or ethnicity of the person involved (Merton 1957). For example, incompetence exhibited by racial or ethnic minorities is often attributed to their race or ethnicity. In contrast, incompetence exhibited by an advantaged race or ethnicity is almost always treated as an individual shortcoming.

Discrimination

In contrast to prejudice, **discrimination** is not an attitude but a behavior. It includes intentional or unintentional unequal treatment of individuals or groups based on attributes unrelated to merit, ability, or past performance. Discrimination denies people equal opportunities to achieve socially valued goals such as getting an education, finding employment, accessing health care, and living a long life.

Sociologist Robert K. Merton explored the relationship between prejudice (the attitude) and discrimination (the behavior). He argued that knowing whether people are prejudiced does not help to predict whether they will discriminate. Merton employed a four-part typology to show that both nonprejudiced people (who believe in equal opportunity) can discriminate and prejudiced people can refrain from discrimination.

---

prejudice A rigid and usually unfavorable judgment about an out-group that does not change in the face of contradictory evidence and that applies to anyone who shares the distinguishing characteristics of the out-group.

stereotypes Inaccurate generalizations about people who belong to an out-group. Stereotypes "give the illusion that one knows the other" and has the right to construct images of the other (Crapanzano 1985, pp. 271–272).

selective perception The process in which prejudiced people notice only those things that support the stereotypes they hold about an out-group.

discrimination An intentional or unintentional act of unequal treatment of individuals or groups based on attributes unrelated to merit, ability, or past performance. Discrimination denies people equal opportunities to achieve socially valued goals such as getting an education, finding employment, accessing health care, and living a long life. It is a behavior, not an attitude.

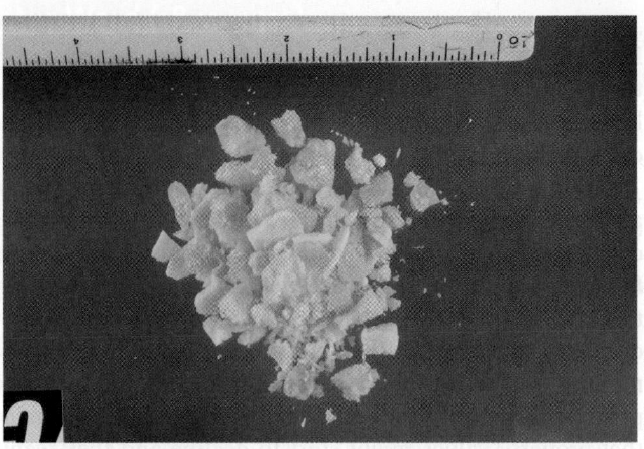

Crack (right) and powder cocaine (left) have the same physiological and psychotropic effects but are handled very differently for sentencing purposes. On average, sentences for crack offenses are three to six times longer than those for offenses involving equal amounts of powder. Approximately 85 percent of defendants convicted of crack offenses in federal court are black, whereas 78 percent of defendants in powder cocaine cases are white; thus, the severe sentences are imposed "primarily upon black offenders" (Kimbrough v. United States 2007).

rights movement. In Brazil, one response was the black consciousness movement.

The Civil Rights Movement

Shortly after slavery was abolished in the United States, a state-sanctioned system of racial discrimination, known as Jim Crow was put into place. Under Jim Crow laws, blacks (and other minorities) were denied the right to vote and sit on juries; subjected to racial segregation (separate and unequal facilities); disadvantaged with regard to employment opportunities; and subjected to widespread, systematic discrimination, including violence against person and property. In many states, whites and nonwhites were denied the right to marry (see Table 9.5). Examples of such laws include the following:

- No person or corporation shall require any white female nurse to nurse in wards or rooms in hospitals, either public or private, in which Negro men are placed. (Alabama)
- It shall be unlawful for colored people to frequent any park owned or maintained by the city for the benefit, use, and enjoyment of white persons . . . and unlawful for any white person to frequent any park owned or maintained by the city for the use and benefit of colored persons. (Georgia)
- All marriages of white persons with Negroes, mulattos, Mongolians, or Malaya hereafter contracted in the State of Wyoming are and shall be illegal and void. (Wyoming)

The civil rights movement was a response to such systematic discrimination, not just in the South but also across the nation. The civil rights movement was waged by hundreds of thousands of people, many of whom put their lives on the line, demanding integrated schools, decent housing, and an end to racial bias. In the popular imagination, the civil rights movement involved

TABLE 9.5 Number and Percentage of U.S. Population Classified as Black and Mulatto, 1910

In the context of the United States between 1850 and 1910, unless people with white and African ancestries could pass as white, they were classified as mulatto, but still considered "Negro" or black and never "white." Although laws in the United States prohibited marriage and sexual relationships between whites and nonwhites, especially blacks, that does not mean sexual relationships were absent. Evidence of intermixing before 1967 (the year miscegenation laws were abolished) is reflected in U.S. census data from 1910. Notice that, in that year 20.9 percent of the "Negro" population was considered "mulatto," and yet, despite mixed ancestry, "mulattos" were still part of the "Negro" racial category, not the "white" category.

	"Negro" Population	Number and Percentage of Total "Negro" Population			
Census Year	Total	Black	Mulatto	% Black	% Mulatto
1910	9,827,763	7,777,077	2,050,688	79.1	20.9

Source: Data from U.S. Bureau of the Census 1910.

confronting white supremacists such as the Ku Klux Klan. However, activists also confronted institutional discrimination as embodied in local, state, and federal agencies, judicial systems, and legislative bodies, including police, the National Guard, judges, and all-white citizens' town/city councils. Most notably, police departments, especially in the South, arrested civil rights activists on false or trumped-up charges, and all-white juries found whites who murdered blacks not guilty. Some officials, such as Alabama Governor George Wallace, used the National Guard and state police to prevent school integration.

The black churches played a key role in the civil rights movement. In an atmosphere of profound discrimination and inequality, the black church had become not just a place to worship but also served as a community clearinghouse, a credit union, a support group, and center of political activism (National Park Service 2009). In fact, churches were the context for the emergence of key civil rights leaders such as reverends Martin Luther King Jr. and Fred Shuttlesworth and the emergence of key organizations such as the Southern Christian Leadership Conference (SCLC), the Student Nonviolent Coordinating Committee (SNCC) and the National Association for the Advancement of Colored People (NAACP).

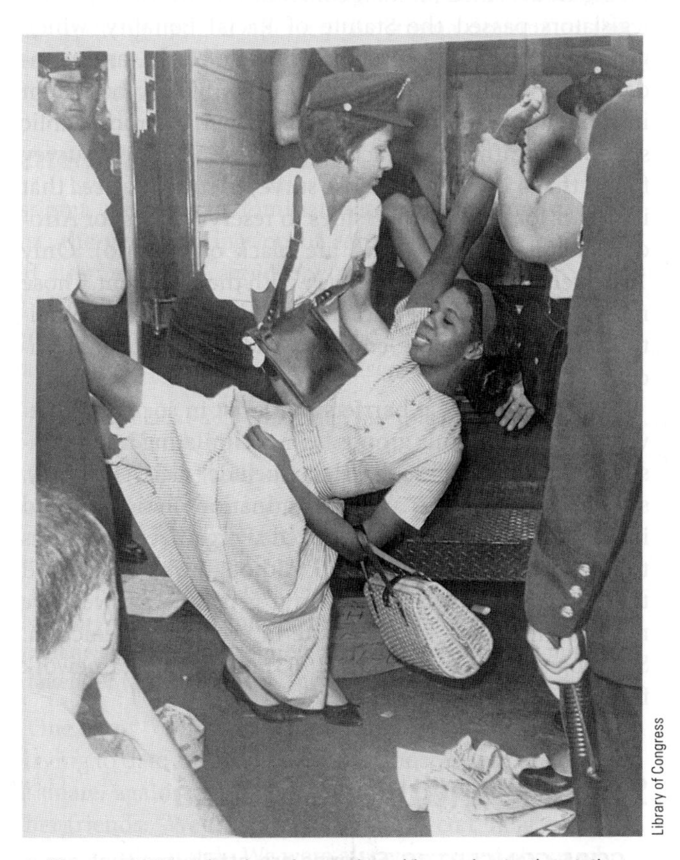

Library of Congress

College and high school students played key roles in the civil rights movements. They participated in bus boycotts, sit-ins, freedom rides, and school integration.

The Selma-to-Montgomery March was an especially significant moment for the civil rights movement, because it was the first event of the activist movement to be televised. What Americans saw created enough outrage to lend national support to the movement. The march began on what is known as Bloody Sunday, March 7, 1965, when state and local law enforcement agents stopped 600 demonstrators six blocks into the march and attacked them with clubs and tear gas. Civil rights leaders sought protection from the courts to march and it was granted. The escalating intensity of this movement pushed the federal government to become involved. Most notably, President John F. Kennedy used the power of his office to enforce desegregation in schools and public facilities. Attorney General Robert Kennedy filed suits against at least four states to secure the right to vote for blacks. When Congress passed the Civil Rights Act of 1964 and the Voting Rights Act of 1965, President Lyndon Johnson signed them into law knowing that it might cost him the next presidential election and severely weaken the Democratic Party's chance of winning in the next election cycle. Of course, federal and Supreme Court judges also played key roles in ruling against segregation and discrimination (National Park Service 2009).

The civil rights movement reached its most organized phase in the late 1950s and 1960s. It encompassed other related movements as well, including the American Indian Movement, La Raza Unida (the Unified Race), the antiwar movement (Vietnam), and the women's movement. After decades of struggle and resistance, that discrimination supported by law was overturned with the ratification of the Civil Rights Act of 1964, the Voting Rights Act of 1965, and the Fair Housing Act of 1968.

Black Consciousness Movement in Brazil

In the case of Brazil, we can say that the black consciousness movement came of age on May 11, 1988, two days before that country was to celebrate the 100th anniversary of the abolition of Brazilian slavery. On this day, 5,000 people marched through downtown Rio de Janeiro led by Frei Davi the leader of the Commission of Black Religious, Seminarians, and Priests. Davi yelled, "They say the good white masters gave us our freedom! Nonsense!" He reminded the participants and crowd that Brazilian blacks were still enslaved. Brazil's "racial democracy is a lie!" (Davi 1992).

This event was pivotal in pushing Brazilians to see that their country was a racial democracy. According to official ideology, race was deemed unimportant because Brazilians had a tradition of encouraging intermarriage and sexual relations across race. In addition, unlike the United States, Brazil had never created laws that enforced segregation. Brazil's traditions led many in and outside of Brazil to hold the country as a model of race relations. When the UN commissioned a series

A second way the stigmatized respond involves devoting a great deal of time and effort to overcoming stereotypes or to appearing as if they are in full control of everything around them. The stigmatized may try to be perfect—to always be in a good mood, to be extra friendly, to outperform everyone else, or to master an activity not expected of them. This response is common among plumbers, electricians, building contractors, and other service workers who must make house calls in other race neighborhoods. One black contractor interviewed in the *New York Times* maintained that when doing business in white neighborhoods, he tries to "Avoid working at night or showing up at a job early in the morning. Never linger inside houses or gaze at a resident's possessions. And always keep your tools at hand to allay suspicion" (p. A1).

As a third response to being stigmatized, people may use their subordinate status for secondary gains, including personal profit, or as "an excuse for ill success that has come [their] way for other reasons" (Goffman 1963, p. 10). If an employee accuses an employer of discrimination and threatens to file a lawsuit when he or she has been justly sanctioned for poor work, then that person is using stigma for secondary gains. Keep in mind, however, that the person can use stigma this way only because discrimination is commonplace in the larger society.

A fourth response to being stigmatized is to view discrimination as a blessing in disguise, especially for its ability to build character or for what it teaches about life and humanity. Finally, a stigmatized person can condemn all the normals and view them negatively.

Online Poll

If given a chance to indicate your race on a survey, with how many official races will you identify?

- ○ One race
- ○ Two races
- ○ Three or more races

To see how other students responded to these questions, go to **www.cengagebrain.com.**

Ester seng, Courtesy of Joan Ferrante

Summary of CORE CONCEPTS

The photo shows Brazilian students holding the flag of their country. In this chapter, we learned that Brazilian identity is built around the idea that everyone is multiracial and no racial group is distinct from others. This is in sharp contrast to the United States, where races are viewed as distinct categories. We used sociological concepts and theories to think about how ideas of race are constructed and the effects these categories have on life chances, race relations, and identity.

CORE CONCEPT 1 Sociologists define race as human-constructed categories that assume great social importance.

The shortcomings of racial and ethnic categories become evident when we imagine trying to classify the world's 7 billion people. This chapter identifies at lease three shortcomings: (1) There are no sharp dividing lines that distinguishes one racial category from an another. (2) Boundaries between races can never be fixed and clear, if only because males and females of any alleged race can produce offspring together. (3) Most classification schemes force diverse people who vary in ethnicity, language, generation, social class, or physical appearance under one umbrella category.

CORE CONCEPT 2 Most societies, if not all, have created racial categories and rules for placing people into those categories. Keep in mind that, when subjected to scrutiny, category schemes rarely, if ever, make sense.

The U.S. official system recognizes five race categories plus a sixth category, "other" race. Those categories are American Indian or Alaska Native, Asian, Black or African American, Native Hawaiian or Other Pacific Islander, and White. Until the 2000 census, the U.S. system of racial classification required individuals to identity with only one category. Now a person can identify with two or more categories. Only about 2 percent have done so.

The Brazilian Census Bureau employs five categories: white, brown, black, yellow, and indigenous. Only about one percent is considered Asian indigenous. Brazilians tend to use one of six popular terms; one-third use the term *moreno*, a term not employed by the census. The black consciousness movement seeks to divide populations into two categories, white and Negro.

CORE CONCEPT 3 Like race, when sociologists study ethnicity, they are interested in studying the processes by which people make ethnicity important (or not).

An ethnic group consists of people within a larger society (such as country) who possess a group consciousnesses because they share or believe they share a common ancestry, a place of birth, a history, a key experience, or some other distinctive social traits they have defined as the "essence of their peoplehood" (Schermerhorn 1978, p. 12). To capture ethnicity in all its complexity, sociologists employ a number of key concepts, including selective forgetting, ethnic renewal, involuntary ethnicity, and dominant ethnic groups. Dominant ethnic groups are the most advantaged ethnic groups. In Brazil and the United States, those dominant groups are of European ancestries.

CORE CONCEPT 4 The racial and ethnic categories to which people belong are a product of three interrelated factors: chance, context, and choice.

Chance is something not subject to human will, choice, or effort. We do *not* choose our biological parents, nor can we control the physical characteristics we inherit from them. Context is the social setting in which racial and ethnic categories are recognized, created, and challenged. Choice is the act of choosing from a range of possible behaviors or appearances. Individual choices are constrained by chance and context. Most blacks as well as many whites living in the United States would be classified as brown in Brazil. Likewise, everyone in Brazil is considered multiracial.

CORE CONCEPT 5 Every country in the world has people living within its political boundaries who are immigrants and were born elsewhere. Often considerations of race and ethnicity figure into immigration policies.

No government welcomes just anyone; often race, ethnicity, and skills figure into the probability that one will gain admission to a country legally or will have to enter illegally. Historically, race and ethnicity considerations have played a major role in U.S. and Brazilian immigration policies.

CORE CONCEPT 6 Minority groups are subpopulations within a society that are systematically excluded (whether consciously or unconsciously) from full participation in society and denied equal opportunities to access power, prestige, and wealth.

Membership in minority groups is involuntary, and a minority may be the numerical majority in a society. Minority groups are excluded from full participation in the larger society. That is, they do not enjoy the taken-for-granted advantages and immunities a dominant group enjoys. Often minority groups are subjected to social and spatial isolation. Minority status overshadows any accomplishments. These characteristics of minority group status apply especially to the situation of involuntary minorities.

CORE CONCEPT 7 Assimilation is a process by which ethnic and racial distinctions between groups disappear because one group is absorbed into another group's culture or because two cultures blend to form a new culture.

There are two types of assimilation: absorption and melting pot. In the first type, two or more racial and ethnic group minorities adapt to the point where they are completely "absorbed" into the dominant culture. In practice, this type of assimilation is difficult if only because disadvantaged racial and ethnic groups experience de jure or de facto segregation. The segregation may be spatial or hierarchical. The second type of assimilation is melting pot where racial and ethnic groups accept many new behaviors and values from one another. This exchange produces a new blended cultural system.

CORE CONCEPT 8 Racism is the belief that genetic or biologically based differences explain and even justify inequalities that exist between advantaged and disadvantaged racial and ethnic groups.

Those alleged genetic differences—as manifested in skin color, hair texture, and eye shape—are deemed so significant that they are used to explain racial and ethnic differences as they relate to crime, graduation, teen pregnancy, and marriage rates. From a racist point of view, there is no other possible explanation for these inequalities. Racism has probably always existed. Modern racism, however, emerged as a way to justify European exploitation of people and resources in Africa, Asia, and the Americas. More precisely, the exploitation was justified by scientific racism. In Brazil, eugenicists used scientific racism to justify policies aimed at "whitening" the population. In the United States, scientific racism justified laws aimed at preserving racial purity.

CORE CONCEPT 9 Prejudice, stereotyping, and discrimination are tools that preserve the system that benefits advantaged race and ethnic groups.

Both prejudiced and nonprejudiced people can discriminate. In contrast to prejudice, discrimination is not an attitude but a behavior. It is aimed at denying people equal opportunities to achieve socially valued goals (such as education, employment, health care, and long life) or blocking their access to valued goods and services. Prejudice does not necessarily predict discriminatory conduct.

CORE CONCEPT 10 Stigmas are attributes that are so deeply discrediting that they come to dominate social experiences and interactions.

A stigma is a physical trait or other attribute that is deeply discrediting. Sociologists examine social encounters between those deemed normal and stigmatized, known as mixed contacts. A stigma can come to dominate those contacts in many ways. For example, normals and stigmatized often take measures to avoid one another. The stigmatized respond in a variety of ways. They may attempt to "correct" the source of stigma or view it as a blessing in disguise.

Resources on the Internet

Login to CengageBrain.com to access the resources your instructor requires. For this book, you can access:

Sociology CourseMate

Access an integrated eBook, chapter-specific interactive learning tools, including flash cards, quizzes, videos, and more in your Sociology CourseMate.

Take a pretest for this chapter and receive a personalized study plan based on your results that will identify the topics you need to review and direct you to online resources to help you master those topics. Then take a posttest to help you determine the concepts you have mastered and what you will need to work on.

CourseReader

CourseReader for Sociology is an online reader providing access to readings, and audio and video selections to accompany your course materials.

Visit **www.cengagebrain.com** to access your account and purchase materials.

Key Terms

absorption assimilation 229
assimilation 229
chance 222
choice 223
context 223
discrimination 233
dominant ethnic group 222
ethnic group 220
ethnic renewal 220
ethnicity 220
hidden ethnicity 222
individual discrimination 235
institutionalized discrimination 235
involuntary ethnicity 220
involuntary minorities 229
melting pot assimilation 231
minority groups 226
mixed contacts 239
nonprejudiced discriminators (fair-weather liberals) 235
nonprejudiced nondiscriminators (all-weather liberals) 235
prejudice 232
prejudiced discriminators (active bigots) 235
prejudiced nondiscriminators (timid bigots) 235
race 214
racial common sense 215
reify 216
scientific racism 231
segregation 230
selective forgetting 220
selective perception 233
spatial segregation 230
stereotypes 233
stigma 238

GENDER

10 With Emphasis on AMERICAN SAMOA

When sociologists study gender they seek to understand the processes by which people learn behaviors and appearances expected of someone of their sex. They also are very interested in people who conform and defy expectations and how others react when they do. The two people pictured are American Samoan fa'afafines (pronounced fah-ah-fuh-fee-nay), someone not considered biologically female but who has taken on the "way of women" in dress, mannerism, appearance, and role. The two pictured are also international fa'afafine boxing champions. Fa'afafines are very diverse in the way they present themselves. The fa'afafines who live in cities tend to be more visible and dress flamboyantly, often performing in what are thought of as American-style drag shows. Those who live in rural areas or who are from devout Christian families are more discreet and tend not to flaunt sexuality (Tok Blong Pacifik 2008).

Why Focus On AMERICAN SAMOA?

In this chapter, the topic of gender is paired with American Samoa. Most Americans probably do not realize that American Samoa has been a territory of the United States since 1899, when the Samoan Islands were partitioned between Germany and the United States. Many Americans probably do know of an American Samoan who plays for their favorite NFL or Division I college football team, as at least 200 Samoans play at those levels. That number is truly amazing, given that the entire population of American Samoa is about 65,000, hardly large enough to fill a Super Bowl stadium (CBS 60 Minutes 2010).

In contrast to the rugged image of the football player, Samoans recognize a third gender that they call fa'afafine, people not considered biologically female but who have taken on the "way of women" in dress, mannerism, appearance, and role. The most visible fa'afafines dress as famous female celebrities such as Britney Spears, Madonna, or Kelly Clarkson. Their success in Samoan society comes from giving the most stunningly accurate imitation of these women. There is even a Miss Island Queen contest (since 1979) judged by very high-profile men and women outside the transgender community including U.S. military officers, writers, the reigning Miss American Samoa, and musicians (Matà afa 2006). One year, a second-grade teacher won. The winners are expected to engage in a range of community service projects that often involve the elderly, the local Red Cross chapter, and hospitals (Matà afa 2009).

Sociologists are fascinated by a society where both football players and fa'afafines are commonplace. They look for social forces that support the existence of both in such a tiny population. In addition, sociologists draw lessons about gender that go beyond American Samoa and inform a general understanding of gender.

Online Poll

If you were expecting a baby, would you want to know the sex of the baby in advance of its birth?

○ Yes

○ No

To see how other students responded to these questions, go to **www.cengagebrain.com.**

Distinguishing Sex and Gender

In this chapter, we begin by examining a crucial distinction sociologists make between *sex* and *gender*. Although many people use those two words interchangeably, the two terms have very different meanings.

Sex as a Biological Concept

CORE CONCEPT 1 Sex is a biological concept, whereas gender is a social construct.

Sex is a biological distinction determined by primary sex characteristics or the anatomical traits essential to reproduction. Most cultures classify people in two categories—male and female—based largely on what are considered to be clear anatomical distinctions. Anatomical sex is not something that is clear-cut, however, if only because some

sex A biological concept based on primary sex characteristics.

Phil Walter/Getty Images

- If I stayed in the construction program, I would have to fight the belief that men are the only real construction workers.
- My career options would narrow.
- I would have to be conscious of the way I sit.

The women in the class believed that as men they would have to worry about asking women out and about whether their major was appropriate. They also believed, however, that they would make more money, be less emotional, and be taken more seriously. Some of their responses follow:

- I would worry about whether a woman would say yes if I asked her out.
- I would earn more money than my female counterpart in my chosen profession.
- People would take me more seriously and not attribute my emotions to PMS.
- My dad would expect me to be an athlete.
- I'd have to remain cool when under stress and not show my emotions.
- I think that I would change my major from "undecided" to a major in construction technology.

These comments show the extent to which life is organized and constrained in gendered ways. They reveal that decisions about how early to get up in the morning, which subjects to study, whether to show emotion, how to sit, and whether to encourage a child's athletic development are influenced by a society's gender ideals rather than by criteria such as self-fulfillment, interest, ability, or personal comfort. For example, one study found that almost 40 percent of women will wear uncomfortable shoes to be fashionable; 17 percent of men admit to buying the wrong size shoes, probably larger than needed (Daily Mail 2009).

coka/Shutterstock

Many women admit to wearing uncomfortable shows that are damaging to foot and body health just to be in fashion.

College students are influenced by gender ideals if they choose a major they believe is appropriate to their sex. Consider that 82 percent of bachelor's degrees in computer and information sciences are awarded to men, whereas 91 percent of bachelor's degrees in library sciences are awarded to females. Other majors dominated by women include education, health professions, family and consumer sciences/human sciences, and public administration/services (almost 80 percent of all bachelor's degrees awarded in these fields go to women) (*Chronicle of Higher Education* 2010).

To this point, we have shown that not everyone fits easily into two biological categories, that gender is a social construction, and that gender ideals are impossible to realize and maintain. So, it should be no real surprise that there are some societies such as American Samoa that accept a third gender. In Samoa, that third option is *fa'afafine*.

A Third Option

CORE CONCEPT 2 While all societies distinguish between male and female, some also recognize a third gender.

Jeannette Mageo (1992) begins her article "Male Transvestitism and Cultural Change in Samoa" by describing the guests attending a wedding shower in Samoa. Of approximately 40 "women," 6 were *fa'afafine*—people who are not biologically female but who have taken on the "way of women" in dress, mannerism, appearance, and role. Those who study *fa'afafine* maintain that to understand this third gender, we must set aside any cultural preconceptions we have about being male, female, gay, or transsexual (Fraser 2002).

During that wedding shower, the *fa'afafine* staged a beauty contest in which each sang and danced a love song. Such beauty contests are well known in Samoa, and the winner "is sometimes the 'girl' who gives the most stunningly accurate imitation of real girls, such that even Samoans would be at a loss to tell the difference; sometimes the winner is the most brilliant comic" (Mageo 1998, p. 213). Often the *fa'afafine* imitate popular foreign female vocalists, such as Whitney Houston, Britney Spears, Madonna, or Kelly Clarkson.

Mageo believes that *fa'afafines were* not common in pre-Christian Samoa (before 1830). If so, she believes that early Christian missionaries, preoccupied with documenting sexual habits of the Samoans would have mentioned it in their written accounts of that society. How, then, did assuming the role of *fa'afafine* become commonplace among males, especially in urban areas? Mageo (1992, 1998) argues that *fa'afafine* could not have become commonplace unless something about Samoan society supported gender

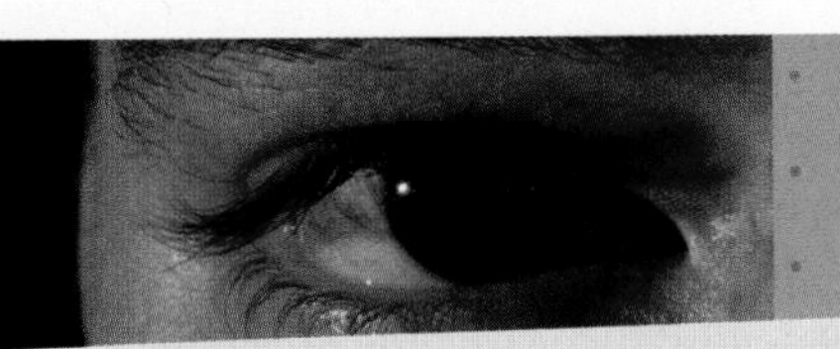

The Sociological Imagination Compliance and Resistance to Gender Ideals

If we simply think about the men and women we encounter every day, we quickly realize that most people fall short of society's gender ideals. This fact does not stop most people from using these ideals to evaluate their own and others' behavior and appearance. Some accounts of resistance, compliance, and the accompanying inner conflicts and social strains speak to the importance of gender ideals in shaping our lives.

Learning about and Accepting Gender Ideals. In high school, I began to realize how important my appearance and behavior were to being accepted. In the bathrooms, the juniors and seniors showed freshmen how to throw up after lunch in order to stay thin. Anorexia and bulimia were widespread, and we were very ruthless in our criticism of others. I began to practice walking with a sway in my hips and to flirt with my eyes lowered and head tilted. I started to wear makeup to accentuate my feminine features. I began to giggle, something I never really did growing up. Gender ideals have had their effects upon me. I work as an amateur model and actress, banking on the male–female distinction.

Attempting to Change the Behaviors that Deviate from Gender Ideals. There is something that drove me crazy. The feeling I experienced and the warmth I felt toward this woman were extraordinary, but so wrong. I didn't have anywhere to turn and no one to talk to. I did not want to tell anyone who or what I had done. Society said it was wrong to feel and act this way, but why did I feel so good and think about her so much? I nearly drove myself crazy and ran as far as I could. I joined the army to punish myself. Every day I hated what I had done and I wanted to hurt myself to make it go away. I hoped that by surrounding myself with men, I would forget what I had done and find some man that would help make me into a "normal" heterosexual woman.

Challenging Those Who Do Not Conform to Gender Ideals/ Refusing to Conform to Gender Ideals. I went on a blind date and met a girl who seemed too good to be true. Everything went well for the next three weeks. We hadn't been very intimate, so when she asked me to come over to her house and stay with her, I jumped at the opportunity. I entered her house to find candles lit and soft music playing. We began to kiss, and I noticed that she hadn't shaved her legs in a while. Disgusting, but under the circumstances I tried to block it out. As we got more involved I was surprised to find a lot of hair under her arms. I quickly pulled away. She asked what was wrong, and after I told her she laughed. Needless to say I felt like I was going to throw up. I put my clothes on and quickly left. On the way home I kept thinking how sick it made me that she didn't shave. I decided that I would never talk to her again. Sometime later she called me to explain. She told me that she didn't think women should have to shave and that she hadn't shaved for at least a year. She believed I should respect and even like her more for making a stand. Making a stand toward what, I thought? Anyway, I told her that I thought it was disgusting and I told her I didn't want to see her again.

blurring. She notes that "on a personal level Samoans do not distinguish sharply between men and women, boys and girls." For example, "boys and girls take equal pride in their skills in fights; pre-Christian personal names are often not marked for gender, and outside school little boys and girls still wear much the same clothing" (1998, p. 451).

Once Samoan boys reach the age of 5 or 6, they begin spending the majority of their time in the company of other boys; at this point, they are prohibited from flirting with girls. At the same time, "close and physically affectionate relations with same-sex people are established practices. In Samoa, as in much of the Pacific, boys may walk about hand-in-hand or with an arm draped around their comrade, and so may girls" (Mageo 1992, p. 452).

There are other historical factors that support the existence of *fa'afafine*. When the Christian missionaries observed girls engaged in sexualized forms of entertainment, they pressured the Samoans to change this practice and the girls to abandon this role. Mageo makes the case that this change opened the doors for *fa'afafine* to stand-in for the girls.

Another factor that may account for the widespread emergence of the *fa'afafine* relates to changes in the opportunities open to men in Samoa. Specifically, these changes are connected to the gradual decline in the importance of the *augama*, an organization of men without titles. At one time, the *augama* was considered the "strength of the village" (Mead 1928, p. 34), serving "as a village police force or an army reserve" (Mageo 1992, p. 444). It took responsibility for the heavy work, whether that be "on the plantation, or fishing, [or] cooking for the chiefs" (Mead 1928, p. 34).

The rise of compulsory education, the shift away from an agriculture-based to a wage-based economy, and the introduction of labor-saving technology contributed to the decline of the *augama*. That decline removed an important source of status for Samoan males. That loss of status has been compounded by an unemployment rate of 29.8 percent in American Samoa (U.S. Central Intelligence Agency 2011). Moreover, when we consider that the size of the Samoan labor force is just a little over 17,000 and that the largest employers are a tuna cannery (employing 2,282) and the Samoan government (employing 6,052), we can see the blows to status that many males might experience (U.S. Department of Interior 2010). This economic reality has left the average Samoan male without a clear sense of place. For some men, becoming a *fa'afafine* offered them an opportunity to step out of their lowered status and assume the status of well-known female impersonators. Other options for Samoan men include joining the military or becoming football players for the United States, Canadian, or European league.

In evaluating Mageo's research, keep in mind that she was writing about the social forces that push some fa'afafines to assume highly visible roles such as female impersonators. In reality, the fa'afafine population is very diverse. Keep in mind that Western labels—such as gay, transvestite, drag queen, or even transgendered—do not capture traditional Samoan conceptions of fa'afafine and their place in society. It is important to point out that fa'afafine do not define themselves as women in men's bodies. In fact, their mothers or other women close to the family decided they were fa'afafine based on the way they behaved or because there were not enough girls in the family to help with "women's work." In general, fa'afafine typically do not have children, and they often become teachers. They are thought to be devoted to family, caring for aging parents, keeping house, and babysitting (Tok Blong Pacifik 2008).

How do fa'afafines express themselves sexually? The answer is not a simple one. Many who consider themselves women "enter into clandestine, short-term relationships with men who see themselves as straight. Some fa'afafine, motivated by social pressure and the wish for children, leave their feminine identity behind and marry women, but many others don't. Occasionally they live openly with male partners" (Tok Blong Pacifik 2008).

Christina Douglas (9th Mission Support Command)

Economic opportunities are limited in American Samoa. After one of two tuna canneries closed in 2009, affecting one-third of all Samoan workers, many turned to the U.S. military. At this town hall meeting held in Pago Pago, the capital of American Samoa, attendees check with U.S. military personnel about plans to increase the number of Army Reserve positions in American Samoa.

sexual orientation "An enduring pattern of emotional, romantic, and/or sexual attractions to men, women, or both sexes. Sexual orientation also refers to a person's sense of identity based on those attractions, related behaviors, and membership in a community of others who share those attractions" (APA 2009).

Sexuality

We are bombarded daily with messages about sexuality. They may come from sex education classes that warn of the health dangers of unprotected sexual activities, from fairy tales, from song lyrics, or from commercials, movies, and news events. Other messages come from those close to us: friends who come out to us or parents who kidded us when we were toddlers about having a boyfriend or girlfriend. Finally, messages come from observing the treatment of people around us. For example, we notice uncomfortable reactions toward women who breast-feed their babies in public and toward men who appear feminine or women who appear masculine; we take notice of the boy and girl everyone wants to date or to not date; we take note of reactions and facial expressions when someone says they are from San Francisco.

CORE CONCEPT 3 Sexuality encompasses all the ways people experience and express themselves as sexual beings. The study of sexuality considers the range of sexual expressions and the social activities, behaviors, and thoughts that generate sexual sensations.

Sexuality is not an easy subject to present for several reasons. First, for most of us, sex/sexuality education focused on the dangerous consequences of sexuality (sexually transmitted diseases) uninformed by any discussions of what to make of sexual excitement, sexual attraction, or of the commercial uses of sexuality to sell products from cars to music. Second, people who have had difficult sexual experiences—those molested or raped as children, men who cannot achieve erections, and women and men who have been sexually assaulted—may be uncomfortable with the topic (Davis 2005). Third, it is very difficult to discuss human sexuality in all its dimensions when heterosexuality and all that it entails is presented as the norm and other expressions of sexuality are considered deviant and in need of fixing.

Sexual orientation is an expression of sexuality. Although sociologists are interested in the topic of sexual orientation, the American Sociological Association does not issue a statement about what sexual orientation means. According to the American Psychiatric Association (2009), sexual orientation refers to "an enduring pattern of emotional, romantic, and/or sexual attractions to men, women, or both sexes. Sexual orientation also refers to a person's sense of identity based on those attractions, related behaviors, and membership in a community of others who share those attractions." The word *enduring* suggests that one encounter does not make someone gay or lesbian. This caveat speaks to the fact that many people have experienced at least one same-sex sexual encounter at some point in their lives. Results from the most recent Centers for Disease Control (2002) survey found that 1 in 9 women and 1 in 18 men ages 15 to 44 have had a sexual experience with someone of the same sex.

Sexual orientation falls along a continuum, with its endpoints being exclusive attraction to the other sex and exclusive attraction to the same sex. In the United States, we tend to think of sexual orientation as falling into three distinct categories: heterosexual (attractions to those of the other sex), gay/lesbian (attractions to those of one's own sex), and bisexual (attractions to both men and women). It is important to realize that there are other labels that cultures apply to expressions of human sexualities (APA 2009).

Sexual orientation should not be confused with other related and intertwining terms, including the following:

gender identity—the awareness of being a man or woman, of being neither, or something in between (gender identity also involves the ways one chooses to hide or express that identity)

transgender—the label applied to those who feel that their inner sense of being a man or woman does not match their anatomical sex, so they behave and/or dress to actualize their gender identity (see Working for Change: "Transgender Day of Remembrance"). The *Diagnostic and Statistical Manual of Mental Disorders*, a reference book used by mental health practitioners, estimates that 1 in 30,000 people born male and 1 in 100,000 people born female have a "gender identity disorder" (Barton 2005). A revised edition of this manual is scheduled for release in 2012. Many practitioners are calling for the removal of gender identity disorder as a mental illness (Ault and Brzuzy 2009).

People enact sexual orientation in relationships with others. Thus, according to the APA (2009), "sexual orientation is closely tied to the intimate personal relationships that meet deeply felt needs for love, attachment, and intimacy." Based on what we know to date, the core attractions that emerge in middle childhood through early adolescence prior to sexual experiences are the foundation of adult sexual orientation. The experiences of coming to terms with sexual orientation vary. People can be aware of their sexual orientation even if they are celibate or have yet to engage in sexual activity. Others come to label their orientation after a sexual experience with a same-sex and/or other-sex partner. Still others ignore, suppress, or resist pulls toward those of the same sex because of widespread social disapproval (APA 2009).

Regardless of sexual orientation, every group has established **sexual scripts**, responses and behaviors that people learn, in much the same way that actors learn lines for a play, to guide them in sexual activities and encounters. The sexual scripts of the dominant culture call for behaviors and responses that support heterosexual ideals. Other sexual scripts, constructed by those in lesbian, gay, bisexual, and transgender and questioning (LGBTQ) communities, are dismissed as deviant. Even those who resist or reject the heterosexual scripts know that script and know they are ideally expected to follow them. As a result, those outside the heterosexual community often feel like they must account for deviating from the script. In this regard, most of my students have acknowledged knowing a same-sex person for whom they felt deep affection and attraction and who made them feel alive (descriptions we typically reserve for male–female attractions). Likewise, most if not all these students account for these feelings by emphasizing that they should in no way be interpreted as sexual.

Chris Caldeira

The Human Rights Campaign is a grass roots, not-for-profit, organization that claims one million supporters. HRC works to promote equal rights and a safe environment where LBGT can live an open, honest and safe life in their homes, workplaces and communities.

Socialization

CORE CONCEPT 4 Gender expectations are learned and culturally imposed through a variety of social mechanisms, including socialization, situational constraints, and commercialization of gender ideals.

Masculinity and femininity are not innate; children learn these characteristics. Once a child is labeled male or female, everyone who comes in contact with the child begins to treat him or her as such. With encouragement from others, children learn to talk, walk, and move in gendered

sexual scripts Responses and behaviors that people learn, in much the same way that actors learn lines for a play, to guide them in sexual activities and encounters.

Working for Change Transgender Day of Remembrance

Activist, writer and graphic designer Gwendolyn Ann Smith launched the Transgender Day of Remembrance (TDR) on November 20, 1999, as a tribute to Rita Hester who died in 1998. Hester was an African American transgendered female who was found stabbed to death at her apartment in Allston, Massachusetts. Although the crime against her was never solved, it was widely believed to be a hate crime. This small Internet project is snowballing into a worldwide movement to recognize the rights of the transgendered. Now on November 20 each year, Rita Hester's death, the deaths of other transgender people, and those who have suffered from gender-based intolerance and discrimination are remembered. According to the organization's website, the event's purpose is to raise public awareness about hate crimes against transgender people, stories that the mainstream media typically doesn't cover (St. Pierre 2007). "Day of Remembrance publicly mourns and honors the lives of our brothers and sisters who might otherwise be forgotten. Through the vigil, we express love and respect for our people in the face of national indifference and hatred. Day of Remembrance reminds non-transgender people that we are their sons, daughters, parents, friends, and lovers" (St. Pierre 2007).

Transgender people constitute a group that has endured discrimination and prejudice. In the United States, there are 315 documented cases of murders and suicides of transgender people that occurred between 1970 and 2010. For the most part, freedom of gender identity is not a recognized right, and politics and ideology play an important part in the way transgendered people are perceived and treated. But there have been some recent successes. On October 28, 2009, President Barack Obama signed the Matthew Shepard and James Byrd, Jr. Hate Crimes Prevention Act. In 1998, three men targeted Shepard, a University of Wyoming student, as gay. They tortured him, tied him to a fence post, and left him to die. In that same year, known white supremacists chained Byrd, an African American man, by his ankles to the back of a pickup truck and then dragged him for miles along a Jasper, Texas, road to his eventual death. Among other things, this federal law extends 1969 hate crime laws to cover crimes that target victims because of their actual or perceived gender, sexual orientation, gender identity, or disability (Robinson 2010).

In 2010, the annual Transgender Day of Remembrance marked its 12th anniversary. There were events in 200 cities in 20 different countries, including Malaysia, Minsk, Netherlands, New Zealand, Philippines, Poland, Switzerland, Germany, Greece, Israel, Italy, Ireland, Sweden, Australia, Canada, England, Finland, France, Scotland, and the United States (St. Pierre 2007). Each event is unique, but candlelit vigils are the most common. In Dublin, the St. Stephen's Green Unitarian Church honored the lives of 180 fellow Irish transgendered who died as a result of anti-transgender hate crimes with a candle-lighting ceremony and Irish song (Cedar Lounge Revolution 2010). In Santa Rosa, Australia, people gathered in the downtown courthouse square to light candles and remember the dead. Flyers listed the names of transgendered people who were murdered or committed suicide (Coffey 2010). In Vancouver, Canada, 120 people participated in a march from the Carnegie Community Center to SFU Harbour Center where they viewed the documentary *Isn't It Obvious*, listened to several readings, and sang songs to honor the dead (Starlight 2010). The Chicago Transgender Coalition sponsored two events. One, the Night of Fallen Stars, featured young transgendered speakers and a film about anti-transgender hate crimes and transgender people who have committed suicide. On the second night, 100 people gathered for a candlelit vigil. Although this was the first public celebration in Chicago, it is expected to occur annually (Sosin 2010). All events have the same mission, to honor and remember those transgendered who have suffered. If awareness and participation continue to grow, then the Transgender Day of Remembrance may become a celebration that reaches well beyond transgender and small sympathetic communities.

Source: Written by Brooke Goerman, Northern Kentucky University, Class of 2011.

ways (Lorber 2005). They also learn dominant gender roles, the behavior, and activities expected of someone who is male or female. These expectations channel children's energies in directions defined as sex-appropriate. As children learn to look and behave like boys or girls, most reproduce and perpetuate their society's version of how the two sexes should be. When children fail to behave in gender-appropriate ways, their character becomes suspect (Lorber 2005). At the minimum, people call girls who violate the rules tomboys and boys who do so sissies.

Gender Socialization

The gender socialization process may be direct or indirect. It is indirect when children learn gender expectations by observing others' words and behavior, such as the jokes, comments, and stories they hear about men and women or portrayals of men and women they see in magazines, books, and on television (Raag and Rackliff 1998). Socialization is direct when significant others intentionally convey the societal expectations to children.

Terra Shultz, Courtesy of Joan Ferrante

If this little boy (center) showed up for a birthday party at your home as a princess, what would you say to him? Do you think other parents might get upset? How might other parents treat his mother?

Agents of socialization are the significant people, groups, and institutions that act to shape our gender identity—whether we identify as male, female, or something in between. Agents of socialization include family, classmates, peers, teachers, religious leaders, popular culture, and mass media. Child development specialist Beverly Fagot and her colleagues (1985) observed how preschool teachers shape gender identity. Specifically, the researchers focused on how toddlers, ages 12 and 24 months, in a play group interacted and communicated with one another and how teachers responded to the children's attempts to communicate. Fagot found no differences in the interaction styles of 12-month-old boys and girls: All of the children communicated by gestures, gentle touches, whining, crying, and screaming.

The teachers, however, interacted with them in gender-specific ways. They were more likely to respond to girls who communicated in gentle, "feminine" ways and to boys who communicated in assertive, "masculine" ways. That is, the teachers tended to ignore girls' assertive acts but respond to boys' assertive acts. Thus, by the time these toddlers were 2, they communicated in very different ways.

Fagot's research was conducted more than 25 years ago. A more recent study found that early childhood teachers are more accepting of girls' cross-gender behaviors and explorations than they are of boys'. According to this research, teachers believe that boys who behave like "sissies" are at greater risk of growing up to be homosexual and psychologically ill adjusted than are girls who behave like "tomboys." This finding suggests that although American society has expanded the range of behaviors and appearances deemed acceptable for girls, it has not extended the range for boys in the same way (Cahill and Adams 1997).

Children's toys and celebrated images of males and females figure prominently in the socialization process, along with the ways in which adults treat children. Barbie® dolls, for example, have been marketed since 1959, with the purpose of inspiring little girls "to think about what they wanted to be when they grew up." The dolls are available in 67 countries. An estimated 95 percent of girls between ages 3 and 11 in the United States have Barbie dolls, which come in several different skin shades and 45 nationalities (Mattel 2011). Then there are the estimated 26,000 Disney princess items on the market and merchandise for girls that "seems to come in only one color: pink jewelry boxes, pink vanity mirrors, pink telephones, pink hair dryers, pink fur stoles" (Paul 2011).

For boys, G.I. Joe was launched in 1964, as the first action figure toy on the market, and it was followed by a long line of action figures, including Transformers™, Micronauts™, Star Wars™, Power Rangers™, X-Men™, Street Fighter™, Bronze Bombers, and Mortal Kombat™. The popularity of these toys has generated comic books, motion pictures, and cartoons, and they appear on school supplies, video games, card games, lunch boxes, posters, and party supplies (Hasbro Toys 2011, Son 1998).

Learning to be male or female involves learning norms governing the way males and females present themselves; that includes learning the sex-appropriate norms governing body language. According to women's studies professor Janet Lee Mills (1985), norms governing male body language suggest power, dominance, and high status, whereas norms governing female body language suggest submissiveness, subordination, vulnerability, and low status. Mills argues that these norms are learned and that people give them little thought until someone breaks them, at which point everyone focuses on the rule breaker.

Mills describes how men hold power by taking an authoritative stance—often with a wide stance, one or both

Terra Shultz, Courtesy of Joan Ferrante

Parents and others buy children toys that channel their interests and behavior in masculine and feminine directions. Would you encourage this little boy to continue showing an equal amount of interest in his toy truck and his princess dress?

hands in pocket, straight posture, and head high. Women often suggest a submissive stance when they smile, cant their head, hold arms and hands close to their bodies, and assume an unstable stance. When talking to men who are shorter, women tend to slip into a scrunched up posture.

We might conclude that if we change socialization experiences, behavior will change accordingly. The Christian missionaries assigned to Samoa must have recognized this principle, as they sought to "destroy most of the social institutions that guided young Samoans through childhood to adulthood" (Cote 1997, p. 7). Among other things, these missionaries attempted to end the practice of tattooing, and they targeted the *aualuma*, a group of unmarried adolescent girls who "lived together," "supported one another emotionally," and carried out village work projects. As part of the missionaries' efforts, unmarried girls, instead of going to live with the *aualuma*, were brought to live with the pastors and their wives (Cote 1997, p. 8). By introducing mass education, the missionaries also changed the role of the *augama* (the organization of younger and older men without titles). Instead of learning skills as members of the *augama*, boys studied inside the classroom, which prepared them to work for wages instead of for the village as a whole.

Socialization of Samoan Boys

There are about 4,600 high school students (grades 9 to 12) in American Samoa. Of this number, about 400 males graduate from high school each year. Each year, about 46 graduates leave the Pacific island to play football in the United States at the collegiate level (Syken 2003). That figure translates into about one in every nine high school graduates. What socialization mechanisms channel young Samoan males' energies into a sport that may take them 4,150 miles or more from home?

First, for young Samoan boys to play football, the sport must be available in their society. Football was introduced to the island in 1969, after a U.S. government official decided that the public schools should field football teams. Second, the celebrity status of successful Samoan football players highlights for young Samoan males the rewards of pursuing a football career. For example, 120 football players of Samoan descent played at the University of Hawaii and Arizona State between 1997 and 2000, under Coach Dick Tomey. Today, 200 players of Samoan descent play on Division I college teams and about 20 play in the NFL. Others play in the Canadian Football League and NFL Europe (U.S. House of Representatives 2004, Ferguson 2005). Samoan youth hear and read praise from college and NFL coaches about Samoan players, which fuels their interest in the game and channels their choices toward football and not some other sport, such as golf or tennis:

- "They're so physical. Even in scrimmages they go all out." (Busch 2003)
- "There are no athletes that are, in my estimation, more competitive, more athletic, or more family-oriented, or who fit into a team concept as well as Samoan athletes. The more we could get on our team, the better I felt." (Tomey 2003)
- "Why not use what God gave you? You wouldn't mind putting a golf club in their hands, but you have to be realistic. I don't see a [Samoan] Tiger Woods out there. We're going to use what we know." (Malauulu 2003)

Third, high school football is very popular in American Samoa. The six high school football teams use the same 5,000-seat stadium to host all of their games. The schools play each other twice each season and meet again for playoffs. In addition to the fans who attend games, other fans listen to the games on the island's only radio station or watch them on TV (Saslow 2007, Ferguson 2005).

Finally, American Samoan males have relatively few career opportunities. Consider that one in every 50 people moves away each year. Surely this high rate is tied in part to the limited opportunities. As a case in point, consider that one-third of all Samoan jobs are connected to tuna fishing and processing. The importance of this industry to the economy is such that tuna represents 93 percent of American Samoa's total exports (U.S. Central Intelligence Agency 2011). In an interview three high school athletes reflected on football and their desire to leave the island that they call the "rock," a 54-square-mile volcanic island about 4,500 miles from mainland United States and approximately 2,300 miles from Hawaii (Saslow 2007):

- "The pride of the whole high school is football. That's the most important sport now . . . what ever part of the island has the best team, that part of the island celebrates all year."

BRIAN SNYDER/Reuters/Landov

Troy Polamalu is an NFL football player of Samoan ethnicity. He believes that football allows him to carry out his Samoan heritage in the way he runs, hits, and otherwise plays the game (CBS News 2010).

- "If you want to go the mainland, then you have to play football or go into the military. After high school everybody wants to leave, at least for awhile."
- "Everyone down here is just looking to get off 'the Rock.' That's what we call the island. It's just a rock in the middle of the ocean and we don't ever see anything else."

Commercialization of Gender Ideals

Commercialization of gender ideals is the process of introducing products into the market using advertising campaigns that promise consumers they can achieve masculine and feminine ideals if they buy and use the products. Keep in mind that sales depend on people buying a product. One way to convince people to buy products is to play on their insecurities about whether they measure up to gender ideals. Of course, achieving gender ideals requires a great deal of effort, and retailers offer products that can help in that effort. If we consider the strategies corporations employ to increase profits, we can see how commercialization of gender ideals supports profit-making goals. Those profit-making strategies include creating new products that consumers "need" to buy and creating new markets.

Creating New Products The list of products is endless, especially for women. From hair dye to toenail polish, products are available to improve almost every female body part or body function (from the top of the head to the tip of the toe). There are vaginal moisturizers, chin gyms (a mouthpiece that includes a miniature weight-lifting system to help those who use it avoid or lose a double chin), Botox, wrinkle creams, hair removers, and artificial fingernails. Relatively new products on the market for men include body washes, the erectile dysfunction drugs Viagra, Levitra, and Cialis, which are being advertised even to men who do not have the medical condition for which these drugs should be prescribed. Millions of men have tried these drugs to date, and the manufacturers hope to attract millions more "by suggesting that if men cannot have an erection 'on demand,' if they 'fail' even once, they are candidates for these drugs" (Tuller 2004).

Creating New Markets The female market is saturated with products. From a marketing perspective, the amount of discretionary money that female consumers have to spend on cosmetic products may have reached its limit. Thus, marketers must search for a new market—and one new market appears to be males. The problem for marketers is how to sell men products that have traditionally been viewed as "feminine." One strategy is to "masculinize" feminine products by selling them a "body wash that's not for sissies" or a revitalizing face cream that is "more evolved," playing on a hierarchy that puts men at the top of the evolutionary chain.

Structural Constraints

Structural constraints are the established and customary rules, policies, and day-to-day practices that affect a person's life chances. One example is the structural constraints that push men and women into jobs that correspond with society's ideals regarding sex-appropriate work. Women are pushed into work roles that emphasize personal relationships and nurturing skills or that pertain to family-oriented and "feminine" products and services. Men are pushed into jobs that emphasize decision making and control and that pertain to machines and "masculine" products and services.

Table 10.1 shows the top ten female-dominated occupations held by women who work full-time. Because women are 47.4 percent of the full-time labor force, we can see that they are very overrepresented in the secretaries/administrative assistant category. Notice that more than 3.0 million women in the United States work as secretaries or administrative assistants, filling 96.8 percent of such positions. Other occupational categories in which women make up 90 percent or more of the workforce include receptionists, bookkeepers, and child care workers. In Table 10.2, notice that almost 3 million men in the United States work as truck drivers. Men are 52.6 percent of the full-time labor force yet they are 94.2 percent of truck drivers. Other occupational categories dominated by men include construction laborers, carpenters, grounds maintenance workers, and electricians.

Sociologists consider the ways occupations deemed appropriate for one sex or the other channel behavior in stereotypically male and female directions. The point is that it is not the day care worker per se that is feminine; it is the skills needed for the job that makes the day care worker behave in ways we associate with femininity. Presumably anyone holding the job of day care worker will display "feminine" characteristics. Likewise, the skills we require of truck drivers means anyone holding that job behaves in ways we have come to associate with masculinity. This suggests that a person does not have to be an anatomical woman to be a day care worker or an anatomical man to be a truck driver, but they do have to possess the necessary skills.

commercialization of gender ideals The process of introducing products to the market by using advertising and sales campaigns that promise consumers they will achieve masculine and feminine ideals if they buy and use the products.

structural constraints The established and customary rules, policies, and day-to-day practices that affect a person's life chances.

Global Comparisons Countries that Allow Gays and Lesbians to Serve in the Military Openly

Since 1993, the Department of Defense discharged an estimated 13,000 servicemen and servicewomen for homosexuality under the "don't ask, don't tell" policy (Bumiller 2010, Stolberg 2010). Gay men and lesbians in 161 different occupations, including foreign-language specialists and infantrymen, have been discharged on the grounds that homosexuals create an unacceptable risk to military effectiveness (Online Newshour 2007). Yet little if any scientific evidence supports this policy. In fact, until 2010, the year Defense Secretary Robert Gates declared his intention to roll back this policy, whenever Pentagon researchers (with no links to the gay and lesbian communities and with no ax to grind) found evidence that runs contrary to this policy, high-ranking military officials have generally refused to release it or directed researchers to rewrite their reports. A 2007 Zogby poll found that 75 percent of U.S. soldiers returning from Iraq and Afghanistan indicated that they were comfortable interacting with gay colleagues; only 5 percent indicated that they were extremely uncomfortable (Shalikashvili 2007).

People who oppose the presence of gay men and lesbians in the military stereotype them as sexual predators just waiting to pounce on heterosexuals while they shower, undress, or sleep. Opponents seem to believe that *any* same-sex person is attractive to a gay man or lesbian. But as one gay ex-midshipman noted, "Heterosexual men have an annoying habit of overestimating their own attractiveness" (Schmalz 1993, p. B1). Supporters of gay men and lesbians in the military point out that almost 75 percent of troops say that they usually shower alone; only 8 percent say they usually take showers where others can see them (Zogby 2007). To date, 24 countries allow gays to serve openly. Those countries shown on the map are Australia, Austria, Belgium, Canada, Czech Republic, Denmark, Estonia, Finland, France, Germany, Ireland, Israel, Italy, Lithuania, Luxembourg, Netherlands, New Zealand, Norway, Slovenia, South Africa, Spain, Sweden, Switzerland, and the United Kingdom (Sosa 2010).

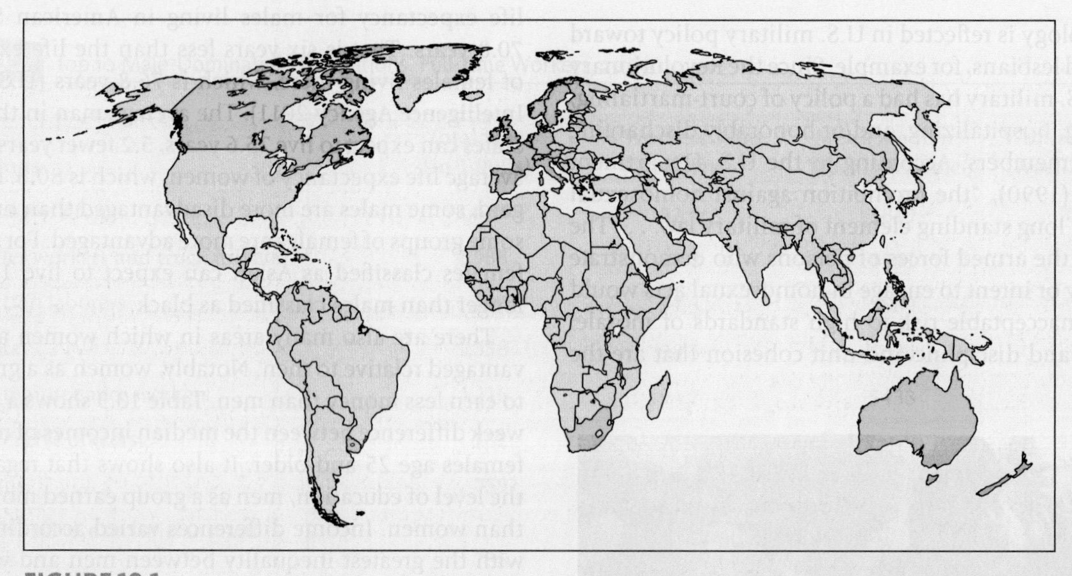

FIGURE 10.1
Source: Data from Palm Center (2010)

TABLE 10.3 Median Weekly Income of Full-Time Wage and Salary Workers Age 25 and Older by Education Level, 2010

	Overall	Less Than High School	High School Graduate	4-Year Degree	Graduate/ Professional Degree
Female	$709	$392	$539	$910	$1,126
Male	$878	$479	$718	$1,196	$1,559
Difference in Weekly Wages	$169	$87	$179	$286	$433

Source: U.S. Bureau of Labor Statistics (2010)

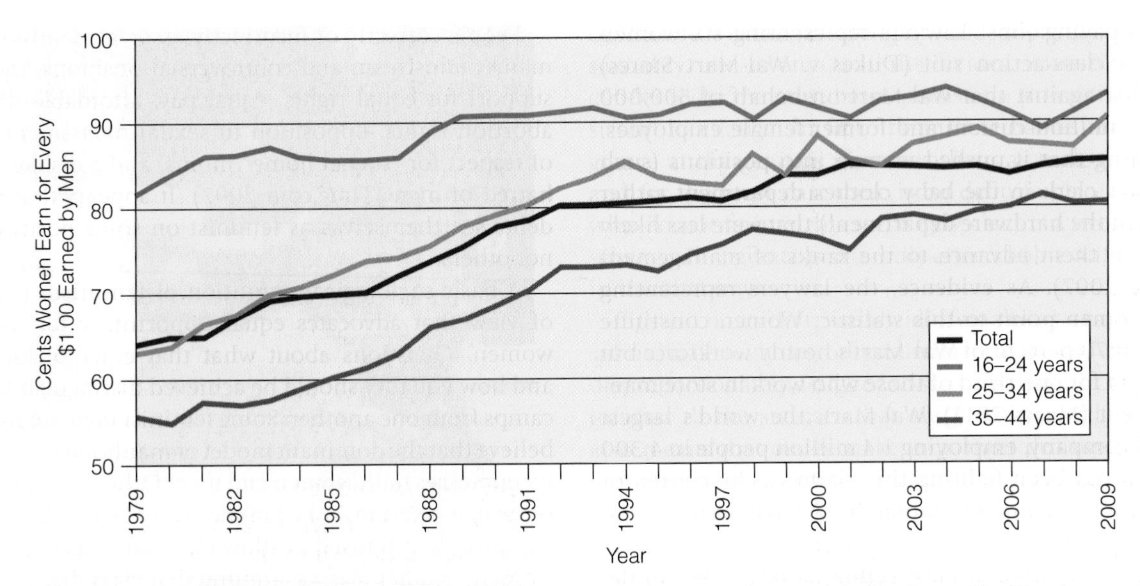

FIGURE 10.2 Cents Full-Time Female Wage and Salary Workers Earn for Every Dollar Earned by Men in 2009 dollars, 1979–2009
This graph shows year-by-year variations in money women earned relative to men. Which age group as a whole has the highest earning relative to men? Which age group has the lowest earning relative to men?
Source: U.S. Bureau of Labor Statistics (2010)

38 percent of what the average man did (Madrick 2004). Economists Stephen J. Rose and Heidi Hartman (2004) compared men's and women's total earnings between 1983 and 1998 and found that the average woman earned $273,592 whereas the average man earned $722,693.

Male–female income differences vary by occupation. Tables 10.1 and 10.2 show the top ten male- and female-dominated occupations and the percentage of males and females in each. Notice that for the ten occupations in which women dominated, their median weekly earnings ranged from a low of $364 (for child care workers) to a high of $1,035 (for registered nurses). Likewise, in the top ten male-dominated occupations, men's median weekly earnings ranged from a low of $433 (for grounds maintenance worker) to a high of $2,217 (for chief executive officers). Note that for all occupations considered in Tables 10.1 and 10.2, the median weekly income was always greater for males than for females, even when males worked in occupations that are female dominated.

There are many possible explanations for this income difference. They include the following:

- Women are disproportionately employed in lower-paying, lower-status occupations.
- Women choose or are forced into lower-paying positions that are considered sex-appropriate, such as teacher, secretary, and caregiver. Female-dominated occupations (day care worker) are valued less and thus pay than male-dominated occupations (such as auto mechanics and construction laborers).
- Women choose or are forced into positions that offer fewer, more-flexible hours to meet caregiving responsibilities.
- Women choose or are forced into lower-paying subspecialties within high-paying professions (for example, women tend to be divorce lawyers rather than corporate lawyers, and pediatricians rather than heart surgeons).
- Employers underinvest in the careers of childbearing-age women because they assume the women will eventually leave to raise children.
- Women leave the labor market to take care of children and elderly parents, and then they reenter it.
- Women choose or are forced into occupations that will not require them to relocate, work in unpleasant environments (such as mines), or take on hazardous assignments—three activities associated with higher pay (Sahadi 2006).
- Employers still view women's salary needs as less important than men's and pay women accordingly. Unfortunately, women's earnings are often considered supplemental to men's—earnings that can be used to buy "extras"—when in reality many women are heads of households.
- When negotiating for salaries, women underestimate their worth to employers and ask for less than their male counterparts.
- Some employers steer males and females into sex-appropriate assignments and offer them different training opportunities and chances to move into

privileges them. African Americans who possess eloquent analyses of racism often persist in viewing poor white women as symbols of white power. . . . In essence, each group identifies the type of oppression with which it feels most comfortable as being fundamental and classifies all other types as being of lesser importance" (p. 459).

Collins believes that each of us must come to terms with the multiple ways the categories to which we belong shape our lives and relationships through institutionalized practices and symbolic representations. She maintains that each of us carries around the cumulative effects of these multiple structures of oppression and that we enforce that system in the personal choices we make about who to include or exclude in our lives and in the errors in judgment that we make about others who are not like us. One way to assess the extent to which you have been affected by these systems is to ask yourself: "Who are your close friends? Who are the people with whom you share your hopes, dreams, vulnerabilities, fears, and victories? How much are they like you?"

Online Poll

Have you ever decided not to pursue some activity or experience because you believed it was appropriate for someone of the other sex?

- ○ Yes
- ○ No

To see how other students responded to these questions, go to **www.cengagebrain.com**.

Molly Hayden, USAG Grafenwoehr

Summary of CORE CONCEPTS

When sociologists study gender, they focus on differences between males and females that are socially constructed and they focus on those behaviors and appearances that depart from gender ideals. To illustrate some of these important concepts, this chapter considers the case of fa'afafines and male football players in American Samoa.

CORE CONCEPT 1 Sex is a biological concept, whereas gender is a social construct.

Most cultures place people in two categories—male and female—based largely on what are considered to be clear anatomical distinctions. Biological sex is not a clear-cut category, however, if only because some babies are born intersexed. In addition, a person's primary sex characteristics may not match his or her sex chromosomes. The existence of transsexuals also challenges the practice of separating everyone into the two biological categories male and female. Gender is the socially created and learned distinctions that specify the ideal physical, behavioral, and mental and emotional traits characteristic of males and females and that make them masculine and feminine. Gender ideals are socially constructed standards that shape practically every aspect of life—influencing, among other things, how people dress, how they express emotions, and what occupations they choose.

CORE CONCEPT 2 While all societies distinguish between male and female, some also recognize a third gender.

The existence of *fa'afafine*—people who are not biologically female but who take on the "way of women" in dress, mannerism, appearance, and role in American Samoa and other Pacific island areas—challenges the two-gender classification scheme. Jeannette Mageo argues that *fa'afafine* could not have become commonplace unless something about Samoan society supports gender blurring.

CORE CONCEPT 3 Sexuality encompasses all the ways people experience and express themselves as sexual beings. The study of sexuality considers the range of sexual expressions and the social activities, behaviors, and thoughts that generate sexual sensations.

Sexual orientation is an expression of sexuality. It also refers to a person's enduring sense of identity, which can be based on to whom one is attracted, their behaviors, and membership in a group that shares similar attractions. Sexual orientation should not be confused with other related and intertwining terms that shape the experiences of sexuality and sexual orientation, including biological sex, gender role, gender identity, and transgender.

CORE CONCEPT 4 Gender expectations are learned and culturally imposed through a variety of social mechanisms, including socialization, situational constraints, and commercialization of gender ideals.

Socialization theorists argue that an undetermined yet significant portion of male–female differences is the product of the ways in which males and females are socialized. Another powerful mechanism for conveying gender expectations is the commercialization of gender ideals—the process of introducing products into the market by using advertising and sales campaigns that promise consumers they will achieve masculine and feminine ideals if they buy and use the products. Finally, structural constraints—the established and customary rules, policies, and day-to-day practices that affect a person's life chances—channel people's behavior in desired directions. Structural constraints push men and women into jobs that correspond with society's ideals for sex-appropriate work.

CORE CONCEPT 5 Sexism is the belief that one sex—and by extension, one gender—is superior to another, and that this superiority justifies inequalities between sexes.

Sexism revolves around four notions: (1) People can be placed into two categories: male and female; (2) a close correspondence exists between a person's primary sex characteristics and characteristics such as emotional activity, body language, personality, intelligence, the expression of sexual desire, and athletic capability; (3) primary sex characteristics are so significant that they explain and determine behavior and the social, economic, and political inequalities that exist between the sexes; and (4) people who behave in ways that depart from ideals of masculinity or femininity are considered deviant, in need of fixing, and subject to negative sanctions ranging from ridicule to physical violence.

CORE CONCEPT 6 When sociologists study inequality between males and females, they seek to identify the social factors that put one sex at a disadvantage relative to the other.

Social inequality exists between men and women when one category relative to the other (1) faces greater risks to physical and emotional well-being; (2) possesses a disproportionate share of income and other valued resources; and/or (3) is accorded more opportunities to succeed. In its most basic sense, the feminist perspective advocates equality between men and women. Sociologists take a feminist perspective when they emphasize in their teaching and research the following kinds of themes: the right to bodily integrity and autonomy, access to safe contraceptives, the right to choose the terms of pregnancy, access to quality prenatal care, protection from violence inside and outside the home, freedom from sexual harassment, equal pay for equal work, workplace rights to maternity and other caregiving leaves, and the inescapable interconnections between sex, gender, social class, race, culture, and religion.

CORE CONCEPT 7 Worldwide, females as a category are subordinate to males.

Of all the divides, gender seems to be the most basic, persistent, prevalent, and resistant to change. Worldwide, we can easily document that females as a category are subordinate to males. There are few, if any, countries in the world in which women hold political control or a majority share of power that allows them as a group to determine national policies. There also appear to be no societies in which women dominate the highest-paying, highest-status jobs, nor is there a society in which full-time working women earn the equivalent of male salaries. But why is gender inequality everywhere? One answer can be found in cognitive mechanisms that support, sustain, and justify the gender inequalities. Those narratives maintain that men and women are different by divine design and/or nature, and that inequalities between them are natural.

CORE CONCEPT 8 A gender category does not stand alone; it intersects with other socially significant and constructed categories, including social class, age, religious affiliation, sexual orientation, ethnicity, disability status, and nation.

The concept of intersectionality helps us to see that no social category is homogeneous; the categories to which someone belongs place them in a complex system of domination and subordination, and the effects of the categories a person occupies cannot simply be added together to obtain some grand total. Each of us derives varying amounts of penalty and privilege from the multiple systems of oppression that frame our lives.

Resources on the Internet

Login to CengageBrain.com to access the resources your instructor requires. For this book, you can access:

Sociology CourseMate

Access an integrated eBook, chapter-specific interactive learning tools, including flash cards, quizzes, videos, and more in your Sociology CourseMate.

Take a pretest for this chapter and receive a personalized study plan based on your results that will identify the topics you need to review and direct you to online resources to help you master those topics. Then take a posttest to help you determine the concepts you have mastered and what you will need to work on.

CourseReader

CourseReader for Sociology is an online reader providing access to readings, and audio and video selections to accompany your course materials.

Visit **www.cengagebrain.com** to access your account and purchase materials.

Key Terms

commercialization of gender ideals 257
feminism 262
femininity 247
gender 247
gender polarization 249
intersexed 246
intersectionality 266
masculinity 247
penalties 266
privilege 266
secondary sex characteristics 246
sex 245
sexual orientation 252
sexual scripts 253
sexism 259
structural constraints 257

Keith Farley

CORE CONCEPT 1 The ongoing agricultural, industrial, and information revolutions have profoundly shaped the world's economic systems.

A society's economic system is the social institution that coordinates human activity to produce, distribute, and consume goods and services. Three major, ongoing revolutions have shaped the world's economic systems: agricultural, industrial, and information revolutions. We considered how these revolutions shaped six types of societies: hunting and gathering, pastoral, horticultural, agrarian, industrial, and postindustrial.

CORE CONCEPT 2 The world's economic systems fall along a continuum whose endpoints are capitalism and socialism in most pure forms.

We can classify the economies of the world as falling somewhere along a continuum that has ideal forms of capitalism and socialism as its extremes. Ideally, capitalism is an economic system in which the means of production are privately owned. It is profit-driven, free of government interference, and governed by the laws of supply and demand. Socialism is an economic system in which the raw materials and the means of producing and distributing goods and services are collectively owned. Socialists believe that government or some worker- or community-oriented organization should play the central role in regulating economic activity on behalf of the people as a whole. Societies do not fall neatly into socialist or capitalist. The concept *welfare state* is a term that applies to economic systems that are a hybrid of the two.

CORE CONCEPT 3 Capitalists' responses to economic downturns and stagnation have driven a 500-year-plus economic expansion, which has facilitated interconnections between local, regional, and national economies.

World system theory focuses on the forces underlying the development of economic transactions that transcend national boundaries. In particular, world system theorists identify profit-generating strategies that have caused capitalism to dominate and facilitate a global network of economic relationships. As a result, capitalism has spread steadily throughout the globe. In addition, every country of the world has come to play one of three different and unequal roles in the global economy: core, peripheral, or semiperipheral.

CORE CONCEPT 4 The United States qualifies as a core economy; India represents an example of a semiperipheral economy.

The United States and India have the largest and fifth largest economies in the world, respectively. When measured by per capita GDP, the United States has the tenth largest economy, and India has the 163rd largest. Both countries depend on oil from foreign sources. The Indian economy is dominated by agriculture, whereas the U.S. economy is dominated by the information and service sector. The U.S. workforce is 153 million compared to India's workforce of 470 million—436 of which work in the informal sector. India is a destination for many IT and customer service jobs once done by Americans.

CORE CONCEPT 5 When people believe that power differences are legitimate, those with power possess authority.

Authority is legitimate power—power that people believe is just and proper. A leader has authority to the extent that people view him or her as being entitled to give orders. Authority can be of three types: (1) traditional, (2) charismatic, and (3) legal-rational, which derives from a system of impersonal rules that formally specify the qualifications for occupying a powerful position.

CORE CONCEPT 6 Government is an organizational structure that directs and coordinates human activities in the name of a country or some other territory, such as a city, county, or state. That structure may be democratic, authoritarian, totalitarian, or theocratic.

No matter the form of government or the scope of its jurisdiction, all make laws, formal rules that move people to behave in specified ways or to refrain from behaving in some specified way. Laws are created by those in power and enforced by regulatory institutions such as police, military, or other bodies. Laws govern almost every area of life. When sociologists study laws, they consider questions of fairness and justice in their creation, interpretation, enforcement, and consequences.

CORE CONCEPT 7 There are two major models of power: the power elite and pluralistic models. The two models help us to evaluate whether an elite few hold the power in society or whether power is dispersed among competing interest groups.

The power elite consists of people whose positions in the military, the government, and the largest corporations are so high that their decisions affect the lives of millions, even billions, of people. For the most part, the source of this power is legal-rational. In writing about the power elite in the United States, sociologist C. Wright Mills focused on those who occupy the highest positions in the leading institutions: the military, the 200 largest corporations, and the government. Pluralist models view politics as a plurality of special interest groups competing, compromising, forming alliances, and negotiating with each other for power. A pluralist model views power as something dispersed among special interest groups.

CORE CONCEPT 8 Empire, imperialism, hegemony, and militarism are concepts that apply to political entities such as governments that can exercise their will over other political entities.

A government is an empire when it controls the political, economic, and cultural development of a group of countries. An imperialistic power controls foreign entities through military force or through political policies and economic pressure. Imperialists justify their control by claiming that they have cultural, political, or economic superiority and that they are using it for the greater good of all humankind. Hegemony is a process by which a power formalizes its dominance over foreign entities—using established institutions, bureaucratic structures, the mass media, and military or police force. A militaristic power believes that military strength, and the willingness to use it, is the source of national and even global security.

Resources on the Internet

Login to CengageBrain.com to access the resources your instructor requires. For this book, you can access:

Access an integrated eBook, chapter-specific interactive learning tools, including flash cards, quizzes, videos, and more in your Sociology CourseMate.

Take a pretest for this chapter and receive a personalized study plan based on your results that will identify the topics you need to review and direct you to online resources to help you master those topics. Then take a posttest to help you determine the concepts you have mastered and what you will need to work on.

CourseReader

CourseReader for Sociology is an online reader providing access to readings, and audio and video selections to accompany your course materials.

Visit **www.cengagebrain.com** to access your account and purchase materials.

Key Terms

527 group 296
authoritarian government 293
authority 289
capitalism 279
charismatic authority 290
colonization 276
democracy 291
domestication 274
economic system 274
empire 296
goods 274
government 291
hegemony 297
imperialistic power 297
insurgents 298
laws 294
legal-rational authority 290
militaristic power 297
pluralist model 295
pluralist models of power 295
political system 289
postindustrial society 278
power 289
power elite 294
primary sector (of the economy) 287
secondary sector (of the economy) 287
services 274
socialism 280
surplus wealth 274
tertiary sector (of the economy) 287
theocracy 294
totalitarianism 293
traditional authority 289
welfare state 281

FAMILY

12 With Emphasis on JAPAN

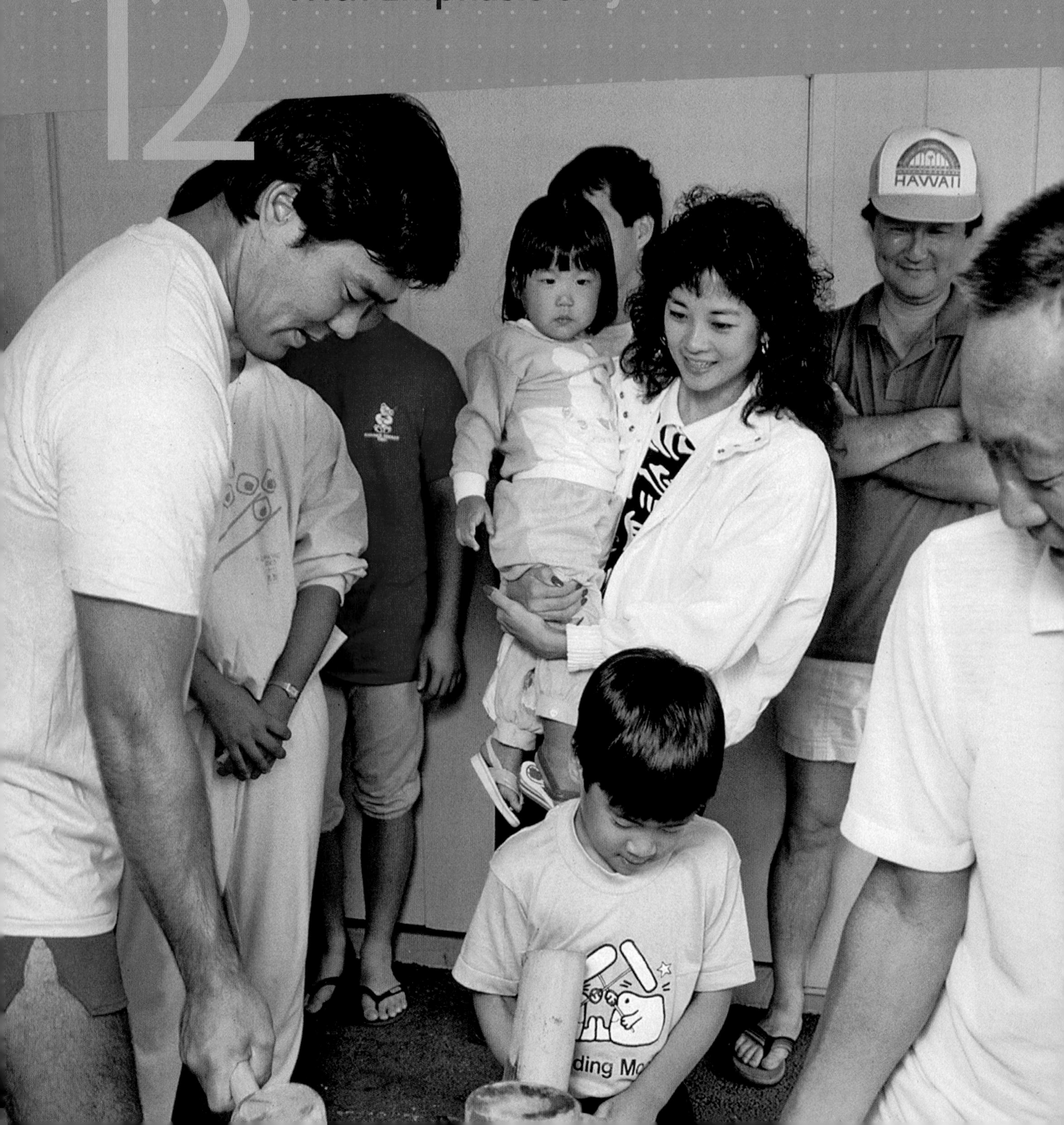

When sociologists study the family, they focus on the many factors that affect its structure and composition. In Japan, an aging population and low fertility rate are dramatically shaping the structure of family life. That is, people age 65 and older are increasing relative to other age groups. In addition, the number of children born each year is not enough to replace the population as its members die. For example, in 2010, there were 925,000 births and 1.3 million deaths in Japan (U.S. Central Intelligence Agency 2011).

Why Focus On JAPAN?

The family is an ever-changing entity. Millions of seemingly personal decisions influence the variety of family arrangements that exist in any country—decisions about whether to (1) have children, and if so, how many to have, when to have them, and how to space them; (2) marry, and if so, when; (3) work for pay; and (4) become a caregiver to dependent relatives. As we will see, these decisions are shaped and constrained by several larger forces, including average life expectancy, employment opportunities, and social norms.

In this chapter, we compare family life in Japan and in the United States. Based on indicators that seem to be associated with family well-being and stability, Japan appears to do better than the United States. The country has lower infant mortality than the United States; it has a lower teen birth rate (4.6 live births per 1,000 females age 15 to 19 versus 41.9 births per 1,000 females age 15 to 19), and a much smaller percentage of single-parent households. In addition, Japan has much fewer reported cases of domestic and child abuse than the United States (see Figure 12.1).

On the other hand, people in the United States seem more "optimistic" about having children than people in Japan; the United States has a higher **total fertility rate**; the average number of children that a woman bears in her lifetime is 2.1 versus Japan's rate, which is 1.2. The United States has a lower abortion rate, a higher teen birth rate, and a greater percentage of births to unmarried mothers (see Figure 12.1).

Both the United States and Japan have an **aging population**, one where the percentage of those age 65 and older is increasing relative to younger age groups. Currently, 13.0 percent of the U.S. population and 22.8 percent of the Japanese population are 65 years old or older. In 2020, 30 percent of Japan's population will be 65 and older, compared to 17.6 percent of the U.S. population (U.S. Bureau of the Census 2011).

Japan's low total fertility rate, combined with its long life expectancy and low immigration rate, means that the country has one of the oldest populations in the world. The low fertility rate is a major national concern, and it has prompted a variety of responses, including condemning young people for being selfish, delivering urgent appeals to couples to have babies, and initiating policies that make it easier on women to pursue a career and raise children (Yoshida 2008).

In the United States, the decline in the number of households occupied by married parents and their children has grabbed headlines along with a high teen birth rate and high percentage of single-parent households. In this chapter, we seek to understand the major social forces shaping family life in such different ways in the United States and Japan.

Online Poll

How many generations are a part of your family?

- Two generations
- Three generations
- Four generations
- Five generations
- More than five generations

To see how other students responded to these questions, go to **www.cengagebrain.com.**

Defining Family

CORE CONCEPT 1 **An amazing variety of family arrangements exists in the world. This variety makes it difficult to define family.**

total fertility rate The average number of children that a woman bears in her lifetime.

aging population A population in which the percentage that is age 65 and older is increasing relative to other age groups.

	Japan	United States
Reported cases of domestic abuse (per 100,000 population age 15 and older)	19	524
Reported cases of child abuse (per 100,000 population age 0–19)	203	7,142
Abortion rate (per every 1,000 live births)	370	279
Teen birth rate (per 1,000 females age 15-19)	4.6	41.9
Total fertility rate (per woman)	1.2	2.1
Percentage of births to unmarried women	2.1	40.6

FIGURE 12.1 Selected Indicators of Family Structure, Composition, and Well-Being: Japan and United States

Sources: Data from NationMaster.com 2011; U.S. Bureau of Census (2009, 2011a, 2011b); Lah, 2011; Population Reference Bureau 2009; U.S. Department of Health and Human Services 2009

Family is a social institution that binds people together through blood, marriage, law, and/or social norms. Family members are generally expected to care for and support each other. This definition is very general because even though everyone is born into a family, there is no structure common to the amazing variety of family arrangements that exist worldwide. That variety is captured in the numerous norms that specify how two or more people can constitute a family. Among other things, these norms govern who can marry, the number of partners people can have, and the ways people trace their ancestry (see Table 12.1). In light of this variability, we should not be surprised that when people think of family, they often emphasize different dimensions, such as kinship, ideal members, or legal ties.

Kinship

Definitions of *family* that emphasize kinship view the family as comprising members who are linked together by blood, marriage, or adoption. Based on this definition, the size of any given person's family network is beyond comprehension, because one person has an astronomical number of living and deceased kin. Keep in mind that to calculate the number of a person's relatives, one would have to count primary kin (mother, father, sister, brother), secondary kin (mother's mother, mother's father, sister's son, brother's daughter), tertiary kin (brother's daughter's son, mother's sister's son), and beyond (brother's daughter's son's son).

Given that it is virtually impossible to keep track of even one's living relatives, let alone maintain social relationships with them, every society finds ways to exclude some kin from their idea of family. For example, some societies trace family lineage through the maternal or the paternal side only. Selective forgetting and remembering is

Missy Gish

Most of the people in this photo are connected to one another by blood or marriage. Yet the family shown represents only a fraction of all those who share the status of relative.

family A social institution that binds people together through blood, marriage, law, and/or social norms. Family members are generally expected to care for and support each other.

TABLE 12.1 Norms Governing Family Structure and Composition

Number of Marriage Partners	
Monogamy	One husband, one wife
Serial monogamy	Two or more successive spouses
Polygamy	Multiple spouses at one time
Polygyny	One husband, multiple wives at one time
Polyandry	One wife, multiple husbands at one time
Choice of Spouse	
Arranged	Parents select their children's marriage partners (sometimes in collaboration with children)
Romantic	Self-selected partner based on love
Endogamy	Marriage within one's social group
Exogamy	Marriage outside one's social group
Homogamy	Marriage to a partner whose social characteristics—such as class, religion, and level of education—are similar to one's own
Authority	
Patriarchal	Male dominated
Matriarchal	Female dominated
Egalitarian	Equal authority between sexes
Descent	
Patrilineal	Traced through father's lineage
Matrilineal	Traced through mother's lineage
Bilateral	Traced through both mother's and father's lineage

Household Type	
Nuclear	Husband, wife, and their immediate children
Extended	Three or more generations living together under one roof
Single-parent	Mother or father living with children
Nonfamily household	People who share the same residence but are not considered family
Domestic partnership	People committed to each other and sharing a domestic life but not joined in marriage or civil union
Civil union	A legally recognized partnership providing same-sex couples with some of the rights, benefits, and responsibilities associated with marriage
Family Residence	
Patrilocal	Wife living with or near husband's family
Matrilocal	Husband living with or near wife's family
Neolocal	Wife and husband live apart from their parents

another way of excluding some kin. That is, people make conscious or unconscious decisions about which kin they will acknowledge as family and which they will "forget" to mention to their children (Waters 1990).

Membership

Some popular definitions of *family* include specific ideas about who should count as family. In this regard, one of the broadest definitions of *family* is "anyone who lives under one roof and expresses love and solidarity" (Aguilar 1999). Organizations such as the World Congress of Families (2011) argue that the ideal or natural family is one "centered around the voluntary union of a man and a woman in a lifelong covenant of marriage," that is, welcoming of children. An **ideal** is a standard against which real cases can be compared. If we simply think about the living and procreation arrangements we observe every day, we quickly realize that many do not match that ideal. This fact does not stop people from using their ideal to judge their own and others' living arrangements, and from labeling families that do not fit their ideal as nontraditional, dysfunctional, immoral, or at-risk (Cornell 1990).

ideal A standard against which real cases can be compared.

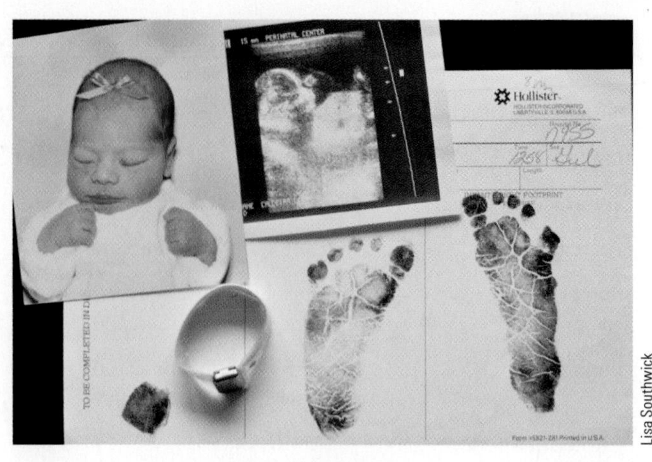
Lisa Southwick

Families function to replace the members of society who die, by providing a legally or socially sanctioned arrangement that can bring new members into the world.

Legal Recognition

When legal recognition is the defining criterion, a family is defined as two or more people whose living and/or procreation arrangements are recognized *under the law* as constituting a family. U.S. federal laws define *spouse* as "a person of the opposite sex who is a husband or wife" and *marriage* as "a legal union between one man and one woman as husband and wife" (Defense of Marriage Act 1996, section 7). Legal recognition of family and marriage arrangements means that laws enforce any awarded benefits, responsibilities, and rights. Some state and local governments within the United States formally recognize family relationships that are in violation of federal law such as same-sex marriages, domestic partnerships, and civil unions. This recognition acknowledges that many people (including gay, lesbian, and heterosexual couples) form lasting, committed, caring, and faithful relationships—and that they are entitled to the legal protections, benefits, and responsibilities associated with marriage. Without these benefits and protections, such people might suffer numerous obstacles and hardships. As a case in point, the U.S. General Accounting Office (2004) identified 1,138 federal statutory provisions in United States Legal Code "in which marital status is a factor in determining or receiving benefits, rights, and privileges."

Functionalist View of Family Life

CORE CONCEPT 2 Family can be defined in terms of the social functions it performs for society.

Because of the debate over what constitutes family, some sociologists argue that *family* should be defined in terms of valued social functions. In this regard, the family performs at least five functions: (1) regulating sexual behavior, (2) replacing the members of society who die, (3) socializing the young, (4) providing care and emotional support, and (5) conferring social status.

Regulating Sexual Behavior Marriage and family systems generate norms (which can take the form of laws) that regulate sexual behavior. These norms may prohibit sex outside of marriage, or they may specify the social characteristics of partners. Such norms may prohibit marriage and sexual relationships between certain relatives (such as first cousins), specified age groups (such as an adult and a minor), and different racial or ethnic groups. Other norms regulate the number of partners (such as monogamy, serial monogamy, and polygamy).

Replacing the Members of Society Who Die All people eventually die. Thus, for humans to survive as a species, we must at least replace those who die. Marriage and family systems provide a socially and legally sanctioned environment into which new members can be born and nurtured.

Socializing the Young Recall that socialization is a learning process that begins immediately after birth. The family is the most significant agent of socialization, because it gives society's youngest members their earliest exposure to relationships and the rules of life.

Providing Care and Emotional Support A family is expected to care for the emotional and physical needs of its members. No matter how old, all humans require meaningful social ties to others. Without such ties, people deteriorate both physically and mentally. The human life cycle is such that we all experience at least one stage of extreme dependency (infancy and childhood). Unless we die suddenly, we are also likely to experience some level of mental and/or physical deterioration in adulthood, accompanied by varying degrees of dependency.

Conferring Social Status We cannot choose our family, the quality of the relationship between our parents, or the economic conditions into which we are born. Among the things we inherit from our parents are a genetic endowment and a social status. For example, the physical characteristics we inherit affect the racial category to which we are assigned. Parents' occupations and incomes are also important predictors of **life chances**—a critical set of

life chances A critical set of potential social advantages, including the chance to survive the first year of life, to live independently in old age, and everything in between.

Global Comparisons Countries with Highest and Lowest Total Fertility Rates

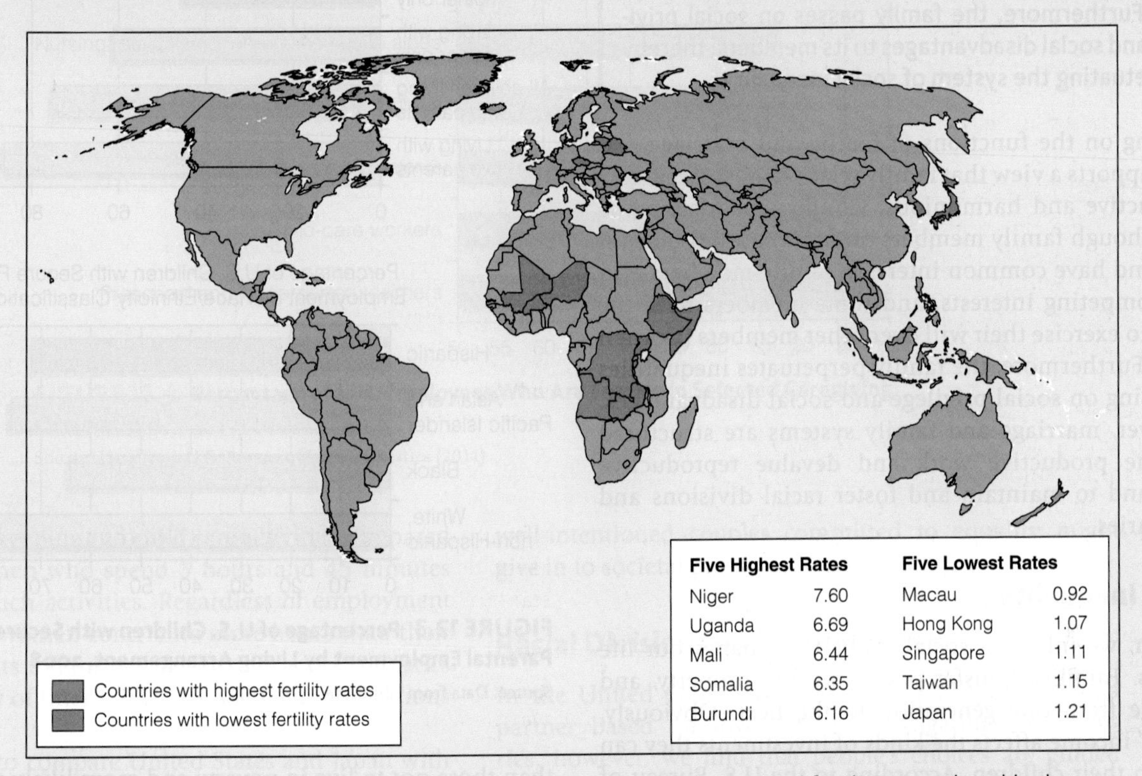

Five Highest Rates		Five Lowest Rates	
Niger	7.60	Macau	0.92
Uganda	6.69	Hong Kong	1.07
Mali	6.44	Singapore	1.11
Somalia	6.35	Taiwan	1.15
Burundi	6.16	Japan	1.21

FIGURE 12.2 The map shows the 10 countries with the highest and lowest total fertility rates; the table shows data for the five countries with highest and lowest rates. Macau has the lowest total fertility rate, at .92; Niger has the highest total fertility rate, at 7.60. That is, the average woman in Macau bears .92 children in her lifetime, whereas the average woman in Niger bears 7.60 children. The total fertility rate for the United States is 2.06. What part of the world has the highest fertility rates? Note that the countries with lowest fertility rates are islands or physically small making it difficult to see them on any map.

Source: Data from World Factbook (2011)

potential social advantages, including the chance to survive the first year of life, to live independently in old age, and everything in between.

Although everyone might agree that families should fulfill these five functions, we must acknowledge that families often fail to achieve one or more of them, as evidenced by the following facts:

- Family systems do not always succeed at regulating sexual activity so that it is confined to a husband and wife or legal partners. Based on tests for organ tissue typing, physicians estimate that between 5 to 20 percent of the donors are genetically unrelated to the men that they believed to be their biological fathers (Anderlik and Rothstein 2002).
- Marriage and family systems do not always succeed in replacing the members of society who die. In at least 96 countries, the total fertility rate is below 2.1—the minimum number of live births needed to replace the members who die (U.S. Central Intelligence Agency 2011). Why a minimum of 2.1 and not 2? Not all babies survive to replace their biological parents. Of course, in societies that have high childhood mortality, the total fertility rate must be more than 2.1. (See Global Comparisons: "Countries with Highest and Lowest Total Fertility Rates.")
- Family members do not always care for one another in positive ways. Domestic and child abuse is a problem in both the United States and Japan. Although the number of *reported* cases of such abuse in Japan is very low compared with the number reported in the United States, many cases go unreported. In one survey taken by the Japanese Cabinet Office's Gender Equality Bureau, 33 percent of women and 17.4 percent of men reported that their partner abused them physically, mentally, or sexually. About 10 percent of women and 2.6 percent of men said they suffered repeated abuse (*Japan Times* 2006).

CORE CONCEPT 3 Family life is not always harmonious. Furthermore, the family passes on social privilege and social disadvantages to its members, thereby perpetuating the system of social inequality.

Conflict theorists argue that the family perpetuates existing inequalities by passing on social privilege and social disadvantages to its members. Moreover, marriage and family systems are structured to value productive work and devalue reproductive work, and to maintain and foster racial divisions and boundaries.

CORE CONCEPT 4 Family structures are not static; they change in response to larger economic, cultural, historical, and social forces. Sociologists track changes in family structure over time and seek to identify triggers of change.

One of the most important changes affecting the structure of families in the United States relates to the percentage of women working in the paid labor force. Between 1900 and 1980, the percentage rose from less than 20 percent to 60 percent, and the percentage of married women in the labor force rose from 15.4 percent to 53.1 percent. Women's entry (and especially married women's entry) into the paid work force can be connected to at least five factors: (1) the rise and fall of the breadwinner system, (2) declines in total fertility, (3) increased life expectancy, (4) higher divorce rates, and (5) increased employment opportunities for women.

In Japan, declining fertility rates, along with an aging population, are having a major impact on family structures. Japan's fertility has declined to 1.2 births per woman, and the population over age 65 is 22.7 percent. The drop in total fertility is intertwined in a number of historical forces, including the fall of the multigenerational household system, the rise of a breadwinner system, the decline in arranged marriages, the rise of the so-called "parasite single," and increased employment opportunities, especially for single women.

CORE CONCEPT 5 Over the past 100 years, fundamental shifts in the economy, a decline in parental authority, the changing status of children, and dramatic increases in life expectancy have changed the family structure.

Four economic arrangements affect the family and relationships among its members: low-technology tribal societies, fortified households, private households, and advanced market economies. Because of economic forces that accompanied industrialization and the rise of private households and advanced market economies, children no longer learned from their parents and other relatives the skills needed to make a living. Parental authority over adult children lost its economic and legal supports.

The technological advances associated with the Industrial Revolution and the shift from an agriculture-based economy to a manufacturing-based altered the status of children from economic assets to liabilities. Increases in life expectancy have also changed the family in fundamental ways. Parents can expect their children to survive infancy and early childhood, children can expect their parents to live a long life, couples can count on a long marriage, and people have more time to settle on a partner, an occupation, and whether to have children.

CORE CONCEPT 6 The aging of the population has no historical precedent. The family must find ways to adapt to this situation and to balance the caregiving needs of the elderly against others who need care.

Finding ways to care for the elderly population is among the greatest challenges. In both the United States and Japan, caregivers are overwhelmingly female. Although most elderly do not live in nursing homes, many do need care with daily activities, such as bathing, walking, dressing, and eating. There are others who need care besides the elderly, and the family must find ways to balance the needs of its elderly members against others who also need care.

Resources on the Internet

Login to CengageBrain.com to access the resources your instructor requires. For this book, you can access:

Sociology CourseMate

Access an integrated eBook, chapter-specific interactive learning tools, including flash cards, quizzes, videos, and more in your Sociology CourseMate.

Take a pretest for this chapter and receive a personalized study plan based on your results that will identify the topics you need to review and direct you to online resources to help you master those topics. Then take a posttest to help you determine the concepts you have mastered and what you will need to work on.

CourseReader

CourseReader for Sociology is an online reader providing access to readings, and audio and video selections to accompany your course materials.

Visit **www.cengagebrain.com** to access your account and purchase materials.

Key Terms

aging population 305
disability 328
endogamy 311
exogamy 311
family 306
fortified households 321
ideal 307
impairment 328
life chances 308
low-technology tribal societies 320
productive work 310
reproductive work 310
secure parental employment 310
total fertility rate 305
tyranny of the normal 328

EDUCATION

13 With Emphasis on THE EUROPEAN UNION

This chapter emphasizes formal education or the ways in which societies structure the educational experience. We consider how educational experiences shape successes and failures among advantaged and disadvantaged populations. As one example, European college students experience an advantage over their American counterparts in that taxpayers cover a greater share of the tuition costs. In light of the economic downturn, that advantage is in jeopardy. Here, students in France protest proposed cuts to higher education.

Why Focus On THE EUROPEAN UNION?

The European Union (EU) is an economic and political alliance that began in 1952, with six member countries. Today, that alliance includes 27 member countries. We focus on the European Union in this chapter for several reasons. First, the EU is investing heavily in education and research to boost its international competitiveness and to ensure that Europeans have the skills necessary to thrive in the twenty-first century. The EU is also offering scholarships to attract the world's super-scholars, and has opened its higher education institutions to the rest of the world, thereby challenging the United States' dominance as a host country to international students (European Commission 2011, Lee 2004).

Second, the U.S. Department of Education routinely compares its students and education system with foreign, especially European, counterparts on a host of attributes, including teachers' salaries, reading scores, scientific literacy, per capita spending on education, and access to educational opportunities. This comparative analysis allows an assessment of U.S. strengths and weaknesses relative to the world's wealthier economies.

Online Poll

How much debt do you expect to acquire over the course of your college career?

- No debt
- Under $5,000
- $5,000 to $9,999
- $10,000 to $14,999
- $15,000 to $19,999
- $20,000 to $29,999
- More than $30,000

To see how other students responded to these questions, go to **www.cengagebrain.com.**

Education

CORE CONCEPT 1 In the broadest sense education includes the formal and informal experiences that train, discipline, and shape the mental and physical potentials of the maturing person.

An experience that educates may be as commonplace as reading a sweater label and noticing that the sweater was made in China or as intentional as performing a scientific experiment to learn how genetic makeup can be modified deliberately through the use of viruses. In view of this definition and the wide range of experiences it encompasses, we can say that education begins when people are born and ends when they die.

Sociologists make a distinction between informal and formal education. **Informal education** occurs in a spontaneous, unplanned way. Experiences that educate informally occur naturally; they are not designed by someone to stimulate specific thoughts or to impart specific skills. Informal education takes place when a child puts her hand inside a puppet and then works to perfect the timing

education In the broadest sense, the experiences that train, discipline, and shape the mental and physical potentials of the maturing person.

informal education Education that occurs in a spontaneous, unplanned way.

Chris Caldeira

Formal education occurs when someone designs the educating experience with an outcome in mind, as when city officials display a wrecked van to teach the dangers of drinking and driving.

between the words she speaks for the puppet and the movement of the puppet's mouth.

Formal education is a purposeful, planned effort to impart specific skills or information. Formal education, then, is a systematic process (for example, military boot camp, on-the-job training, or smoking cessation classes) in which someone designs the educating experiences. We tend to think of formal education as an enriching, liberating, or positive experience, but it can be an impoverishing and narrowing experience (such as indoctrination or brainwashing). In any case, formal education is considered a success when those being instructed internalize the skills, thoughts, and information that those designing the experience seek to impart.

This chapter is concerned with a specific kind of formal education: schooling. **Schooling** is a program of formal, systematic instruction that takes place primarily in classrooms but also includes extracurricular activities and out-of-classroom assignments. In its ideal sense, "education must make the child cover in a few years the enormous distance traveled by mankind in many centuries" (Durkheim 1961, p. 862). More realistically, schooling represents the means by which instructors pass on the values, knowledge, and skills that they or others have defined as important for success in the world. What constitutes an ideal education—in terms of learning objectives, material, and instructional techniques—varies according to time and place.

formal education A purposeful, planned effort aimed at imparting specific skills and information.

schooling A program of formal, systematic instruction that takes place primarily in classrooms but also includes extracurricular activities and out-of-classroom assignments.

Social Functions of Education

CORE CONCEPT 2 Schools perform a number of important social functions that, ideally, contribute to the smooth operation of society.

These functions include transmitting skills, facilitating personal growth, contributing basic and applied research, integrating diverse populations, and screening and selecting the most qualified students for what are considered the most socially important careers.

Transmitting Skills Schools exist to teach children the skills they need to adapt to their environment. To ensure that this end is achieved, society reminds teachers "constantly of the ideas, the sentiments that must be impressed" on students. Educators must pass on a sufficient "community of ideas and sentiments without which there is no society." These ideas and sentiments may relate to instilling a love of country, training a skilled labor force, or encouraging civic engagement. Without some agreement, "the whole nation would be divided and would break down into an incoherent multitude of little fragments in conflict with one another" (Durkheim 1968, pp. 79, 81).

Facilitating Personal Growth Education can be a liberating experience that releases students from the blinders imposed by the accident of birth into a particular family, culture, religion, society, and time in history. Education can broaden students' horizons, making them aware of the conditioning influences around them and encouraging them to think independently of authority.

Contributing to Basic and Applied Research Universities employ faculty whose job descriptions require them to do basic and applied research (in addition to teaching and sometimes in lieu of teaching). The following examples of university-based research centers or institutes suggest that universities add to society's knowledge base and influence policy in a variety of fields: the Center for Aging Research at Indiana University, the Artificial Intelligence Center at the University of Georgia, the Institute for the Study of Planet Earth at the University of Arizona, and the Institute for Drug and Alcohol Studies at Virginia Commonwealth University.

Integrating Diverse Populations Schools function to integrate (for example, to Americanize or Europeanize) people of different ethnic, racial, religious, and family backgrounds. In the United States, schools play a significant role in what is known as the melting-pot process. Recall that the "peopling of America is one of the great dramas in all of human history" (Sowell 1981, p. 3). Among other things, it involved the conquest of the native peoples, the annexation of Mexican territory along with many of its inhabitants (who lived in what is now New Mexico, Utah, Nevada, Arizona, California, and parts of Colorado and Texas), and the voluntary and involuntary immigration of millions of people from practically every country in the world. Early American school reformers—primarily those of Protestant and British backgrounds—saw public education as the vehicle for Americanizing a culturally and linguistically diverse population, instilling a sense of national unity and purpose, and training a competent workforce.

The European Union is relying on its schools to facilitate smooth relationships and interactions among 492.4 million people in 27 member states speaking 23 official languages (U.S. Central Intelligence Agency 2011). Among other things, the EU Commission recommends that schools prepare EU residents to be conversationally proficient in at least two languages beyond the mother tongue. (See Figure 13.1).

Screening and Selecting Schools use tests and grades to evaluate students and reward them accordingly by conferring or withholding degrees, assigning students to academic tracks, rejecting or admitting students into programs, and issuing grades. Thus the schools channel students toward different career paths. Ideally, they channel the best skilled into the most desirable and important careers and the least skilled into careers believed to require no special talents.

Solving Social Problems Societies use education-based programs to address a variety of social problems, including parents' absence from the home, racial inequality, drug and alcohol addictions, malnutrition, teenage pregnancy, sexually transmitted diseases, and illiteracy. Although all countries support education-based programs that address social problems, the United States is probably unique in the emphasis it places on education as the *primary* solution to many problems, including childhood obesity, illegal drug use, poverty, hunger, unwanted pregnancy, and so on.

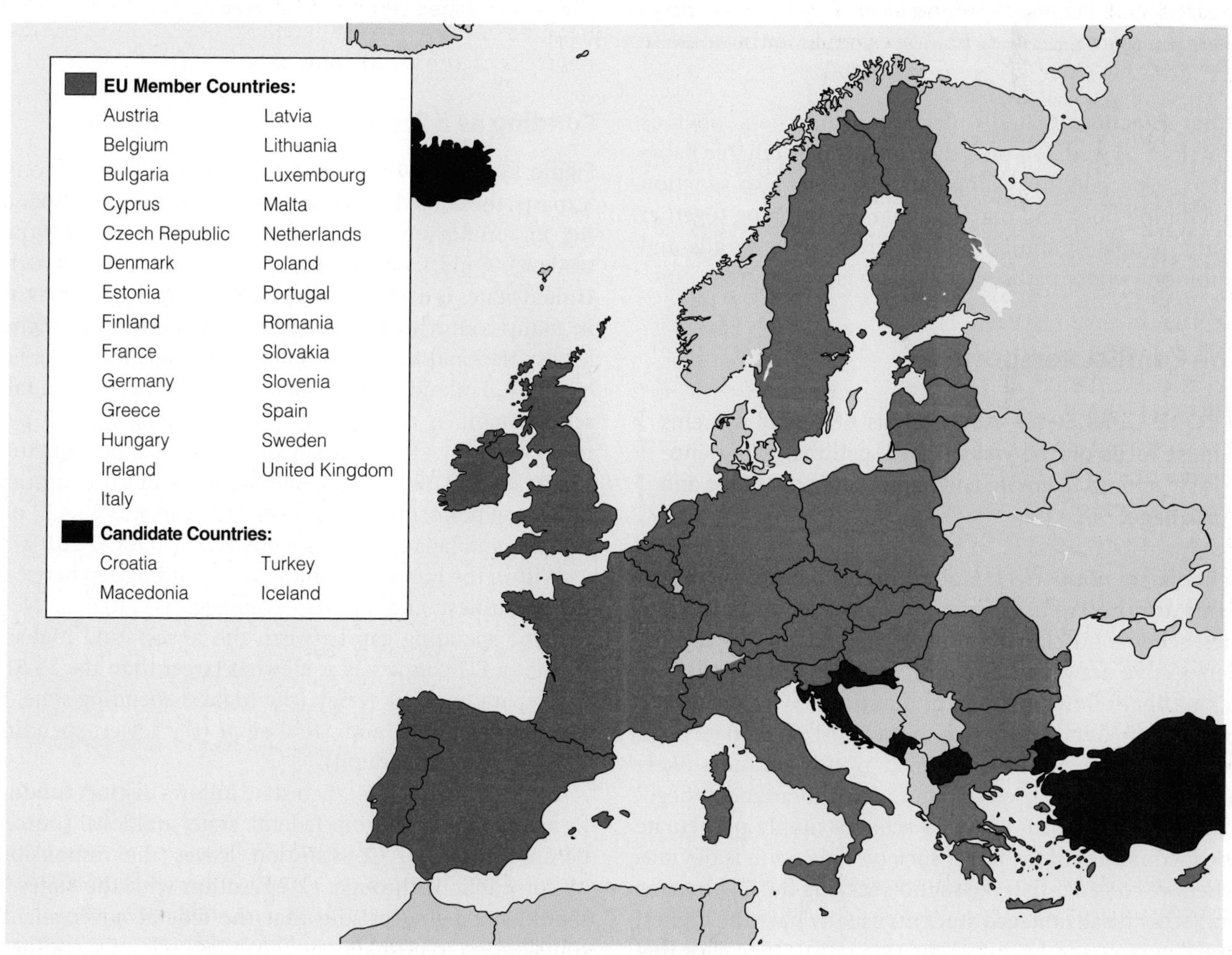

FIGURE 13.1 Map of European Union Member and Candidate Countries
Source: Data from World Factbook

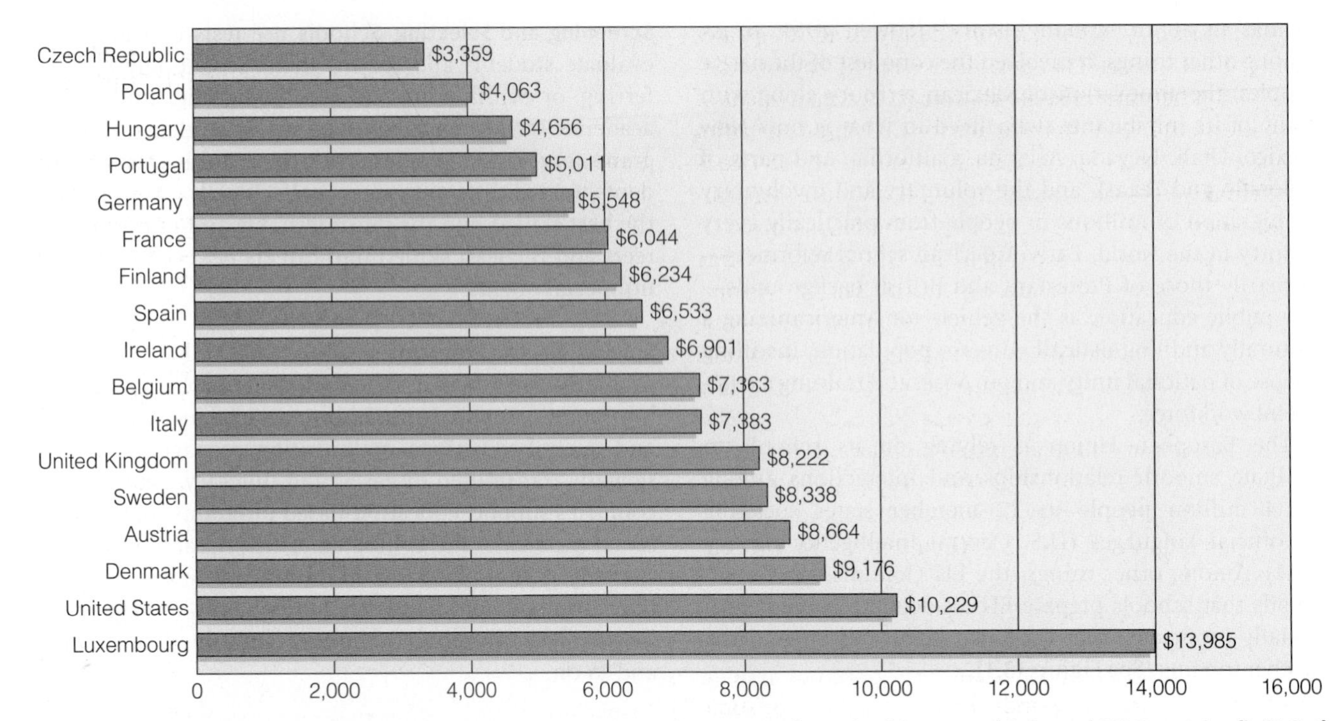

FIGURE 13.2 Per Pupil Spending on Primary and Secondary Education in the United States and Selected EU Countries (in U.S. $)

Source: Data from Organization for Economic Cooperation and Development (OECD) 2010a.

Other Functions Schools perform other, less obvious functions as well. For one, they function as reliable babysitters, especially for young children. They also function as a dating pool and marriage market, bringing together young people of similar and different backgrounds and ambitions whose paths might otherwise never cross.

The Conflict Perspective

CORE CONCEPT 3 Any analysis of school systems must focus on the ways the educational experience is structured to create and perpetuate advantage and privilege.

Schools are not perfect: Not all minds are liberated; students drop out, refuse to attend, or graduate with deficiencies; schools misclassify students; and so on. The conflict perspective draws our attention to issues of inequality by asking the following kinds of questions: Who writes the curriculum? Who has access to the most up-to-date computer or athletic facilities? Which groups are most likely to drop out of high school, and which to attend college? Who studies abroad? How do schools simply perpetuate the inequalities of the larger society? This point is obvious when we consider that the schools serving the low-income and other disadvantaged students usually have the highest dropout rates and lowest graduation rates. One factor that contributes to this inequality is, of course, differences in funding.

Funding as a Broad Measure of Inequality

Figure 13.2 shows that, when compared with its EU counterparts, the United States ranks second in per-pupil spending for primary and secondary education ($13,985 per student). Only Luxembourg spends more. Although the United States is a world leader in spending, students living in many countries that spend far less perform at higher levels (National Center for Education Statistics 2011). For example, Finland, which spends $6,234 per elementary school pupil, is consistently ranked among the top performers on the OECD survey of educational performance that tests 500,000 15-year-old students in 70 countries. Other top performers are Korea, which spends $5,437 per student, and Japan, which spends $7,247 (OECD 2010a).

Within the European Union, the spending gap between the wealthiest and poorest countries is $10,626. Note that the spending gap between the lowest- and highest-spending EU country is somewhat larger than the $9,515 gap separating New Jersey (the highest-spending state, at $16,163 per pupil) and Mississippi (the lowest-spending state, at $6,648 per pupil).

In the United States, each state follows distinct funding formulas that draw from federal, state, and local sources. Because the U.S. Constitution leaves the responsibility for public K through 12 education with the states, it should not be surprising that the federal government's contribution to primary and secondary school spending is 10.8 percent of the total cost. The federal contribution includes funding for the Head Start program, the School

Lunch Program, and the Race to the Top fund. The states contribute 45.6 percent, and 43.6 percent comes from local and other intermediate sources (U.S. Department of Education 2011, National Center for Education Statistics 2011). Such a heavy reliance on state and local revenue is problematic, because less wealthy states and local communities generate less tax revenue than do wealthier ones.

One way to evaluate areas in which U.S. school systems fall short or succeed at educating their students is by comparing those systems with EU education systems, using criteria that sociologists have identified as critical to profiling and evaluating education systems. Most of the data for our comparative analysis comes from the Organization for Economic Cooperation and Development (OECD) education-related reports. We begin this comparative analysis by comparing the percentages of the populations in the United States and selected EU countries that are considered functionally illiterate.

Illiteracy In the most general and basic sense, **illiteracy** is the inability to understand and use a symbol system, whether it is based on sounds, letters, numbers, or some other type of symbol. In the United States (as in all countries), some degree of illiteracy has always existed, but conceptions of what people need to know to be considered literate have varied over time. At one time, Americans were considered literate if they could sign their names and read the Bible. At another time, completing the fourth grade made someone literate. The National Literacy Act of 1991 defines literacy as "an individual's ability to read, write, and speak English and compute and solve problems at levels of proficiency necessary to function on the job and in society, to achieve one's goals, and to develop one's knowledge and potential" (U.S. Department of Education 1993, p. 3).

This point suggests that illiteracy is a product of one's environment. Today, people may be considered functionally illiterate if they cannot use a computer, read a map to find a destination, make change for a customer, read traffic signs, follow instructions to assemble an appliance, fill out a job application, and comprehend the language others are speaking.

Among other things, the OECD (2010a) report focuses on three kinds of literacy: reading, mathematical, and scientific. It seeks to determine how successful school systems are at developing these literacies. For example, in the area of mathematical literacy—defined as "an individual's capacity to identify and understand the role that mathematics plays in the world, to make well-founded judgments, and to use and engage with mathematics in ways that meet the needs of that individual's life as a constructive, concerned, and reflective citizen" (p. 72)—the OECD classifies 28.1 percent of U.S. students as illiterate. In comparison to selected EU countries listed in Table 13.1, the United States has the greatest percentage of 15-years-old students classified as illiterate, followed by Italy.

TABLE 13.1 Percentage of 15-year-olds in the United States and EU Countries Whom the OECD Classified as Mathematically Illiterate

Students who take the OECD math test are classified into one of seven levels: below 1, level 1, level 2, level 3, level 4, level 5, and level 6. Those who are classified as level 1 or below are universally considered illiterate. Individual countries, however, have different criteria for mathematical literacy. To be considered literate in the Netherlands, for example, students must be classified at level 4 or better (de Lange 2006). The percentages in the table show percentages classified as below level 1 or level 1. Which country has the smallest percentage of students classified as illiterate? Which has the largest percentage?

Country	% Illiterate
Finland	4.1
Netherlands	13.0
Denmark	13.6
Hungary	15.0
Germany	15.4
Ireland	15.5
Czech Republic	15.6
Sweden	16.4
Poland	17.0
Belgium	17.0
Spain	19.6
Slovakia	20.0
Austria	20.0
France	21.2
Luxembourg	22.1
Portugal	24.3
Italy	25.3
United States	28.1

Source: Data from OECD 2010.

Foreign Language Illiteracy If we focus on languages, of which there may be as many as 9,000, we can see that the potential number of illiteracies is overwhelming, as people cannot possibly be literate in every language. If a person speaks, writes, and reads in only one language, by

illiteracy The inability to understand and use a symbol system, whether it is based on sounds, letters, numbers, pictographs, or some other type of symbol.

Christopher Brown

This is one of about 500 U.S. students who choose to study abroad in Egypt each year (Lindsey and Labi 2011). Great Britain is the number one choice for U.S. students.

definition he or she is illiterate in as many as 8,999 languages. For most people, such a profound level of illiteracy rarely presents a problem until they find themselves in a setting (such as a war or a business negotiation) that puts them at a disadvantage for not knowing the language others around them are speaking. Both United States and European Union leaders acknowledge the societal benefits that accompany literacy in a language other than the mother tongue. In fact, the European Commission put forth an action plan, calling for every citizen to "have a good command of two foreign languages together with the mother tongue" (Binder 2006). For its part, the U.S. Department of Education (2006) laments that other nations "have an edge in foreign language instruction, a key to improved national security and global understanding."

In European countries, foreign language instruction is compulsory and can start as early as age 5. In the United States, by contrast, 44 percent of high school students study a foreign language (usually for about two years) and only 8 percent of undergraduates take foreign language courses. Study abroad is one way to immerse oneself in a foreign language and culture. Yet, even when Americans study abroad, many of the most popular destinations are English-speaking (International Institute of Education 2010; see No Borders, No Boundaries: "Study Abroad Destinations").

Because of the European Union's greater emphasis on foreign language instruction, 50 percent of people in EU countries indicate that they can speak at least two languages well enough to carry on a conversation. In the United States, however, only 18 percent report an ability to speak a language other than English. It is very likely that the United States is the only country in the world where it is possible to complete high school and college without studying a foreign language. In fact, only ten states require foreign language study as a prerequisite for high school graduation. To exacerbate the situation, upon entering college, most students do not continue on with language studies. When American students do study languages, they are predominantly European languages—most notably Spanish, French, or German—rather than languages named as critical to U.S. national security such as Arabic and Mandarin Chinese, Urdu and Farsi, Pashto and Dari (U.S. Department of Education 2010).

Education critic Daniel Resnick (1990) argues that the absence of serious foreign language instruction contributes to the parochial nature of American schooling. The almost exclusive attention paid to a single language has deprived students of the opportunity to appreciate the connection between language and culture and to see that language is a tool that enables them to think about the world. According to Resnick, the focus on a single language "has cut students off from the pluralism of world culture and denied them a sense of powerfulness in approaching societies very different from their own" (1990, p. 25; see Global Comparisons: "The Legacy of European Colonization on Language Instruction.").

Chris Caldeira

Look at the writing on this sign. If you can only read the top line and not the bottom line, then you are illiterate in the symbol system known as English.

The Availability of College

Only a handful of countries in the world give a significant share of the population the opportunity to attend college. The United States and EU countries represent places in the world where the college-educated constitute

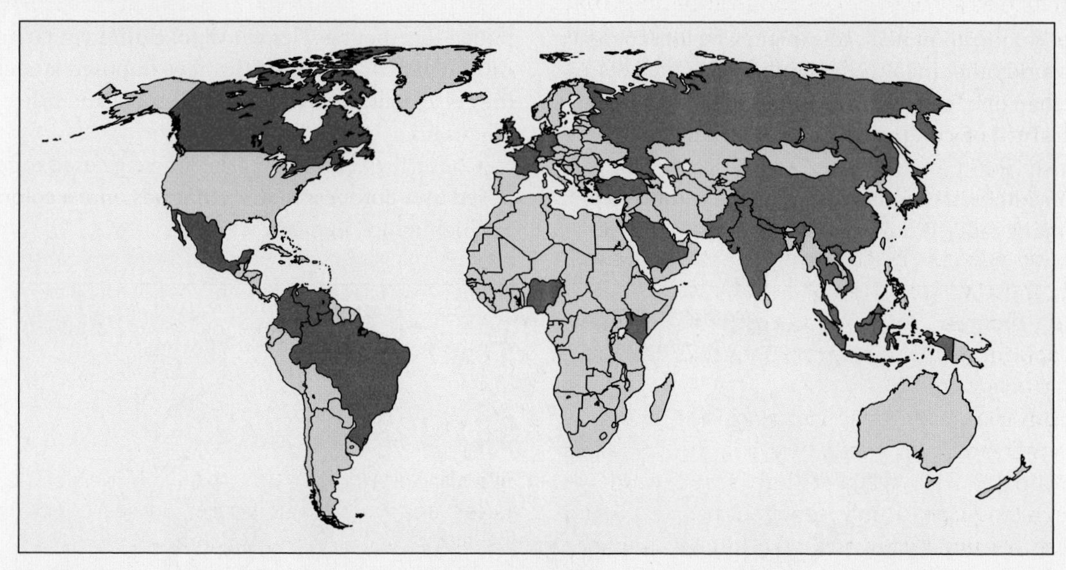

FIGURE 13.3 Top 25 Countries (Highlighted in Red) Sending Students to the United States, 2009–2010 School Year

Over 500,000 international students attend college in the United States. Figure 13.3 shows the top 25 countries sending students to the United States in 2009. China ranked number 1, sending 127,628, followed by India, which sent 104,897, and South Korea, which sent 72,153. Russia, the country ranked 25th, sent 4,827.

Source: Data from International Institute of Education (2010)

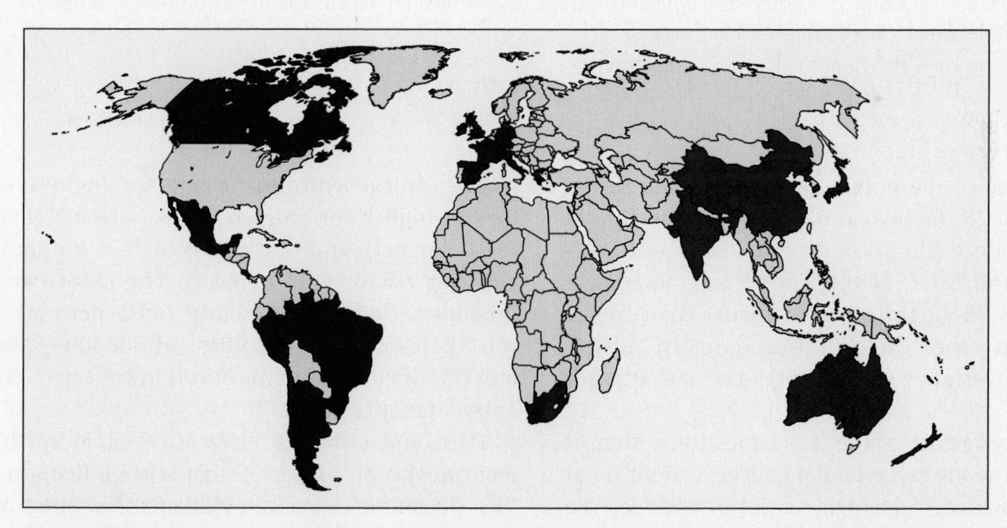

FIGURE 13.4 Top 25 Countries (Highlighted in Blue) to Which the United States Sent Students, 2009–2010 School year

Approximately 260,000 U.S. students studied in a foreign country in 2009. Of the 25 most popular destinations, 11 are EU countries. In fact, 141,953 or 55 percent of American students who study abroad each year go to EU countries. The number one destination was the United Kingdom, with 31,342 students or 12 percent of all students who studied abroad, followed by Italy (27,362) and Spain (24,169). The 25th most popular destination was South Korea (2,062).

Source: Data from International Institute of Education (2010)

Global Comparisons The Legacy of European Colonization on Language Instruction

The legacy of colonization helps to explain why most people around the world other than those born in the United States speak more than one language. Beginning as early as 1492, Europeans learned of, conquered, or colonized much of North America, South America, Asia, and Africa. In doing so, Europeans imposed their languages on the colonized peoples. In many cases, Europeans created countries that pulled together peoples who spoke different languages. Although people learned the new, imposed languages, they also continued to speak their own and other native languages.

Countries where the *official* language used today was imposed by a European nation that was once a colonial power are highlighted in green.

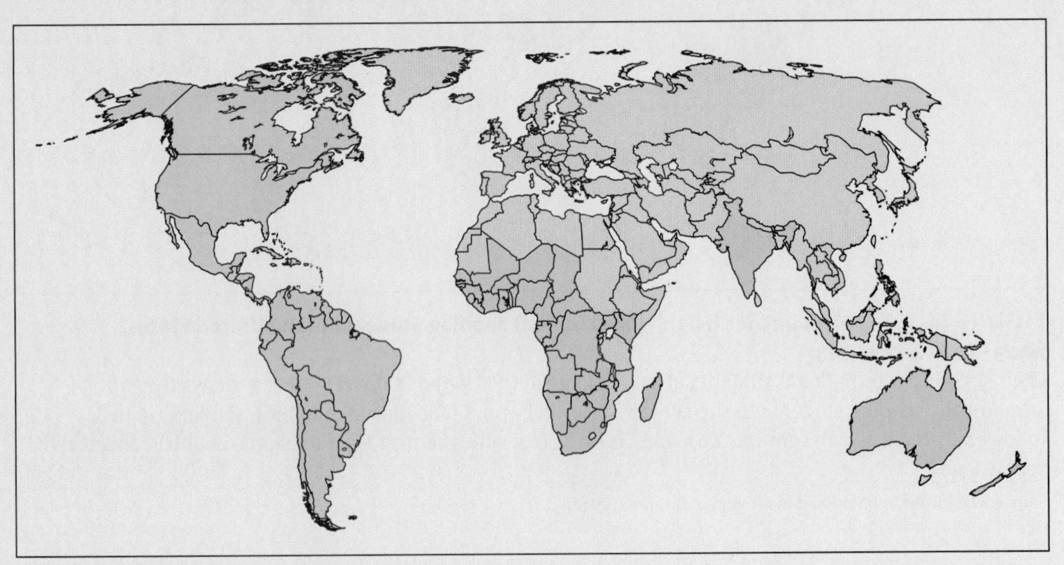

FIGURE 13.5 Countries where the *official* language was imposed by a European nation that was once a colonial power are highlighted in green.

Source: Data from U.S. Central Intelligence Agency (2011)

a significant share of the population. Figure 13.6 shows that 32 percent of 25- to 34-year-olds in the United States have at least a college education. In some EU countries—Denmark, Finland, the Netherlands, and Sweden—the percentage is about the same or higher than in the United States. In other EU countries—notably, Austria (13 percent) and Germany (17 percent)—the percentage is much smaller.

One distinctive feature of the U.S. education system is that, in theory, anyone can attend a college, even if he or she has not graduated from high school or has received a general equivalency diploma (GED). A U.S. Department of Education study found that 400,000 students—2 percent (1 in 50) of all college students—did not have a high school diploma or GED (Arenson 2006). Among those who do graduate from high school, almost 70.1 percent enroll in college the following year (U.S. Bureau of Labor Statistics 2010). The share of *all* 18-year-olds who enroll in college is about 40 percent (U.S. Department of Education 2010).

The college enrollment rate for high school graduates is higher for women (73.8 percent) than for men (66.0 percent), and that rate is highest for graduates classified as Asian (92.2 percent). The same percentages of graduates classified as white (69.2 percent) and black (68.7 percent) go on to college. About 60 percent of graduates classified as Hispanic enroll in college (U.S. Bureau of Labor Statistics 2010).

Race and ethnicity have an even more dramatic effect on who drops out of high school. Keep in mind that 26.1 percent of American students who enter ninth grade in any given year do not graduate from high school four years later. But that dropout rate varies for students classified as Asian (21 percent), white (24 percent), Native American (43 percent), black (45 percent), and Hispanic (47 percent) (U.S. Department of Education 2010).

As another indicator of the relatively easy access to college found in the United States, consider that many U.S. four-year colleges and universities accept students regardless of deficiencies or poor grades in high school coursework, or

FIGURE 13.6 Postsecondary Education Statistics, 2009

The chart shows that 32[1] percent of Americans between the ages of 25 and 34 have the equivalent of a college degree. The table also shows that, in the United States, one-third (31.6 percent) of college costs are subsidized with public expenditures. In some EU countries, 80 percent or more of college costs are subsidized with public expenditures. Over the past five or so years, some European governments have reduced the public (taxpayer) contribution. As one example, in 2003, the UK taxpayer paid for 67.7 percent of the costs of college. That percentage has been reduced to 35.8. Other countries that have reduced public contributions by at least 10 percent since are Austria and Slovak Republic. Recently France has also reduced its contribution, but that reduction is not reflected in this table, which shows 2009 data, the most current data available.

Country	Per student cost of tertiary education (in US$)	% of 25- to 34-year-olds who have obtained the equivalent of at least a college degree	Percentage of public expenditure relative to total cost of postsecondary education
Austria	12,845	13	85.4
Belgium	11,860	23	90.3
Czech Republic	6,826	18	83.8
Denmark	15,890	35	96.5
Finland	12,983	33	95.7
France	10,657	24	84.8
Germany	NA	17	84.7
Ireland	10,540	31	85.4
Italy	5,531	20	69.6
Netherlands	11,246	38	72.4
Portugal	NA	23	70.0
Slovak Republic	4,153	18	76.2
Spain	9,740	26	79.0
Sweden	15,774	32	89.3
United Kingdom	5,352	31	35.8
United States	20,154	32	31.6

[1] Twenty percent of American college students attend four-year institutions where they pay less than $6,000 per year to attend school. 13.2 percent attend four-year institutions where they pay $30,000 or more per year in tuition and fees.

Sources: Data from OECD 2010a, U.S. Department of Education 2010, College Board 2010.

low scores received on the ACT or SAT. In fact, ACT scores suggest that only 25 percent of students taking them are prepared for college-level work (Lewin 2005). Almost 80 percent of colleges and universities offer remedial courses for students who lack the skills needed to do college-level work. An estimated 28 percent of entering freshmen take one or more remedial courses in reading, writing, or mathematics (National Center for Education Statistics 2004).

Although the United States makes it easier for people to enroll in college in spite of academic deficiencies, it places more of the funding burden on the private sector. Compared with its EU counterparts, the United States ranks last in the percentage of postsecondary education costs paid by public funds or taxpayer dollars. In the 16 EU countries for which we have data, with the exception of the UK and Italy, at least 70 percent of postsecondary education costs is paid by public funds. In the United States, that share is 31.6 percent (OECD 2010a). The U.S. system relies more heavily on private sources—such as scholarships, tuition reimbursement from employers, and bank loans—to offset personal costs. As a result, 66 percent of U.S. college students borrow money to pay for college. Upon graduation, the average debt burden is $23,186 (Chalker 2009). Given this high level of private investment, one should not be surprised to find that American students (and their parents) are preoccupied with the return on this investment.

Despite the debt burden, the U.S. way of funding higher education may have some advantages. Claude Allègre, a former French education minister and reformer, argues that the U.S. university system drives American prosperity

Missy Gish

Unlike European counterparts, American students have come to expect more amenities of their universities because they pay more. Most U.S. colleges and universities offer apartment living, gyms, food courts, smart classrooms, and services such as Early Alert, a support program that helps students form an action plan when they encounter obstacles that interfere with college success.

and that the French government simply does not invest in higher education. It promises a college education to any high school graduate who passes the baccalaureate exam, but the government does not deliver the facilities, which are considered to be run-down, crowded, and uninspiring (Sciolino 2006). Americans, on the other hand, expect more because they are paying more: "comfortable student residences, gyms with professional exercise equipment, better food of all kinds, more counselors, . . . more high-tech classrooms and campuses that are spectacularly handsome" (Chace 2006).

Credential Society

Sociologist Randall Collins (1961) is associated with the classic statement outlining the credential society. Collins points out that it is difficult to assess the role education plays in occupational success because there have been few, if any, systematic studies of how much of a particular job's skills are learned on the job, through job training, or acquired through something learned in school. In addition, it is the rare study that examines "what is actually learned in school and how long it is retained" (p. 39). In spite of the untested role education plays in job success, Collins points to the steady increase in educational requirements for employment throughout the last century that has created what he calls a **credential society.**

credential society A situation in which employers use educational credentials as screening devices for sorting through a pool of largely anonymous applicants.

Missy Gish

Imagine that this employee works as a customer service rep at a bank and spends 90 percent of her workday talking to clients, fielding basic questions about their retirement accounts. Her answers follow a script from which she must not deviate. Do you think a college degree is needed for this job?

CORE CONCEPT 4 The credential society is a situation in which employers use educational credentials as screening devices for sorting through a pool of largely anonymous applicants.

Employers have come to view applicants with college degrees as having demonstrated responsibility, consistency, and presumably, basic skills. In addition, employers often require a degree for promotion and advancement, even among those who have an excellent work record and have demonstrated a high level of competence. Likewise, employers use a high school diploma as a screening device for hiring low- and mid-level employees who are viewed (relative to high school dropouts) as people "who have acquired a general respect" for cultural values and styles (Collins 1961, p. 36).

Collins asked what historical factors contributed to the emergence of a credential society in the United States. This is what he discovered:

1. The emergence *cannot* be traced to technological advancements associated with full-scale industrialization because the vast majority of jobs created as a result of industrialization do not require advanced knowledge beyond that of an eighth-grade education.
2. The emergence *can* be traced to a long-standing association between high economic status and advanced degrees. Collins argues that, beginning with the colonial era, the high visibility of a relatively small group of educated elite in high-status positions fueled public

demand in the United States that educational opportunities be available "on a large scale" (p. 37).

3. The emergence of the credential society *can* also be traced to the fact that the United States has always left decisions about what to teach to state and local communities. In addition, the country maintains a separation between church and state (no national church). These two characteristics set the stage for various religious groups to establish their own schools and a very large number of them. Collins argues that it was this large number of schools and colleges in the United States that helps explain, in part, the emergence of the credential society. More specifically, religious rivalry helped produce the Catholic and Lutheran school systems and even the public school system. Collins maintains that white Anglo-Saxon Protestant (WASP) elites founded the public school system in response to large-scale Catholic immigration from Europe. Of course, once the system of public mass education was established, the elite founded private schools for their children as a "means of maintaining cohesion of the elite culture itself" (p. 38). These rivalries set the stage for religious and elite groups to establish universities as well. As a result, Collins believes the opportunities for education at all levels expanded faster in the United States than anywhere else in the world, so that today there are an estimated 4,301 colleges in the United States (National Center for Education Statistics 2008).

Collins argues that it was this large number of schools and colleges in the United States that helped explain the emergence of the credential society. The employer's demand for credentials in turn has fueled and perpetuated a widely held belief that a person must go to college to be successful. The ever-increasing supply of college-educated people, in turn, has made a college degree increasingly a requirement for many jobs. With regard to this demand, surveys show that at least 80 percent of the American public believes that a "young person is best advised to pursue a college education rather than take even a good job out of high school." And 91 percent believe that "employers are less likely to hire people without degrees even though they could do the job" (National Center for Public Policy and Higher Education 2008). In other words, the American public sends the message that a college degree is virtually the only option for obtaining a well-paying job, even though skilled electricians and plumbers, for example, can earn more than many college-educated professionals (Reuters 2006).

The emphasis on a college degree may explain why fewer than 10 percent of American high school students are enrolled in vocational programs and why 60 to 70 percent are enrolled in the college preparatory track. In contrast, 35.6 to 80.7 percent of EU high school students enroll in what those in the United States would call vocational programs (see Figure 13.7). Such programs prepare students for direct entry into a specific occupation (OECD

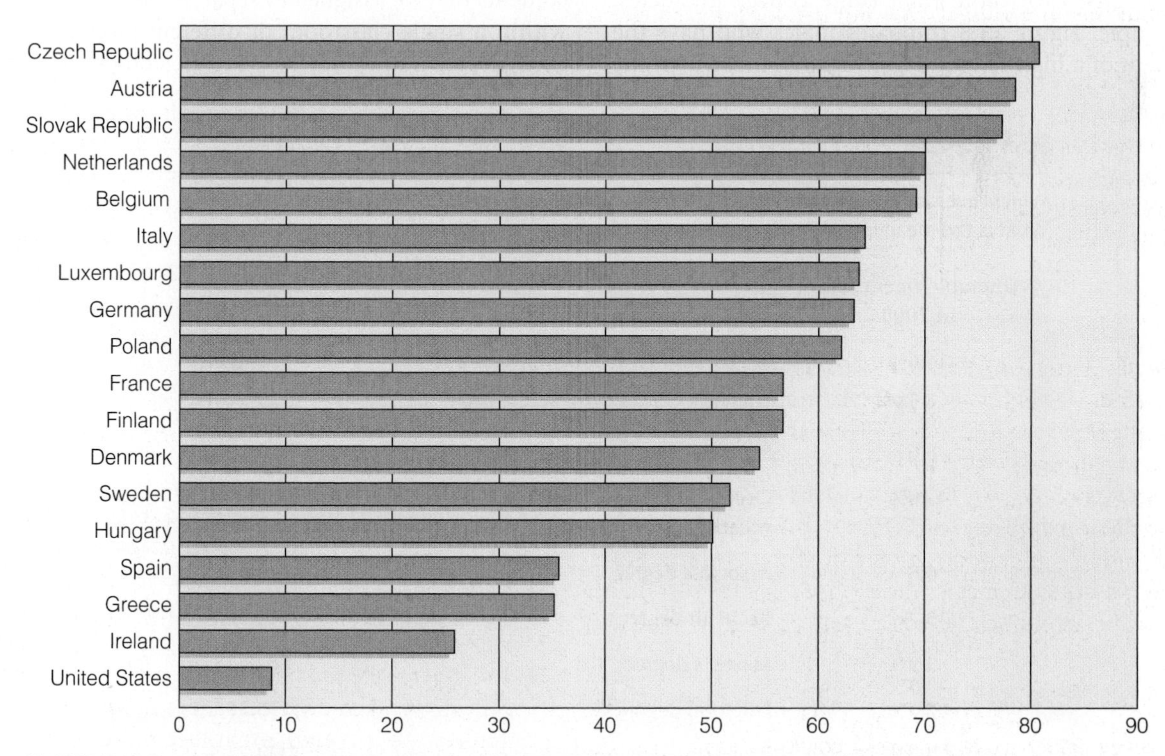

FIGURE 13.7 Percentages of Students in Vocational Programs in Selected EU Countries and the United States

Source: Data from OECD (2010)

2009). In thinking about vocation schooling in the EU, we should not apply our ideas of U.S. vocational education. According to Richard Owen (2000), a Sacramento high school principal who visited ten European countries to learn about their education systems, most vocational programs are equivalent in rigor to U.S. college preparatory programs.

The Promise of Education

Collins has outlined the core elements of credential society. His idea that a college education is not required for many jobs is supported by the 40 percent of college graduates working in the United States who claim that their degree is not needed to do the job they currently hold. Although a college degree is not necessary for many jobs, it is still important to point out that education pays. Table 13.2 shows that income rises and risk of unemployment declines with the level of education obtained.

Sociologist John Reynolds points out that American parents and high school counselors send the message that a college degree is the only option for obtaining a well-paying job, when in fact skilled electricians and plumbers earn more than many college-educated professionals (Reuters 2006). In the United States and elsewhere, the promise that a college education will lead to job opportunities and higher salaries is not always realized on a personal level. To complicate matters, many workers with a high school degree earn more than some college graduates. For example, Figure 13.8 compares males who have the equivalent of a high school degree against those with the equivalent of a 4-year college degree for selected European countries and the United States. Notice that 4 percent of men in Austria with a high school degree earn more than twice the median income compared to 50.3 percent of college educated. In the United States, we see that 11.7 percent of men with a high school degree earn twice the median income. The point is that those without a college degree can be economically successful.

Curriculum

CORE CONCEPT 5 Most, if not all, educational systems sort students into distinct instructional groups according to similarities in past academic performance, performance on standardized tests, or even anticipated performance.

Curriculum encompasses subject content, assessment methods, and activities involved in teaching and learning for a specific course, grade, or degree. When sociologists study a curriculum, they focus on at least two questions: (1) Are students tracked or exposed to different kinds of curricula? (2) As students learn the curriculum, what other lessons do they learn?

Tracking

Under this system, known as tracking or ability grouping, students may be assigned to separate instructional groups within a single classroom or different programs such as college preparatory versus general studies or advanced

TABLE 13.2 Earnings and Unemployment Rate for People 15 and Over, United States

Which levels of education are associated with unemployment rates of 5.2 percent or less? What is the mean income of those with bachelor degrees? With high school, no college?

Unemployment rate in 2009	Level of education completed	Mean (average) earnings in 2009
14.6%	Less than high school diploma	$28,496
9.7%	High school graduate, no college	$40,352
8.6%	Some college, no degree	$46,800
7.0%	Occupational program (trade/ vocational school)	$46,696
6.8%	Associate degree	$48,308
5.2%	Bachelor degree	$71,552
3.9%	Master's degree	$82,628
2.5%	Doctoral degree	$113,308
2.3%	Professional degree	$114,712

Source: Data from College Boards 2011.

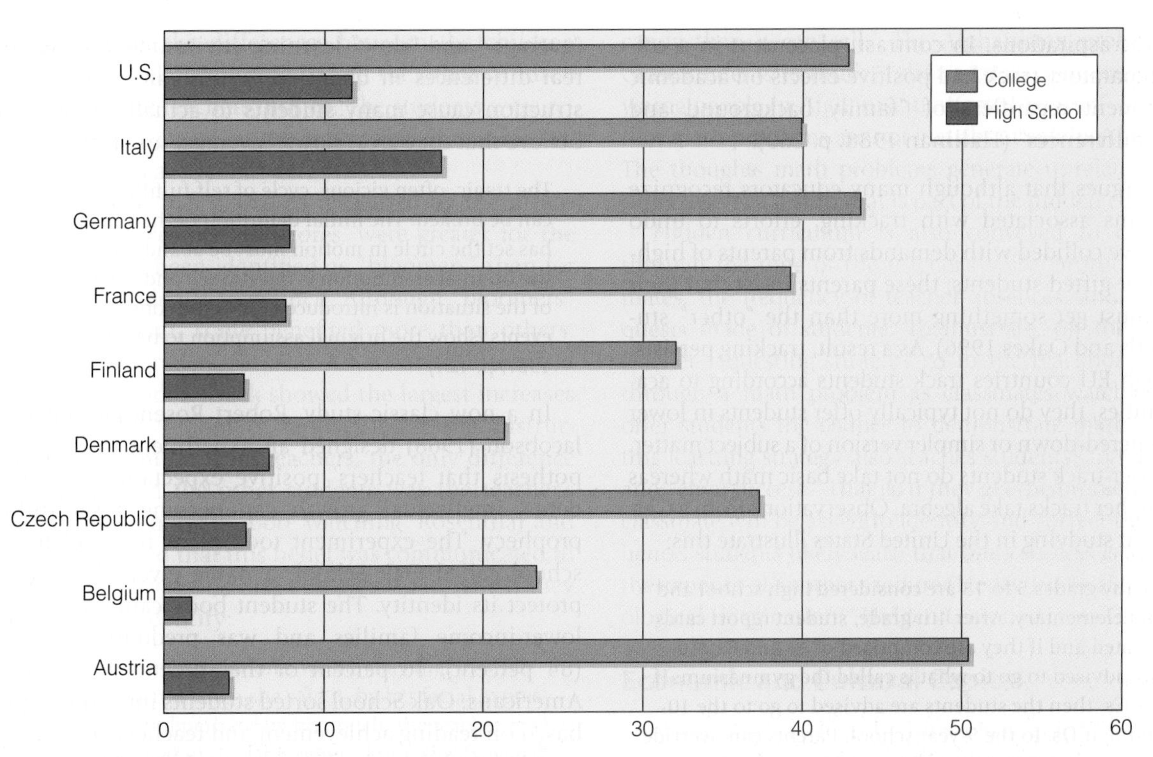

FIGURE 13.8 Percentage of the 25- to 64-Year-Old Male Population Earning Twice the Median Income for Males (College vs. High School Graduates)
This chart shows the percentage of 25- to 64-year-old men who earn salaries at least twice the median income for men. In which country do male high school graduates do best and worst?
Source: Data from OECD (2010b)

placement versus remedial classes. Advocates of tracking offer the following rationales in support of the practice:

- Students learn better when they are grouped with those who learn at the same rate. Slower learners hold back the quickest learners, and slower learners receive the extra time and special attention needed to correct academic deficiencies.
- Slow learners develop more positive attitudes when they do not have to compete with those deemed academically capable.
- Groups of students with similar abilities are easier to teach than groups of students with differing abilities.

There is little evidence to indicate that placing students in remedial or basic courses contributes to their intellectual growth, corrects their academic deficiencies, prepares them for success in higher tracks, or increases their interest in learning. In fact, the special curricula to which advanced-track students are exposed actually helps widen differences between them and students in different tracks (Oakes 1986a, 1986b).

In a now-classic study, sociologist Jeannie Oakes (1985) investigated how tracking affected the academic experiences of 13,719 middle school and high school students in 297 classrooms and 25 schools across the United States. "The schools themselves were different: some were large, some very small; some in the middle of cities; some in nearly uninhabited farm country . . . but the differences in what students experienced each day in these schools stemmed not so much from where they happened to live and which school they happened to attend but rather, from differences within each school" (Oakes 1985, p. 2). Oakes's findings are consistent with the findings of other studies that assess tracking:

Placement: Poor and minority students were placed disproportionately in the lower tracks.
Treatment: The different tracks were not treated as equally valued instructional groups. Clear differences existed in classroom climate and in the quality, content, and quantity of instruction, as reflected in the teachers' attitude, in student–student relationships, and in teacher–student relationships. Low-track students were consistently exposed to inferior instruction—watered-down curricula and endless repetition—and to a more rigid, more emotionally strained classroom climate.
Self-image: Low-track students did not develop positive self-images because they were publicly identified and treated as educational discards, damaged merchandise, or even as unteachable. Among average- and low-track groups, tracking seemed to foster lower self-esteem and promote misbehavior, higher dropout rates, and lower

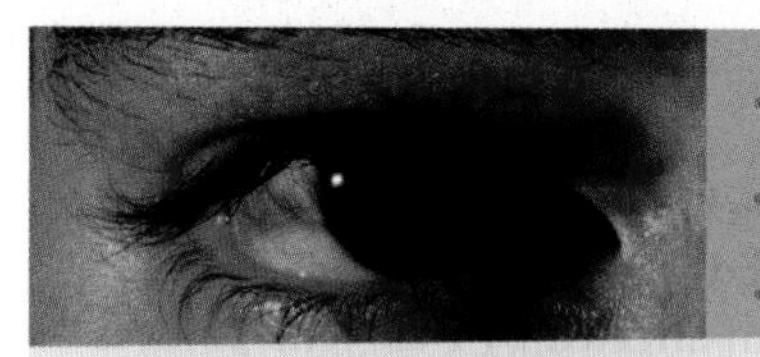

The Sociological Imagination European Students Studying in the United States Comment on American Teachers, Tests, and Study Habits

Three European students attending college in the United States offer insights about the hidden curriculum as it relates to lessons learned about college professors, tests, and study habits. That is, by observing professors—their behavior, tone of voice, body language, interactions with other students—they draw conclusions about the expectations those professors hold. In addition, the number of tests and types of tests convey larger lessons about what type of learning matters and the best ways to study.

College Professors

Germany: American teachers do not seem to ask their students to do as much work as German teachers do. In the United States, the homework is more like busy work such as a scavenger hunt on a website or looking at a catalog to find out what the requirements are for a particular major. These are things that I would do on my own but the students here need to be told to do them.

In Germany, we are taught to be independent and in the United States it is the opposite. In the beginning, it was nice being a college student in the U.S. because I got straight "As" and it helped with my confidence but now I want to be more challenged. In Germany, I was an OK student but in the United States I am a straight-A student. American teachers are very supportive and helpful with keeping students up with the lessons. I feel like I can ask my American teachers anything and they will do whatever they can to help. It's not so much that I would not ask a German teacher something but here I never feel that I ask a stupid question.

Poland: American teachers are more helpful, more approachable, and they seem more equal to students. You would never ask a professor in Poland how they are or call them by their first name. Here you can walk in and talk to a professor about personal things such as what you did over the weekend; that would never happen in Poland. I feel I can ask American teachers for help and I will get it.

Spain: American teachers are more friendly and accessible. They are always asking about personal things like what happened over the weekend. American teachers talk to students like they are kids, not adults. In the United States, professors seem to want to be friends with the students; in Spain, they are just professors, not friends too.

Tests

Germany: In Germany, it was essay only. Here the tests are too easy with multiple choice and true/false. At first it was a good feeling that I was always doing well but now I just want to be more challenged.

Poland: Tests are a different format here. In Poland, they were all essay questions, no multiple choice. We also would have one test at the end of the semester that covered everything. Here we have three or four tests.

Spain: In my entire education in Spain, I have only taken one multiple-choice test. They are not very common. Even essay tests are very easy in the United States. If American students write something even remotely related to the answer, teachers give them some points just for their effort.

Study Habits

Germany: American students are all about memorizing. My roommate memorizes her flash cards the night before the test. At home it is more about understanding the subject, not just a specific date that something happened. We read other books beside the textbook so we can understand the material but American students just study their notes.

Poland: American students spend less time studying than students in Poland. Polish students discuss serious topics for fun. I guess fun is the wrong word. We talk about serious topics because they are serious and need attention. When American students get together to talk, it is more about what was on the headline news or what was going on in class. We don't get into groups and do assignments or have discussions. We are more independent in Poland.

Spain: It's easier in the United States. In Spain, students cannot study two days and expect to do well on a test. . . . In Spain we study almost a month before taking exams and have a week of tests every few months. Even in September when school starts we have a week of exams so we have to study over the summer. In high school, we take ten subjects over the course of a school year and if someone fails two or three exams, he or she has to take the whole year again. American students eat, sleep, and listen to music while they study. I might listen to some music when I study but it would be relaxing music like classical, but not rock.

Sources: Missy Gish, Northern Kentucky University, Class of 2005 (interviewer); Isabell Haage, Class of 2007 (Germany); Anna Nowak, Class of 2005 (Poland); and Cristina Gonzalez, Class of 2004 (Spain).

projected this photograph onto a screen and asked you to comment about it in writing. What would you write? Sociologist Pierre Bourdieu used this line of questioning to document the perceptual schemes that individuals draw upon to think about and react to the world around them. He found that working-class respondents tend to use plain, concrete language to describe the woman's hand (for example, "This man looks like he's got arthritis. His hands are all knotted. I feel sorry seeing that poor old man's hands.") Respondents from more advantaged classes tend use abstract, aesthetic language that transcends the situation of the particular woman pictured: "This photograph is a symbol of toil. It puts me in mind of Flaubert's old servant-woman. . . . It's terrible that work and poverty are so deforming" (Bourdieu 1984). Bourdieu was interested in how these perceptual schemes or points of view come into being. He found that the schemes people draw upon are shaped in large part by their social position. The social position someone occupies depends on the amount of economic and cultural capital upon which they can draw (Appelrouth and Edles 2007).

Economic capital refers to a person's material resources—wealth, land, money. Cultural capital refers to a person's nonmaterial resources, including educational credentials, the kinds of knowledge acquired, social skills, and aesthetic tastes. Both forms of capital are distributed unequally throughout society.

CORE CONCEPT 7 When people locate themselves relative to others, they gain a sense of their place in society, and of what is objectively possible.

According to Bourdieu's theory, high school dropouts come to know and internalize what is objectively possible for someone with their educational credentials. As a result, they are not likely to expect or aspire to a high income. They are also likely to assume that higher incomes are out of reach and to feel that they are just a step away from poverty. Someone with a college or graduate degree, on the other hand, will likely expect to live free of poverty and to earn a relatively high salary. Such assumptions match an objective reality. Figure 13.8 shows that people with college degrees or higher in United States and EU countries have a much better chance of earning twice the median income than those with a high school education.

The **habitus** is the objective reality internalized. That internalized reality becomes the mental filter through which people understand the social world and their place in it. The habitus guides behavior and interpretations of others' actions. Bourdieu believed that the habitus even affects how people physically hold themselves and move about the world (that is, posture, facial expressions, gestures).

The habitus plays a vital role in a process sociologists call **social reproduction**, the perpetuation of unequal relations such that almost everyone, including the disadvantaged, come to view this inequality as normal and legitimate and tend to shrug off or resist calls for change. According to Bourdieu, no institution does more to ensure the reproduction of inequality than education (Appelrouth and Edles 2007). He argues that the system of education is widely misperceived as meritocratic—that grades are awarded according to demonstrated academic abilities and not family connections or class privilege. Upon close scrutiny, however, educational systems actually perpetuate preexisting inequalities. For example, we accept tests as an objective measure of academic achievement. Bourdieu, however, views examinations as one of the most powerfully effective ways of impressing upon students the dominant culture and its values. In fact, tests dominate the educational experience in that students' mental energies are organized around taking them and their grades are largely determined by test scores.

Bourdieu maintains that the purpose of exams cannot just be to measure academic progress. He argues that

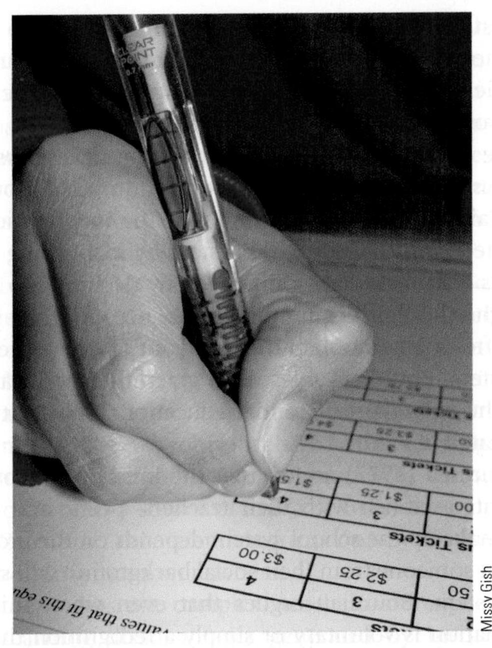
Missy Gish

Bourdieu sees the exam as the "clearest expression of academic values" (1984, p. 142). Tests are treated as a valid measure of knowledge. Thus, test performance is the socially accepted, largely unchallenged way to demonstrate that a specified body of knowledge has been acquired.

habitus Objective reality internalized. That internalized reality becomes the mental filter through which people understand the social world and their place in it.

social reproduction The perpetuation of unequal relations such that almost everyone, including the disadvantaged, come to view this inequality as normal and legitimate and tend to shrug off or resist calls for change.

adolescents were influenced more by the peer group than by their parents. Coleman wondered why the adolescent society penalized academic achievement. He maintained that the manner in which students are taught is a factor. When teachers prescribe assignments and students take tests at a teacher's command, there is no room for creativity, only conformity. Students show their discontent by choosing to become involved in things they can call their own: athletics, dating, clothes, cars, and extracurricular activities. This reaction is inevitable, given the passive roles that students play in the classroom: Athletics is one of the major avenues open to adolescents, especially males, in which they are considered important to the school and the community. Others support this effort, identify with the athletes' successes, and console athletes when they fail. Athletic competition between schools generates an internal cohesion among students as no other school-sponsored event can. For this reason, male athletes gain so much status.

Coleman argues that because athletic achievement is so widely admired, everyone with some athletic ability will try to develop it. In contrast, because academic pursuits go unrewarded relative to athletics, those who have the ability to perform well in school may not be motivated to do the hard work of developing their intellectual potential. This reward structure suggests that the United States does not draw into the competition everyone who has academic potential. Coleman's findings should deliver the message that the peer group represents a powerful influence on learning, but they should not leave the impression that the family and school have no influence.

Online Poll

Who were the <u>most</u> popular male students in the high school you attended?

- ○ Athletes
- ○ Band members
- ○ The top students academically
- ○ Other

Who were the <u>most</u> popular female students in the high school you attended?

- ○ Athletes
- ○ Band members
- ○ The top students academically
- ○ Cheerleaders
- ○ Other

To see how other students responded to these questions, go to **www.cengagebrain.com**.

©Age Fotostock

Summary of CORE CONCEPTS

This chapter emphasizes formal education or the ways in which societies structure the educational experience. We considered how educational experiences in the United States and EU shape successes and failures among advantaged and disadvantaged populations.

CORE CONCEPT 1 In the broadest sense, education includes the formal and informal experiences that train, discipline, and shape the mental and physical potentials of the maturing person.

Sociologists make a distinction between formal and informal education. Informal education occurs in a spontaneous, unplanned way and formal education encompasses a purposeful, planned effort aimed at imparting specific skills or information.

CORE CONCEPT 2 Schools perform a number of important social functions that, ideally, contribute to the smooth operation of society.

Schools perform a number of important functions that serve the needs of society and contribute to its smooth operation. These functions include transmitting skills, liberating minds, facilitating personal change and reflection, performing basic and applied research, integrating diverse populations, solving social problems, and screening and selecting the most qualified students for the most socially important careers. Schools perform other, not so obvious functions; they function as reliable babysitters and as dating pools and marriage markets.

CORE CONCEPT 3 Any analysis of school systems must focus on the ways the educational experience is structured to create and perpetuate advantage and privilege.

The conflict perspective inspires this kind of focus. In this regard, conflict theorists examine a wide range of issues, including differences in school funding, functional illiteracy, and access to college.

CORE CONCEPT 4 The credential society is a situation in which employers use educational credentials as screening devices for sorting through a pool of largely anonymous applicants.

The college degree has become a requirement for jobs that arguably do not need it. Employers have come to view applicants with college degrees as having demonstrated responsibility, consistency, and presumably basic skills. In addition, employers often require a degree for promotion and advancement. Likewise, employers use a high school diploma as a screening device for hiring low- and mid-level employees. Although a college degree is not necessary for many jobs, it is still important to point out that income rises and risk of unemployment declines with the level of education obtained.

CORE CONCEPT 5 Most, if not all, educational systems sort students into distinct instructional groups according to similarities in past academic performance, performance on standardized tests, or even anticipated performance.

The practice of storing students into distinct instructional groups is known as tracking or ability grouping. Advocates of tracking argue that students learn better when they are grouped with those who learn at the same rate; slow learners develop more positive attitudes when they do not have to compete with the more academically capable; and groups of students with similar abilities are easier to teach than students of various abilities. Research suggests that tracking has a positive effect on high-track students, a negative effect on low-track students, and no noticeable effects on middle-track or regular track students.

CORE CONCEPT 6 Schools and teachers everywhere and at all levels of education teach two curricula simultaneously: a formal one and a hidden one.

The content of the various academic subjects—mathematical formulas, science experiments, key terms, and so on—make up the formal curriculum. As teachers deliver the content, they also teach a hidden curriculum that conveys lessons unrelated to the subject matter per se. Hidden curriculum is presented through things like the tone of the teacher's voice, attitudes of classmates, the number of students absent, the frequency of teacher absences, and specific requests made of students.

CORE CONCEPT 7 When people locate themselves relative to others, they gain a sense of their place in society, and of what is objectively possible.

Economic and cultural capital are distributed unequally throughout society. When people locate themselves relative to others, they gain a sense of their place in society, and of what is objectively possible. Habitus is objective reality internalized. We can learn about the inequalities perpetuated by the educational system by identifying the differential educational mortality rate, the rate at which various groups are likely to "voluntarily" drop out of school. The kind of relationships students maintain with their teachers, peers, and others who make up the school system depends on the probability that someone from their social background will survive the system.

CORE CONCEPT 8 Schools are a stage on which critical issues and key concerns are voiced and addressed.

Every government seems to think its education system is failing in major ways. The United States is no exception. Throughout U.S. history, it seems as if public education has always been in a state of crisis and under reform. The ongoing nature of the educational "crises" in the United States and elsewhere suggests that schools are a "stage on which a lot of cultural crises get played out" (Lightfoot 1988, p. 3). Inequality, racial segregation, poverty, chronic boredom, family breakdown, unemployment, and illiteracy are "crises" that transcend the school environment. Yet, we confront them whenever we go into the schools (Lightfoot 1988). Consequently, the schools seem to be both a source of and a solution for our problems.

CORE CONCEPT 9 Not all racial and income groups experience educational success or failure at the same rates. Sociologists seek to understand why this is the case.

Sociologist James Coleman's research, known as the Coleman Report, represents a classic example of how sociologists study inequalities in schools. Coleman determined that the average minority student was likely to come from an economically and educationally disadvantaged household located in a disadvantaged neighborhood. Coleman also found that disadvantaged blacks who had participated in school integration programs scored higher on tests than their disadvantaged counterparts who did not. Clearly, then, the important variable in determining academic success is not race. If race was the factor, test scores would remain unchanged even when black students changed schools. Although family background was identified as the most important factor in the Coleman study, the effects of family background could not be easily separated from the effects of neighborhood and peer groups.

CORE CONCEPT 10 The adolescent society or subculture is a "small society, one that has most of its important interactions within itself, and maintains only a few threads of connection with the outside adult society" (Coleman, Johnstone, and Jonassohn 1961, p. 3).

The emergence of an adolescent society can be traced to industrialization. Around the turn of the twentieth century—the early decades of late industrialization—jobs began to move away from the home and the neighborhood and into factories and office buildings. Parents no longer trained their children, because the skills they knew were becoming obsolete. This shift in training from the family to the school cut adolescents off from the rest of society and forced them to spend most of the day with those of their own age group. Adolescents came to constitute a subculture with great influence over its members and that penalized academic achievement. For adolescent boys, athletic success was extremely important, and for girls, social success with boys is highly valued.

Resources on the Internet

Login to CengageBrain.com to access the resources your instructor requires. For this book, you can access:

Sociology CourseMate

Access an integrated eBook, chapter-specific interactive learning tools, including flash cards, quizzes, videos, and more in your Sociology CourseMate.

Take a pretest for this chapter and receive a personalized study plan based on your results that will identify the topics you need to review and direct you to online resources to help you master those topics. Then take a post-test to help you determine the concepts you have mastered and what you will need to work on.

CourseReader

CourseReader for Sociology is an online reader providing access to readings, and audio and video selections to accompany your course materials.

Visit **www.cengagebrain.com** to access your account and purchase materials.

Key Terms

adolescent society 354
credential society 342
education 333
formal curriculum 347
formal education 334
habitus 349
hidden curriculum 347
illiteracy 337
informal education 333
schooling 334
self-fulfilling prophecy 346
social reproduction 349

RELIGION

14 With Emphasis on THE ISLAMIC REPUBLIC OF AFGHANISTAN

What do you make of this U.S. Air Force sergeant stationed in Afghanistan putting on his taqiyah, a round cap Muslim men wear during worship? He is part of a worship service in Afghanistan marking the end of the Hajj, an annual Muslim pilgrimage to Mecca. When sociologists study religion, they do not study whether God or some other supernatural force exists, whether certain religious beliefs are valid, or whether one religion is better than another. Instead, they focus on the social aspects of religion, such as the characteristics common to all religions and the ways in which people use religion to justify almost any kind of action.

Why Focus On AFGHANISTAN?

On September 20, 2001, nine days after the September 11 terrorist attacks on the United States, then President George W. Bush answered a question many Americans were asking: Who attacked the United States? He identified those who hijacked the commercial aircraft and turned them into "cruise missiles" as belonging to a collection of loosely affiliated terrorist organizations led by Osama bin Laden and known as al-Qaida. Al-Qaida members, believed to be scattered across more than 60 countries, learned the tactics of terrorism in Afghanistan. That night, Bush demanded that the Taliban (an Islamic fundamentalist group that took control of Afghanistan in 1996) close all terrorist training camps, turn over any al-Qaida leaders in Afghanistan, and grant the U.S. government full access to the camps (Bush 2001a). The demands were not met, so on October 7, 2001, the United States began air strikes on Taliban military installations and al-Qaida training camps.

For many Americans, the September 11 attacks represented the most devastating in a long line of attacks by radical Islamists. When radical Islamists become the focus, the attacks are dismissed as simply resulting from the actions of religious fanatics driven by "primitive and irrational" religious conviction. This perspective fails to recognize that "lurking behind every terrorist act is a specific political antecedent. That does not justify either the perpetrator or his political cause. Nevertheless, the fact is that almost all terrorist activity originates from some political conflict and is sustained by it as well" (Brzezinski 2002).

Recognizing political antecedents—more specifically struggles for power—allows us to realize that religious affiliation per se explains little about the causes behind acts considered terrorist. Rather people draw upon religion to justify responses made to political situations. The sociological perspective is useful because it allows us to step back and view in a detached way an often emotionally charged subject. Detachment and objectivity are necessary if we wish to avoid making sweeping generalizations about the nature of religions, such as Islam, that are unfamiliar to many of us.

Online Poll

Is there a mosque in the area in which you live?

○ Yes

○ No

Have you ever visited it?

○ Yes

○ No

To see how other students responded to these questions, go to **www.cengagebrain.com.**

SrA Daryl Knee

What Is Religion?

CORE CONCEPT 1 When sociologists study religion, they are guided by the scientific method and by the assumption that no religions are false.

When sociologists study religion, they do not investigate whether God or some other supernatural force exists, whether certain religious beliefs are valid, or whether one religion is better than another. Sociologists cannot study such questions, because they adhere to the scientific method, which requires them to study only observable and verifiable phenomena. Instead, they investigate the social

Sacramental, Prophetic, and Mystical Religions

In **sacramental religions**, followers seek the sacred in places, objects, and actions believed to house a god or a spirit. These locations may include inanimate objects (such as relics, statues, and crosses), animals, trees or other plants, foods, drink (such as wine and water), places, and certain processes (such as the way people prepare for a hunt or perform a dance). Sacramental religions include many forms of Native American spirituality, none of which are documented in holy books such as the Bible or Koran, and none of which are practiced in man-made churches but rather "in nature, at sacred sites, or in temporary religious structures—such as a tepee or sweat lodge" (Echo-Hawk 1979, p. 280).

In **prophetic religions**, the sacred revolves around items that symbolize historic events or around the lives, teachings, and writings of great people. Sacred books, such as the Christian Bible, the Muslim Qur'an, and the Jewish Tanakh, hold the records of these events and revelations. In the case of historic events, God or some other higher being is believed to be directly involved in the course and outcome of the events (such as a flood, the parting of the Sea of Reeds, or the rise and fall of an empire). In the case of great people, the lives and inspired words of prophets or messengers reveal a higher state of being, "the way," a set of ethical principles, or a code of conduct. Followers then seek to live accordingly. Some of the best-known prophetic religions include Judaism, as revealed to Abraham in Canaan and to Moses at Mount Sinai, Confucianism (founded by Confucius), Christianity (founded by the earliest followers of Jesus Christ), and Islam (founded by Muhammad). The set of ethical principles may include the Ten Commandments of Judaism and Christianity or the Five Pillars of Islam. Muslim tenets include the following:

- The declaration of faith known as the *shahadah* ("There is no god but Allah, and Muhammad is his messenger.")
- Obligatory prayer known as *salah* (performed five times per day)
- Almsgiving (Each year, devout Muslims set aside a percentage of their accumulated wealth to assist the poor and sick.)

Barbara Houghton

The people in this photo are washing clothes and bathing in the river Ganga (also known in the West as the Ganges). The river is considered sacred to Hindus. The millions who live along its banks depend on it to meet daily needs.

- A pilgrimage to the city of Mecca, Saudi Arabia, known as *hajj* (if one is physically and financially able)

In **mystical religions**, followers seek the sacred in states of being that can exclude all awareness of their existence, sensations, thoughts, and surroundings. In such states, mystics become caught up so fully in the transcendental experience that earthly concerns seem to vanish. Direct union with the divine forces of the universe assumes the utmost importance. Not surprisingly, mystics tend to become involved in practices such as fasting or celibacy to separate themselves from worldly attachments.

In addition, they meditate to clear their minds of worldly concerns, "leaving the soul empty and receptive to influences from the divine" (Alston 1972, p. 144).

Sgt. 1st Class Mark Bruce, 8th Theater Sustainment Command

Founded in the sixth and fifth centuries BC by the Buddha, Siddhartha Gautama, Buddhism has an estimated 376 million followers. Buddhism teaches that suffering is an inevitable part of human existence; desires and feelings of self-importance cause suffering; nirvana is achieved through meditation, karma, and righteous actions, thoughts, and attitudes (BBC 2009).

sacramental religions Religions in which the sacred is sought in places, objects, and actions believed to house a god or spirit.

prophetic religions Religions in which the sacred revolves around items that symbolize significant historical events or around the lives, teachings, and writings of great people.

mystical religions Religions in which the sacred is sought in states of being that, at their peak, can exclude all awareness of one's existence, sensations, thoughts, and surroundings.

Buddhism and philosophical Hinduism are two religions that emphasize physical and spiritual discipline as a means of transcending the self and earthly concerns.

Keep in mind that the distinctions between sacramental, prophetic, and mystical religions are not clear-cut. In fact, most religions in each of these categories incorporate or combine elements of the other categories. Consequently, most religions cannot be assigned to a single category.

Beliefs about the Profane

According to Durkheim (1915), the sacred encompasses more than the forces of good: "There are gods of theft and trickery, of lust and war, of sickness and of death" (p. 420). Evil and its various representations are, however, generally portrayed as inferior and subordinate to the forces of good: "In the majority of cases we see the good victorious over evil, life over death, the powers of light over the powers of darkness" (p. 421). Even so, Durkheim considers superordinary evil phenomena to fall within the category of the sacred, because they are endowed with special powers and serve as the objects of rituals (such as confession to rid one of sin, baptism to purify the soul, penance for sins, and exorcism to rid one of evil) designed to overcome or resist the negative influences of such phenomena.

The **profane** encompasses everything that is not considered sacred, including things opposed to the sacred (such as the unholy, the irreverent, and the blasphemous) and that stand apart from the sacred (such as the ordinary, the commonplace, the unconsecrated, and the bodily) (Ebersole 1967).

Contact between the sacred and the profane is viewed as dangerous and sacrilegious. Consequently, people take action to safeguard sacred things by separating them from the profane. For example, some refrain from speaking the name of God when they feel frustrated; others believe that a woman must cover her hair or her face during worship and that a man must remove his hat during worship.

Sgt. Martin Downs

Removing shoes before entering a mosque is an act that separates the profane (ordinary and unconsecrated) from the sacred (holy place of worship).

Rituals

In the religious sense, **rituals** are rules that govern how people behave in the presence of the sacred. These rules may take the form of instructions detailing an appropriate place to engage in worship, the roles of various participants, acceptable dress, and the precise wording of chants, songs, and prayers. Participants engage in rituals with a goal in mind, whether it be to purify the body or soul (as through confession, immersion, fasting, or seclusion), to commemorate an important person or event (as by making a pilgrimage to Mecca or celebrating Passover or the Eucharist), or to transform profane items into sacred items (for example, changing water to holy water and human bones to sacred relics) (Smart 1976, p. 6).

Rituals can be as simple as closing one's eyes to pray or having one's forehead marked with ashes. Alternatively, they can be as elaborate as fasting for three days before entering a sacred place to chant, with head bowed, a particular prayer for forgiveness. Although rituals are often enacted in sacred places, some are codes of conduct aimed at governing the performance of everyday activities, such as sleeping, walking, eating, defecating, washing, and dealing with members of the opposite sex.

According to Durkheim, the nature of the ritual is relatively insignificant. Rather, the important element is that the ritual is shared by a community of worshippers and evokes certain ideas and sentiments that help individuals feel themselves to be part of something larger than themselves.

Community of Worshippers

Durkheim uses the word **church** to designate a group whose members hold the same beliefs regarding the sacred and the profane, who behave in the same way in the presence

profane A term describing everything that is not sacred, including things opposed to the sacred and things that stand apart from the sacred, albeit not in opposition to it.

rituals Rules that govern how people must behave in the presence of the sacred to achieve an acceptable state of being.

church A group whose members hold the same beliefs about the sacred and the profane, who behave in the same way in the presence of the sacred, and who gather in body or spirit at agreed-on times to reaffirm their commitment to those beliefs and practices.

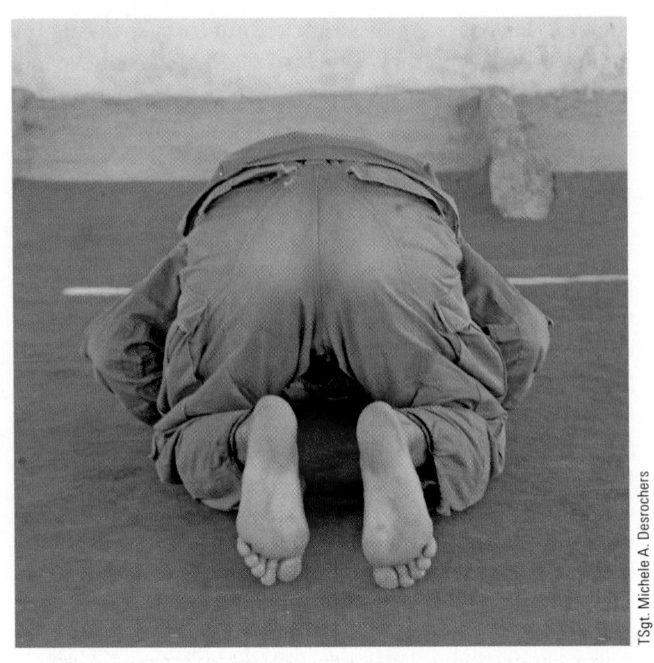

TSgt. Michele A. Desrochers

Kneeling with forehead touching the floor in prayer facing Mecca five times a day constitutes a ritual.

Chris Caldeira

Chris Caldeira

Many religious denominations have churches in the United States. Above is a Hindu temple in San Francisco. Below is a small Christian church located in rural community of the United States.

of the sacred, and who gather in body or spirit at agreed-on times to reaffirm their commitment to those beliefs and practices. Obviously, religious beliefs and practices cannot be unique to an individual; they must be shared by a group of people. If not, then the beliefs and practices would cease to exist when the individual who held them died or if he or she chose to abandon them. The gathering and the sharing create a moral community and allow worshippers to share a common identity. The gathering need not take place in a common setting, however. When people perform a ritual on a given day or at given times of day, the gathering may be spiritual rather than physical.

Sociologists have identified at least five broad types of religious organizations or communities of worshippers: ecclesiae, denominations, sects, established sects, and cults. As with most classification schemes, these categories overlap on some characteristics because the classification criteria for religions are not always clear.

Ecclesiae

An **ecclesia** is a professionally trained religious organization, governed by a hierarchy of leaders, that claims everyone in a society as a member. Membership is not voluntary; it is the law. Consequently, considerable political alignment exists between church and state officials, so that the ecclesia represents the official church of the state. Ecclesiae formerly existed in England (the Church of England [Anglican], which remains the official state church), France (the Roman Catholic Church), and Sweden (the Church of Sweden [Lutheran]). The Afghan constitution signed in 2004 declares the country to be an Islamic republic, makes Islam the official religion, and announces that "no law can be contrary to the sacred religion of Islam." The Afghan government, however, guarantees non-Muslims the right to "perform their religious ceremonies within the limits of the provisions of law" (Feldman 2003).

Individuals are born into ecclesiae, newcomers to a society are converted, and dissenters are often persecuted.

ecclesia A professionally trained religious organization, governed by a hierarchy of leaders, that claims everyone in a society as a member.

Those who do not accept the official religious view tend to emigrate or to occupy a marginal status. The ecclesia claims to be the one true faith and often does not recognize other religions as valid. In its most extreme form, it directly controls all facets of life.

Denominations

A **denomination** is a hierarchical religious organization in a society in which church and state usually remain separate; it is led by a professionally trained clergy. In contrast to an ecclesia, a denomination is one of many religious organizations in society. For the most part, denominations tolerate other religious organizations; they may even collaborate with other such organizations to address problems such as poverty or disaster relief in society. Although membership is considered to be voluntary, most people who belong to denominations did not choose to join them. Rather, they were born to parents who were members. Denominational leaders generally make few demands on the laity (members who are not clergy), and most members participate in limited and specialized ways.

For example, members may choose to send their children to church-operated schools, attend church on Sundays and religious holidays, donate money to the church, or attend church-sponsored functions Although laypeople vary widely in lifestyle, denominations frequently attract people of particular races, ethnicities, and social classes.

Major denominations in the world include Buddhism, Christianity, Confucianism, Hinduism, Islam, Judaism, Shinto, and Taoism. Each is predominant in different areas of the globe. For example, Christianity predominates in Europe, the Americas, and Australia and Oceania; Islam predominates in the Middle East and North Africa; and Hinduism predominates in India.

Sects and Established Sects

A **sect** is a small community of believers led by a lay ministry; it has no formal hierarchy, or official governing body, to oversee its various religious gatherings and activities. Sects are typically composed of people who broke away from a denomination because they came to view it as corrupt. They then created the offshoot in an effort to reform the religion from which they separated.

All of the major religions encompass splinter groups that have sought at one time or another to preserve the integrity of their religion. In Islam, for example, the most pronounced split occurred about 1,300 years ago, approximately 30 years after the death of Muhammad. The split related to Muhammad's successor. The Shia maintained that the successor should be a blood relative of Muhammad; the Sunni believed that the successor should be selected by the community of believers and need not be related to Muhammad by blood. After Muhammad's death, the Sunni (encompassing the great majority of Muslims) accepted Abu Bakr as the caliph (successor). The Shia supported Ali, Mohammad's first cousin and son-in-law, and they called for the overthrow of the existing order and a return to the pure form of Islam. Today, Shiite Islam predominates in the Islamic Republic of Iran (95 percent), whereas Sunni Islam predominates in the Islamic Republic of Pakistan (77 percent) and Afghanistan (80 percent) (U.S. Central Intelligence Agency 2011a).

SrA Daryl Knee

Muslim men at a prayer service in Afghanistan. All work for the International Security Assistance Force in Afghanistan. Muslims are not a monolithic group. In Afghanistan, 20 percent are Shia and 80 percent are Sunni. Muslims also have ties with major ethnic groups such as Pashtun (followers of Sunni Islam) and Hazara (Shiite). The most puritanical are Sunni Taliban whose followers are Pashtun. Obviously, all Pashtuns are not Taliban.

The divisions within Islam have existed for so long that Sunni and Shia have become recognized as **established sects**—groups that have broken from denominations or ecclesiae and have existed long enough to acquire a large following and widespread respectability. As you might expect, divisions can form within established sects, creating splinter groups. Sects exist within the Sunni branch (Wahabis) and the Shiite branch (the Assassins, the Druze).

Similarly, several splits have occurred within Christianity. During a period from about the eleventh century to the

denomination A hierarchical religious organization, led by a professionally trained clergy, in a society in which church and state are usually separate.

sect A small community of believers led by a lay ministry, with no formal hierarchy or official governing body to oversee its various religious gatherings and activities. Sects are typically composed of people who broke away from a denomination because they came to view it as corrupt.

established sects Religious organizations, resembling both denominations and sects, that have left denominations or ecclesiae and have existed long enough to acquire a large following and widespread respectability.

early thirteenth century, for example, the Greek-language Eastern churches (then centering on Constantinople) and the Latin-language Western church (centering on Rome) gradually drifted apart over issues such as the papal claim of supreme authority over all Christian churches in the world. The Protestant churches owe their origins largely to Martin Luther (1483–1546), who also challenged papal authority and protested against many practices of the medieval Roman Catholic Church.

Cults

Generally, **cults** are very small, loosely organized religious groups, usually founded by a charismatic leader who attracts people by virtue of his or her personal qualities. Because the charismatic leader plays such a central role in attracting members, cults often dissolve after the leader dies. Consequently, few cults last long enough to become established religions. Even so, a few manage to survive, as evidenced by the fact that the major world religions began as cults. Because cults form around new and unconventional religious practices, outsiders tend to view them with considerable suspicion.

Cults vary in terms of their purpose and the level of commitment that their leaders demand of converts. They may draw members by focusing on highly specific but eccentric interests, such as astrology, UFOs, or transcendental meditation. Members may be attracted by the promise of companionship, a cure for illness, relief from suffering, or enlightenment. In some cases a cult leader may require members to break all ties with family, friends, and jobs and thus to rely exclusively on the cult to meet all of their needs.

Durkheim's definition of *religion* highlights three essential characteristics: beliefs about the sacred and the profane, rituals, and a community of worshippers. Critics argue that these characteristics are not unique to religious activity. This combination of characteristics, they say, applies to many gatherings (for example, sporting events, graduation ceremonies, reunions, and political rallies) and to many political systems (for example, Marxism, Maoism, and fascism). On the basis of these characteristics alone, it is difficult to distinguish between an assembly of Christians celebrating Christmas, a patriotic group supporting the initiation of a war against another country, and a group of fans eulogizing James Dean or Elvis Presley. In other words, religion is not the only unifying force in society that incorporates the three elements defined by Durkheim as characteristic of religion. Civil religion represents another such force that resembles religion as Durkheim defined it.

cults Very small, loosely organized groups, usually founded by a charismatic leader who attracts people by virtue of his or her personal qualities.

civil religion An institutionalized set of beliefs about a nation's past, present, and future and a corresponding set of rituals. Both the beliefs and the rituals take on a sacred quality and elicit feelings of patriotism. Civil religion forges ties between religion and a nation's needs and political interests.

Chief Photographer's Mate Johnny Bivera

The three essential characteristics of religion also apply to other events such as a football game when people engage in pregame rituals such as praying. The school colors take on a sacred quality, and a community of worshippers gathers to watch the game, often believing "God" is on their side.

Civil Religion

CORE CONCEPT 3 Civil religion is an institutionalized set of beliefs about a nation's past, present, and future and a corresponding set of rituals that take on a sacred quality and elicit feelings of patriotism.

Civil religion forges ties between religion and a nation (Bellah 1992, Hammond 1976, Davis 2002). A nation's values (such as individual freedom and equal opportunity) and rituals (such as parades, fireworks, singing the national anthem, and 21-gun salutes) often assume a sacred quality. Even in the face of division, national beliefs and rituals can inspire awe, respect, and reverence for the country. These sentiments are most evident during times of crisis and war, on national holidays that celebrate important events or people (such as Thanksgiving, July 4th), and in the presence of national monuments or symbols (the flag, the Capitol, the Lincoln Memorial, the Vietnam Memorial).

In times of war, presidents offer a historical and mythological framework that morally justifies the country's involvement in the war and offers the public a vision and an identity. Sociologist Roberta Cole (2002) argues that America's civil religion found its voice in a nineteenth-century political doctrine known as *manifest destiny*. Although the term was first used in 1845, it expressed a long-standing ideology that the United States, by virtue of its moral superiority, was destined to expand across the North American continent to the Pacific Ocean and beyond (Chance 2002). Manifest destiny included the beliefs that the United States had a divine mission to serve as a democratic model to the rest of the world, that the country was a redeemer exerting its good influence upon other nations, and that it represented hope to the rest of the world (Cole 2002). In 1835, Alexis de Tocqueville observed this long-standing belief among Americans that their country was unique:

> For 50 years, it has been constantly repeated to the inhabitants of the United States that they form the only religious, enlightened, and free people. They see that up to now, democratic institutions have prospered among them; they therefore have an immense opinion of themselves, and they are not far from believing that they form a species apart in the human race.

Farm Security Administration and Office of War Information Collection, Library of Congress Prints and Photographs Division

In 1941, the year when these children were reciting the Pledge of Allegiance with hand over heart, they said "one nation indivisible, with liberty and justice for all." They did not say the words "under God" because those words were not part of the oath until the 1950s, during the cold war. At that time, the U.S. government also started stamping "In God We Trust" on its coins.

Civil Religion and the Cold War

The cold war (1945 to 1989) included an arms race, in which the Soviet Union and the United States competed to match and then surpass any advances made by each other in the number and technological quality of nuclear weapons. Although the United States and the Soviet Union fell short of direct, full-scale military engagement, they took part in as many as 120 proxy wars fought in developing countries. In many of these conflicts, the United States and the Soviet Union supported opposing factions by providing weapons, military equipment, combat training, medical supplies, economic aid, and food. Three of the best-known proxy wars were fought in Korea, Vietnam, and Afghanistan (1979 to 1989). Soviet and American leaders justified their direct or indirect intervention on the grounds that it was necessary to contain the spread of the other side's economic and political system, to protect national and global security, and to prevent the other side from shifting the balance of power in favor of its system.

Library of Congress Prints and Photographs Division Washington, D.C.

The painting portrays those who participated in the United States' westward expansion as fulfilling an almost divine mission, represented by the guardian angel-like figure watching over them. Of course, westward expansion was not the peaceful process depicted here.

From 1945 through 1989, the foreign and domestic policies of the United States were largely shaped by cold war dynamics—specifically, a professed desire to *save*, even *redeem*, the world from Soviet influence and the spread of communism. Robert S. McNamara (1989), U.S. Secretary of Defense under Presidents Kennedy and Johnson, remarked that "on occasion after occasion, when confronted with a choice between support of democratic governments and support of anti-Soviet dictatorships, we have turned our backs on our traditional values and have supported the antidemocratic," brutally repressive, and totalitarian regimes (p. 96). President George W. Bush (2003) agreed with McNamara's assessment when he acknowledged in a speech to the British people that

> we must shake off decades of failed policy on the Middle East. Your nation and mine in the past have been willing to make a bargain to tolerate oppression for the sake of stability. Long-standing ties often led us to overlook the faults of local elites. Yet this bargain did not bring stability or make us safe. It merely bought time while problems festered and ideologies of violence took hold.

The United States and Muslims as Cold War Partners

The cold war between the United States and the Soviet Union made Afghanistan a focus of those two countries' conflict (see Figure 14.2). When the Soviet Union invaded Afghanistan in 1979, and put its Afghan supporters in charge, the United States supported Islamic guerrillas, known as the *mujahideen*, by funneling money through Pakistan. At that time, Pakistani president Zia's goals were to turn Pakistan into the leader of the Islamic world and to cultivate an Islamic opposition to Soviet expansion into central Asia. Zia's aims fit well with the United States' cold war goals of containing the Soviet Union. If the United States could show the Soviet Union that the entire Muslim world was its partner, then the United States would indeed be a force to fear (Rashid 2001).

The U.S. Central Intelligence Agency (CIA) worked with its Pakistani equivalent, the Inter-Services Intelligence Agency, on a plan to recruit radical Muslims from all over the world to fight with their Afghan brothers against the Soviet Union. An estimated 35,000 Muslims from 43 countries—primarily in central Asia, North and East Africa, and the Middle East—heeded the call. Thousands more came to Pakistan to study in *madrassas* (Muslim schools) in Pakistan and along the Afghan border (Rashid 2001).

Military training camps staffed with U.S. advisors helped train the guerrillas, and the *madrassas* offered a place for the most radical Muslims in the world to meet, exchange ideas, and learn about Islamic movements in one another's countries. Among those who came to Afghanistan was Osama bin Laden. At that time, the pressing question for the United States, as asked by national security advisor Zbigniew Brzezinski (2002), was, "What was more important in the world view of history? The possible creation of an armed, radical Islamic movement or the fall of the Soviet Empire? A few fired-up Muslims or the liberation of Central Europe and the end of the Cold War?" These Pakistani- and U.S.-supported recruiting and military centers would eventually evolve into al-Qaida ("the base"). In 1989, the year the term *al-Qaida* was first used, Osama bin Laden had taken over as the centers' leader. That same year, Soviet troops withdrew from Afghanistan, leaving behind

> an uneasy coalition of Islamist organizations intent on promoting Islam among all non-Muslim forces. [They also] left behind a legacy of expert and experienced fighters, training camps and logistical facilities, elaborate trans-Islam networks of personal and organizational relationships, a substantial amount of military equipment, . . . and most importantly, a heavy sense of power and self-confidence based on what [they] had achieved, and a driving desire to move on to other victories. (Huntington 2001, p. A12)

To help measure the legacy of U.S.-supported training camps, consider that "key leaders of every major terrorist attack, from New York to France to Saudi Arabia, inevitably turned out to be veterans of the Afghan War" (Mamdani 2004).

Civil Religion and the Gulf War I

In 1990, at the request of the Saudi government, the United States government sent 540,000 troops to the Persian Gulf region after Iraqi troops invaded Kuwait. In a presidential address, George H. W. Bush (1991) fused country and religion together when he described the United States in sacred terms:

> I come to this House of the people to speak to you and all Americans, certain that we stand at a defining hour. Halfway around the world, we are engaged in a great struggle in the skies and on the seas and sands. We know why we're there: We are Americans, part of something larger than ourselves. For two centuries, we've done the hard work of freedom. And tonight, we lead the world in facing down a threat to decency and humanity. . . . Yes, the United States bears a major share of leadership in this effort. Among the nations of the world, only the United States of America has both the moral standing and the means to back it up. We're the only nation on this Earth that could assemble the forces of peace. This is the burden of leadership and the strength that has made America the beacon of freedom in a searching world.

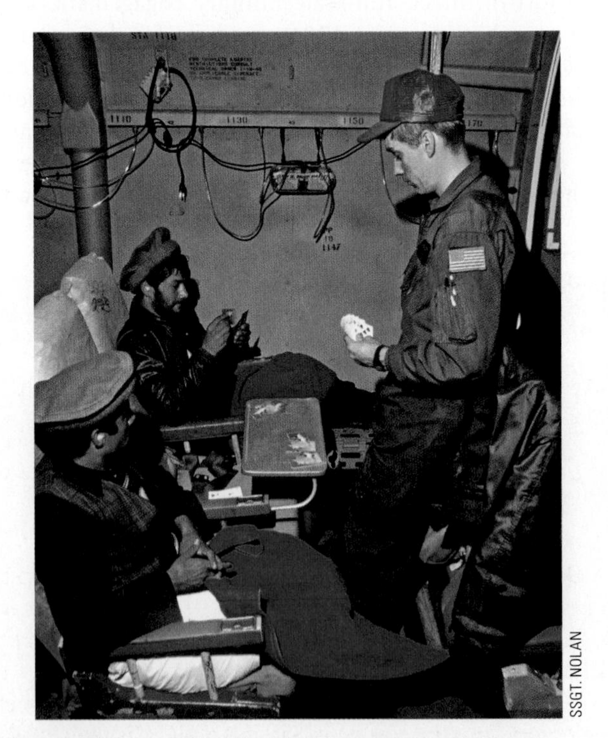

SSGT. NOLAN

The United States offered medical treatment and other assistance to mujahideen who were wounded fighting the Soviet Union. Here, a U.S. medic plays cards with two wounded "freedom fighters," the term the American government used to characterize the anticommunist guerrillas, en route to a U.S. airbase in Germany for specialized medical care.

The Historical Context of Afghanistan

Afghanistan is a mountainous and landlocked county. Its neighbors are Pakistan, Iran, China, and three countries once part of the Soviet Union—Turkmenistan, Uzbekistan, and Tajikistan. Afghanistan connects the Middle East with Central Asia and the India. Afghanistan's borders, especially with Pakistan, have always been in dispute, and great empires and dynasties have fought to control the country, its peoples, and resources, beginning with Alexander the Great in 330 B.C. Here we focus on the nineteenth and twentieth centuries, when Britain invaded Afghanistan twice—the First Anglo-Afghan War of 1838 to 1842 and the Second Anglo-Afghan War of 1878 to 1880—with the goal of limiting Russian influence. Russia invaded in 1979, to support a secular government. The United States supported Afghanistan's military resistance to the Soviets, which was mobilized in large part by proclaiming a "holy war." The Soviet Union eventually pulled out in 1989, leaving a country ravaged and full of rival factions fighting for control. When the Taliban government took power in 1996, it justified many of its new policies on Islamic grounds. Westerners tend to see such policies as simply fanatical and irrational and to overlook the history that led up to them.

In October 2001, the United States and its European allies launched a bombing attack against what was called the epicenter of international terrorism. In October 2011, the campaign marked the tenth-year anniversary.

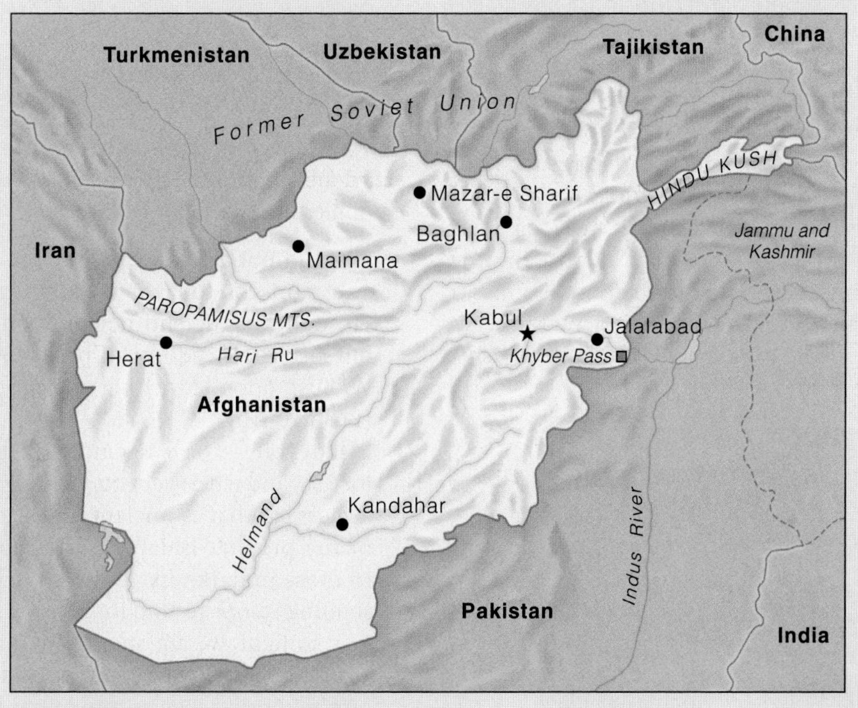

FIGURE 14.2

Source: Map from U.S. Central Intelligence Agency 2011

Osama bin Laden, who had hoped to raise a force composed of veterans of the Afghan War to fight Iraq, was infuriated with the Saudi royal family for calling upon the United States for help. He appealed to Muslim clerics to issue a *fatwa* (ruling or decree) condemning the stationing of non-Muslim troops in Saudi Arabia. His request was denied, and eventually the Saudi royal family, tired of bin Laden's incessant criticism, revoked his citizenship.

According to bin Laden (2001a), the "U.S. knows that I have attacked it, by the grace of God, for more than ten years now. . . . Hostility toward America is a religious duty and we hope to be rewarded for it by God. I am confident that Muslims will be able to end the legend of the so-called superpower that is America." Osama bin Laden claimed credit for the 1993 World Trade Center bombings and attacks on U.S. soldiers in Somalia, the 1998 attacks on U.S. embassies in East Africa, and the 1998 attack on the USS *Cole.* Although not formally claiming credit for the September 11, 2001, attacks on the United States, bin Laden (2001b) condoned them with the following religiously charged words: "Here is America struck by God Almighty in one of its vital organs, so that its greatest buildings are destroyed. Grace and gratitude to God. America has been filled with horror from north to south and east to west, and thanks be to God."

Civil Religion and the War on Terror

On September 20, 2001, President George W. Bush indicated that a global war on terror would begin with air strikes against al-Qaida and Taliban strongholds in Afghanistan. The enemy was larger than Afghanistan, however; it was a "radical network of terrorists" and the governments (as many as 60) that supported them. Bush (2001a) indicated that the war would not end "until every terrorist group has been found, stopped, and defeated."

On March 22, 2003, Bush announced the beginning of Operation Iraqi Freedom. He described the mission as clear: "to disarm Iraq of weapons of mass destruction, to end Saddam Hussein's support of terrorism (it has aided, trained, and harbored terrorists, including operatives of al-Qaida), and to free the Iraqi people." We know now that Iraq had no weapons of mass destruction and no substantiated links to al-Qaida. Our purpose here is *not* to address the question of whether the wars in Iraq and Afghanistan have been just or whether the war effort has succeeded. Rather, we continue to focus on the language that presidents use to justify war and to articulate a national identity in time of war. Usually during such times, the nation assumes a sacred quality and the president projects a moral certitude that some critics liken to "a kind of fundamentalism" and a "dangerous messianic brand of religion, one where self-doubt is minimal" (Hedges 2002). The following are examples of such statements:

TSGT CEDRIC H. RUDISILL, USAF

Five members of al-Qaida hijacked American Airlines Flight 77 and then crashed the aircraft into the Pentagon, killing 64 passengers onboard and 125 people on the ground. The Pentagon was one of three targets on September 11, 2001, the other being the twin towers of the World Trade Center and presumably the White House (which was spared when passengers brought their aircraft down in a Pennsylvania field).

Department of Defense

Consider the legacy of the U.S.-supported training camps in Afghanistan during the cold war: Key leaders behind every major terrorist attack on U.S. interests since then were veterans of the Afghan War.

- "We did not ask for this mission, but we will fulfill it. The name of today's military operation is Enduring Freedom. We defend not only our precious freedoms, but also the freedoms of people everywhere." (George W. Bush upon launching attack on Afghanistan, 2001) (Bush 2001b)
- "All of you . . . have taken up the highest calling of history . . . and wherever you go, carry a message of hope—a message that is ancient and ever new. In the words of the prophet Isaiah, to the captive 'come out,' and to those in darkness, 'be free.'" (George W. Bush upon sending troops to Iraq in 2003)
- ". . . tonight, we are once again reminded that America can do whatever we set our mind to. That is the story of our history . . . Let us remember that we can do these things not just because of wealth or power, but because of who we are: one nation, under God, indivisible, with liberty and justice for all." (Barack Obama on the death of Osama Bin Laden)

The Functionalist Perspective

> **CORE CONCEPT 4 The functionalist perspective maintains that religion serves vital social functions for individuals and for the group.**

Some form of religion appears to have existed for as long as humans have lived (at least 2 million years). In view of this fact, functionalists maintain that religion must serve some vital social functions for individuals and for the group. On the individual level, people embrace religion in the face of uncertainty; they draw on religious doctrine and ritual to comprehend the meaning of life and death and to cope with misfortunes and injustices (such as war, drought, and illness).

Life would be intolerable without reasons for existing or without a higher purpose to justify the trials of existence (Durkheim 1951). Try to imagine, for example, how people might cope with the immense devastation and destruction resulting from decades of war. When Soviet troops invaded Afghanistan in 1979, they attacked civilian populations, burned village crops, killed livestock, used lethal and nonlethal chemical weapons, planted an estimated 10 million mines, and engaged in large-scale high-altitude carpet bombing. "In the countryside it was standard Soviet practice to bombard or even level whole villages suspected of harboring resistance fighters. Sometimes women, children, and old men were rounded up and shot" (Kurian 1992, p. 5).

Even after the Soviets withdrew from Afghanistan in 1989, the civil war continued, as various political parties competed to fill the power vacuum. Table 14.1 summarizes the tragic results of more than 20 years of war in this country. In light of this situation, is it any wonder that Afghan people might turn to religion to cope with the devastation and restore a sense of order out of chaos?

Besides turning to religion in the face of intolerable circumstances, people rely on religious beliefs and rituals to help them achieve a successful outcome (such as the birth of a healthy child or a job promotion) and to gain answers to questions of meaning: How did we get here? Why are we here? What happens to us when we die? According to Durkheim, people who have communicated with their God or with other supernatural forces (however conceived) report that they gain the inner strength and the physical strength to endure and to conquer the trials of existence:

> It is as though [they] were raised above the miseries of the world. . . . Whoever has really practiced a religion knows very well . . . these impressions of joy, of interior peace, of serenity, of enthusiasm, which are, for the believer, an experimental proof of his beliefs. (Durkheim 1915, pp. 416–417)

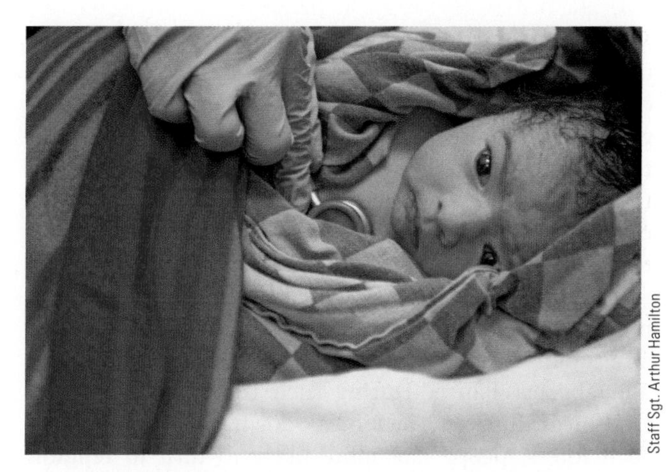

Staff Sgt. Arthur Hamilton

This baby in Afghanistan, born at a hospital staffed with U.S. military doctors, has a 16 percent chance of dying before reaching age 5.

TABLE 14.1 Profile of the Islamic Republic of Afghanistan

Before the United States began its military attacks on Afghanistan in 2001, the country was already devastated. Barnett Rubin (1996) argues that no one paid more for the U.S. cold war victory than did Afghanistan and its people: "Millions of unknown people sacrificed their homes, their land, their cattle, their health, their families, with barely hope of success or reward, at least in this world" (p. 21). Following the Soviet withdrawal, the country experienced a decade of civil war, Taliban rule, and severe drought (1999 to 2001), and then, in 2001 to present, the U.S.-led Afghanistan war. One has to go back to the early 1800s, to find a time when Afghanistan could be considered a unified country (Halloran 2004).

Population	29.8 million
People dependent on food aid	6.0 million
Deaths as a result of war	500,000 military, 1.5 million civilian (since 1979)
Life expectancy at birth	Males: 44.7 years; females: 45.3 years
Malnutrition	49.3% of infants to five-years olds are underweight for their age
Access to drinking water	83% of households have no safe drinking water
Literacy rate	Males: 43.1%; females: 12.6%
Maternal mortality	1,600 per 100,000 pregnancies
Infant mortality	161 babies per 1,000 live births before age 5
Birth rates	39.8 births per 1,000 people

Sources: Data from U.S. Central Intelligence Agency 2011a, United Nations 2009, World Bank 2011

Religion functions in several ways to promote group unity and solidarity. First, shared doctrines and rituals create emotional bonds among believers. Second, religions strive to raise individuals above themselves—to help them achieve a life better than they would lead if left to their own impulses. When believers violate moral codes of conduct, they feel guilt and remorse. Such feelings, in turn, motivate them to make amends. Third, religious rituals reinforce and renew social relationships—thereby binding individuals to a group. Finally, religion functions as a stabilizing force in times of severe social disturbance and abrupt change. During such times, many regulative forces in society may break down. In the absence of such forces, people are more likely to turn to religion in search of a force that will bind them to a group. This tie helps people think less about themselves and more about some common goal whether that goal is to work for peace or to participate more fervently in armed conflict (Durkheim 1951).

That religion functions to meet individual and societal needs, and that people create sacred objects and rituals, led Durkheim to reach a controversial but thought-provoking conclusion: The "something out there" that people worship is actually society.

Society as the Object of Worship

> **CORE CONCEPT 5 The variety of religious responses is endless, because people play a fundamental role in determining what is sacred and how they should act in its presence.**

If we operate under the assumptions that all religions are true in their own fashion and that the variety of religious responses is virtually endless, we find support for Durkheim's conclusion that people create everything encompassed by religion—gods, rites, sacred objects. That is, people play a fundamental role in determining what is sacred and how to act in the presence of the sacred. Consequently, at some level, people worship what they (or their ancestors) have created. This point led Durkheim to conclude that the real object of worship is society itself—a conclusion that many critics cannot accept (Nottingham 1971).

Let us give Durkheim the benefit of the doubt, however, and ask the following: Is there anything about the nature of society that makes it deserving of such worship? In reply to this question, Durkheim maintained that society transcends the individual life, because it frees people from the bondage of nature (as in "nature versus nurture"). How does it accomplish this task? We know from cases of extreme isolation, neglect, and limited social contact that "it is impossible for a person to develop without social interaction" (Mead 1940, p. 135). In addition, studies involving psychologically and socially sound people who experience profound isolation—astronauts orbiting alone in space and prisoners of war placed in solitary confinement—show that when people are deprived of contact with others, they lose a sense of reality and personal identity (Zangwill 1987). The fact that we depend so strongly on society for our humanity supports Durkheim's view of society as "a reality from which everything that matters to us flows" (Durkheim, cited in Pickering 1984, p. 252). Durkheim argued that people create religion that reflects their strongest convictions—those convictions may have to do with the place of women in society, who we can love, and how we treat animals. In fact, Durkheim observed that whenever any group of people has strong conviction, that conviction almost always takes on a religious character. Religious gatherings and affiliations become ways of affirming convictions and mobilizing the group to uphold them, especially when the group is threatened.

A Critique of the Functionalist Perspective of Religion

To claim that religion functions as a strictly integrative force is to ignore the long history of wars between different religious groups and the many internal struggles between factions within the same religious group. For example, although the Afghan mujahideen united to oppose the Soviet occupation and its secular government, many competing factions existed within the mujahideen. After the Soviets withdrew from Afghanistan, the former mujahideen commanders became the major power brokers, and each took control of different cities outside Kabul. At this point, as had happened in the past, the same rugged terrain that made it impossible for the Soviets to gain control over the entire country likewise made it difficult for any one internal

Chris Caldeira

Durkheim believed that the variety of religious responses is endless. This is because people create the churches in which they worship. What clues does this sign offer about the people who are members of this church?

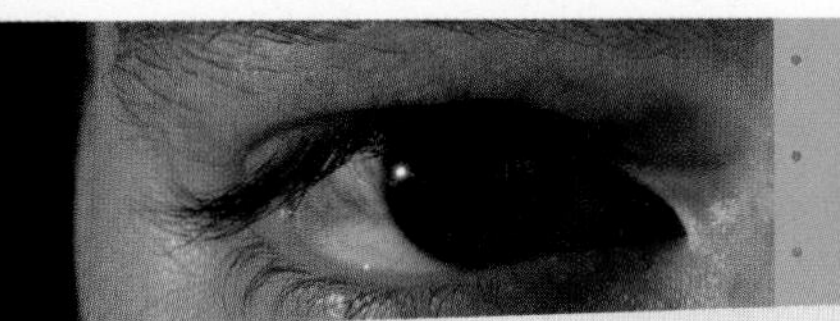

The Sociological Imagination: Personal Images of Jesus

The religious painter and illustrator Warner Sallman (pictured on the right) created the image of Jesus Christ (in the background) in 1940. The number of times this image has been reproduced on "church bulletins, calendars, posters, book marks, prayer cards, tracts, buttons, stickers and stationery" is more than 500 million (Grimes 1994). As a result, many people in the United States have come to think of Jesus's physical appearance as such. Given that Jesus was born in Bethlehem (according to the Christian Bible), a town in the Middle East, is the Sallman image the most accurate representation of how Jesus might have looked? Archaeological evidence suggests that the average man at the time of Jesus was 5 feet, 3 inches tall and weighed approximately 110 pounds (Gibson 2004). Student comments suggest that many accept Sallman's image but others have come to question it:

AP Photo

- Whenever I think about what Jesus looks like, I always see Sallman's image. However, I believe that Jesus cannot look like that because of the geographic region in which he was born. I like to think that maybe God is female.
- I really can't believe that my image of Jesus is a man-made one. Warner Sallman's image has been in my head so long I cannot even comprehend another image.
- It is shocking to me to learn that Jesus probably had dark skin, hair, and facial features, because I have always imagined Jesus to look like Sallman's paintings.
- My image of Jesus used to be that of a Caucasian male, but I remember reading in the Bible that he had "hair like sheep's wool and dark skin."
- The Bible does not give an exact physical description of Jesus, but it does say that he was "unattractive to the eye." People think that he was beautifully pale with long wavy brown hair, when really he was unattractive. Now, no one really knows what he looks like. That's why I belong to a church that's Christian nondenominational and that doesn't display or worship pictures of Jesus.

group to consolidate its power. Tribal elders and religious students, in turn, tried to wrestle control from rebel commanders (U.S. Central Intelligence Agency 2001).

Eventually, the Taliban, with the help of the Pakistani government, rose to power and came to control 90 percent of the country. Its ultimate aim was to establish a pure Islamic state. At first, the Taliban seemed a welcome relief to the chaos of decades of war. Later, its strict interpretation of Islamic laws, which were enforced by amputations and public executions, created widespread resentment among the Afghan people.

This point is that religion is not entirely an integrative force. If it were, then it could not be used as the justification for destroying people who did not follow a particular version of a religion. An Amnesty International (1996c) document, *Afghanistan: Grave Abuses in the Name of Religion*, outlines numerous human rights violations committed by the Taliban in the name of religion. Those abuses included "indiscriminate killings, arbitrary and unacknowledged detention of civilians, physical restrictions on women for reasons of their gender, the beating and ill-treatment of women, children and detainees, deliberate and arbitrary killings, amputations, stoning and executions."

The Conflict Perspective

CORE CONCEPT 6 Conflict theorists focus on ways in which people use religion to repress, constrain, and exploit others.

Scholars who view religion from the conflict perspective focus on how religion turns people's attention away from social and economic inequality. This perspective stems from the work of Karl Marx, who believed that religion was the most humane feature of an inhumane world and that it arose from the tragedies and injustices of human experience. He described religion as the "sigh of the oppressed creature, the heart of a heartless world, and the soul of soulless conditions. It is the opium of the people" (Marx 1976). According to Marx, people need the comfort of religion to make the world bearable and to justify their existence. In this sense, he said, religion is analogous to a sedative.

Even though Marx acknowledged the comforting role of religion, he focused on its repressive, constraining, and exploitative qualities. In particular, he conceptualized religion as an ideology that justifies existing inequities or

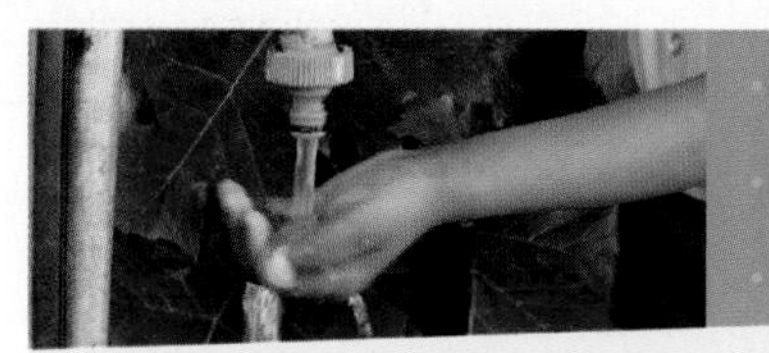

Working for Change Faith-Based Organizations in the United States

Basic statistics suggest that faith-based organizations operate as significant agents of change in American life:

- Faith ministers to the less fortunate. Eighty percent of the more than 300,000 religious congregations in America provide services to those in need.
- Faith shapes lives. Over 90 percent of urban congregations provide social services, ranging from preschool to literacy programs to health clinics.
- Faith shepherds communities. Polls estimate that between 60 and 90 percent of America's congregations provide at least one social service, and about 75 percent of local congregations provide volunteers for social service programs.
- Faith nurtures children. One out of every six child care centers in America is housed in a religious facility. The nation's largest providers of child care services are the Roman Catholic Church and the Southern Baptist Convention.

And perhaps most significantly, faith inspires the faithful to love their neighbors as they'd love themselves. Nearly one-quarter of all Americans volunteer their time and effort through faith-based organizations. Many of America's best ideas—and best results—for helping those in need have come not from the federal government but from grassroots communities, private and faith-based organizations of people who know and care about their neighbors. For years, America's churches and charities have led the way in helping the poor achieve dignity instead of despair, self-sufficiency instead of shame.

Sources: Excerpted from "President Bush's Faith-Based and Community Initiative," www.whitehouse.gov/fbci (August 1, 2004); prepared remarks of Attorney General John Ashcroft, White House Faith-Based and Community Initiative Conference (June 13, 2004).

downplays their importance. In particular, religion is a source of false consciousness in that religious teachings encourage the oppressed to accept the economic, political, and social arrangements that constrain their chances in this life because they are promised compensation for their suffering in the next world. This promise serves to rationalize the political and economic interests of the advantaged social classes.

SSGT JEREMY T. LOCK, USAF

Ironically, in its quest to recruit and support groups that would fight against the Soviet occupation of Afghanistan, the United States supported producers of opium poppies. Thus, as mujahideen insurgents pushed the Soviets out, they ordered peasants to plant opium. U.S. support helped to turn Afghanistan into the world's largest producer of opium and processed heroin. Today, the Afghan economy depends on opium, and profits from its production fund Taliban operations against the United States (Judah 2001, U.S. Central Intelligence Agency 2011).

This kind of ideology led Marx to conclude that religious teaching inhibits protest and revolutionary change. He went so far as to claim that religion would be unnecessary in a truly classless or propertyless society. In the absence of material inequality and exploitation, there are no injustices that cause people to turn to religion. Consider one extreme example of religion-inspired injustice. After the Taliban took control of Afghanistan, they placed, in the name of Islam, severe restrictions on women and the population in general. Women were required to appear covered from head to toe; they had to stay home unless accompanied by a close male relative and could not work outside the home or go to school. Nonreligious schools were closed; music, television, and other forms of entertainment were banned; homosexuals were killed. The Taliban's funding came from two major sources: Osama bin Laden and revenue from the production of opium for heroin (Judah 2001).

A Critique of the Conflict Perspective of Religion

The major criticism leveled at Marx and the conflict perspective of religion is that, contrary to that perspective, religion is not always a sign or tool of oppression. Sometimes religion has been used as a vehicle for protesting or working to change social and economic inequities (see Working for Change: "Faith-Based Organizations in the United States").

Liberation theology represents one such approach to religion. Liberation theologians maintain that Christians have a responsibility to demand social justice for the

PH2 ANTHONY J. PUGLIANI

This Buddhist monk sits outside Kadena Air Base in Afghanistan in protest of U.S. military intervention. He is playing a drum to console the spirits of those who have died in that war.

marginalized peoples of the world, especially landless peasants and the urban poor, and to take an active role at the grassroots level to bring about political and economic justice. Ironically, this interpretation of Christian faith and practice is partly inspired by Marxist thought, in that it advocates raising the consciousness of the poor and teaching them to work together to obtain land and employment and to preserve their cultural identity.

Sociologist J. Milton Yinger (1971) identifies at two interrelated conditions under which religion can become a vehicle of protest or change. In the first condition, a government or other organization fails to deliver on its ideals (such as equal opportunity, justice for all, or the right to bear arms). In the second condition, a society becomes divided along class, ethnic, or sectarian lines. In such cases, disenfranchised groups may form sects or cults and "use seemingly eccentric features of the new religion to symbolize their sense of separation" and to rally their followers to fight against the establishment, or the dominant group (p. 111). In the United States, one religion that emerged in reaction to society's failure to ensure equal opportunity to disenfranchised blacks was the Nation of Islam.

In the 1930s, black nationalist leader Wallace Fard Muhammad founded the Nation of Islam and began preaching in the Temple of Islam in Detroit. (When Fard disappeared in 1934, he was replaced by his chosen successor, Elijah Muhammed.) According to Fard, the white man was the personification of evil, and black people, whose religion had been stripped from them upon enslavement, were Muslim. In addition, he taught that the way out did not entail gaining the "devil's" or white man's approval but by exercising self-discipline and gaining an education. Members received an X to replace their "slave name" (hence, Malcolm X). In the social context of the 1930s, this message was very attractive:

> You're talking about Negroes. You're talking about "niggers," who are the rejected and the despised, meeting in some little, filthy, dingy little [room] upstairs over some beer hall or something, some joint that nobody cares about. Nobody cares about these people. . . . You can pass them on the street and in 1930, if they don't get off the sidewalk, you could have them arrested. That's the level of what was going on. (National Public Radio 1984a)

The Nation of Islam is merely one example of a religious organization working to improve life for African Americans. Historically, African American churches have reached out to millions of black people who have felt excluded from the U.S. political and economic system (Lincoln and Mamiya 1990). For example, African American churches did much to achieve the overall successes of the civil rights movement. Indeed, some observers argue that the movement would have been impossible if the churches had not become involved (Lincoln and Mamiya 1990).

NYWT&S Photograph Collectgion, Library of Congress Prints and Photographs Division

Elijah Muhammad, the man who succeeded the founder of the Nation of Islam, is shown here addressing his followers in 1964, at the height of the civil rights movement. One of his best-known followers is in attendance: Cassius Clay, who later changed his name to Muhammad Ali.

liberation theology A religious movement based on the idea that organized religions have a responsibility to demand social justice for the marginalized peoples of the world, especially landless peasants and the urban poor, and to take an active role at the grassroots level to bring about political and economic justice.

The Interplay between Economics and Religion

Max Weber wanted to understand the role of religious beliefs in the origins and development of **modern capitalism**—an economic system that involves careful calculation of costs of production relative to profits, borrowing and lending money, accumulating all forms of capital, and drawing workers from an unrestricted global labor pool (Robertson 1987).

> **CORE CONCEPT 7 Modern capitalism emerged and flourished in Europe and the United States because Calvinism supplied an ideologically supportive spirit or ethic.**

In his book *The Protestant Ethic and the Spirit of Capitalism*, Weber (1958) asked why modern capitalism emerged and flourished in Europe rather than in China or India (the two dominant world civilizations at the end of the sixteenth century). He also asked why business leaders and capitalists in Europe and the United States were overwhelmingly Protestant.

To answer these questions, Weber studied the major world religions and some of the societies in which these religions were practiced. Based on his comparisons, Weber concluded that a branch of Protestant tradition—Calvinism—supplied a "spirit" or a work ethic that supported profit-oriented behavior and motivations. Unlike other religions that Weber studied, Calvinism emphasized **this-worldly asceticism**—a belief that people are instruments of divine will and that God determines and directs their activities. Consequently, Calvinists glorified God when they worked hard and did not indulge in the fruits of their labor (that is, when they did not use money to eat, drink, or otherwise relax to excess). In contrast, Buddhism, a religion that Weber defined as the Eastern parallel and opposite of Calvinism, "emphasized the basically illusory character of worldly life and regarded release from the contingencies of the everyday world as the highest religious aspiration" (Robertson 1987, p. 7). Calvinists, who believed God to be all powerful and all knowing, emphasized **predestination**—the belief that God has foreordained all things, including the salvation or damnation of individual souls. According to this doctrine, people could do nothing to change their fate and only a small portion of people was destined to attain salvation.

Weber maintained that this-worldly asceticism and predestination created a crisis prompting Calvinists to search for some concrete sign that they were among God's chosen people, destined for salvation. Accumulated wealth became that concrete sign. At the same time, this-worldly asceticism "acted powerfully against the spontaneous enjoyment of possessions; it restricted consumption, especially of luxuries" (Weber 1958, p. 171). Frugal behavior encouraged people to accumulate wealth and make investments—important actions for the success of capitalism.

For Weber, the Protestant ethic was a significant ideological force; it was not the sole cause of capitalism but one force underlying the rise of "*certain aspects* of capitalism" (Aron 1969, p. 204). Unfortunately, many who read Weber's ideas overestimate the importance that he assigned to the Protestant ethic for achieving economic success, drawing a conclusion that Weber himself never reached: The reason that some groups and societies are disadvantaged is simply that they lack this ethic.

In assessing Weber's ideas about the origins of industrial capitalism, keep in mind Weber was not writing about the form of capitalism that exists today, a form that places high value on consumption and self-indulgence. Weber maintained that, once established, capitalism would generate its own norms and become a self-sustaining force. In fact, Weber argued, capitalism came to support a production system "without inner meaning or value and in which men operate almost as mindless cogs" (Turner 1974, p. 155). At that point, religion becomes an increasingly insignificant factor to maintaining the capitalist system.

THE EMPIRE BUILDERS
"Those Christian men to whom God in his infinite wisdom has given control of the property interests of the country."

Library of Congress Prints and Photographs Division

This print, whose setting is New York City's Trinity Church, shows men who were considered empire builders in U.S. history: James J. Hill, Andrew Carnegie, Cornelius Vanderbilt, John D. Rockefeller, J. Pierpont Morgan, Jay Cooke or Edward H. Harriman, and Jay Gould. Notice the caption below the image: "Those Christian men to whom God in his infinite wisdom had given control of the property interests of the country."

modern capitalism An economic system that involves careful calculation of costs of production relative to profits, borrowing and lending money, accumulating all forms of capital, and drawing labor from an unrestricted global labor pool.

this-worldly asceticism A belief that people are instruments of divine will and that God determines and directs their activities.

predestination The belief that God has foreordained all things, including the salvation or damnation of individual souls.

Secularization and Fundamentalism

CORE CONCEPT 8 Secularization and fundamentalism fuel each other's growth in that secularization invites a fundamentalist response.

Some sociologists argue that industrialization and scientific advances that accompany the rise of capitalism cause **secularization**—a process in which religious influences on thoughts and behavior become increasingly irrelevant. Thus, in the face of uncertainty, people are less likely to turn to religion or to a supernatural power to intervene; rather, they rely on scientific explanations and technological interventions. As one example, illness is not a product of God's will. Instead, science explains illnesses, and technology is employed to cure illness.

Secularization invites a fundamentalist response, a belief in the timelessness of sacred writings and a belief that such writings apply to all areas of one's life. Fundamentalists believe sacred principles have been abandoned, and they aim to revive them as the definitive and guiding blueprint for life.

Americans and Europeans tend to associate secularization with an increase in scientific understanding and in technological solutions to everyday problems of living. From a Muslim perspective, secularization is a Western-imposed phenomenon—specifically, a result of exposure to what many people in the Middle East consider the most negative of Western values. This point is illustrated by the following observation by a Muslim student attending college in Great Britain:

> If I did not watch out [while I was in college], I knew that I would be washed away in that culture. In one particular area, of course, was exposure to a society where free sexual relations prevailed. There you are not subject to any control, and you are faced with a very serious challenge, and you have to rely upon your own strength, spiritual strength to stabilize your character and hold fast to your beliefs. (National Public Radio 1984b)

The Complexity of Fundamentalism

In its popular usage, the term *fundamentalism* is applied to a wide array of religious groups around the world, including the Moral Majority in the United States, Orthodox Jews in Israel, and various Islamic groups in the Middle East. Religious groups labeled as fundamentalist are usually portrayed as "fossilized relics . . . living perpetually in a bygone age" (Caplan 1987, p. 5). Americans frequently employ this simplistic analysis to explain events in the Middle East, especially the causes of political turmoil that threatens the interests of the United States (including its demand for oil).

Fundamentalism is a more complex phenomenon than popular conceptions would lead us to believe. It is

SPC LESLIE ANGULO

U.S.-led troops distribute dolls and other toys to orphaned Afghan girls. What does it mean to receive such toys, which challenge Islamic beliefs about modest dress and other aspects of life?

impossible to define a fundamentalist in terms of age, ethnicity, social class, political ideology, or sexual orientation, because this kind of belief appeals to a wide range of people. Perhaps the most important characteristic of fundamentalists is their belief that a relationship with God, Allah, or some other supernatural force provides answers to personal and social problems. In addition, fundamentalists often wish to "bring the wider culture back to its religious roots" (Lechner 1989, p. 51).

Caplan (1987) identifies a number of other traits that seem to characterize fundamentalists. First, fundamentalists emphasize the authority, infallibility, and timeless truth of sacred writings as a "definitive blueprint" for life (p. 19). This characteristic does not mean that a definitive interpretation of sacred writings actually exists. Indeed, any sacred text has as many interpretations as there are groups that claim it as their blueprint. Even members of the same fundamentalist organization may disagree about the true meaning of the texts they follow.

Second, fundamentalists usually conceive of history as a "cosmic struggle between good and evil": a struggle between those dedicated to principles outlined in sacred scriptures and those who digress. To fundamentalists, truth is not a relative; it does not vary across time and place. Instead, truth is unchanging and knowable through the sacred texts.

Third, fundamentalists do not distinguish between the sacred and the profane in their day-to-day lives. Religious principles govern all areas of life, including family, business, and leisure. Religious behavior, in their view, does not just take place in a church, a mosque, or a temple.

Fourth, fundamentalist religious groups emerge for a reason, usually in reaction to a perceived threat or crisis,

secularization A process by which religious influences on thought and behavior are reduced.

whether real or imagined. Consequently, any discussion of a particular fundamentalist group must include some reference to an adversary.

Fifth, one obvious concern for fundamentalists is the need to reverse the trend toward gender equality, which they believe is symptomatic of a declining moral order. In fundamentalist religions, women's rights often become subordinated to ideals that the group considers more important to the well-being of the society, such as the traditional family or the "right to life." Such a priority of ideals is regarded as the correct order of things.

Islamic Fundamentalism

In *The Islamic Threat: Myth or Reality?*, professor of religious studies John L. Esposito (1992) maintains that most Americans' understanding of fundamentalism does not apply very well to contemporary Islam. The term *fundamentalism* has its roots in American Protestantism and the twentieth-century movement that emphasized the literal interpretation of the Bible.

Fundamentalists are portrayed as static, literalist, retrogressive, and extremist. Just as we cannot apply the term *fundamentalism* to all Protestants in the United States, we cannot apply it to the entire Muslim world, especially when we consider that Muslims make up the majority of the population in at least 45 countries. Esposito believes that a more appropriate term is **Islamic revitalism** or *Islamic activism*. The form of Islamic revitalism may vary from one country to another but it involves a disenchantment with, and even rejection of, the West; soul-searching; a quest for greater authenticity; and a conviction that Islam offers a viable alternative to nationalism, socialism, and capitalism (Esposito 1986).

Esposito (1986) believes that Islamic revitalism represents a "response to the failures and crises of authority and legitimacy that have plagued most modern Muslim states" (p. 53). Recall that after World War I, France and Great Britain carved up the Middle East into nation-states, drawing the boundaries to meet the economic and political needs of Western powers. Lebanon, for example, was created in part to establish a Christian tie to the West. For example, Israel was envisioned as a refuge for persecuted Jews when no country seemed to want them; the Kurds received no state; Iraq became virtually landlocked; and resource-rich territories were incorporated into states with very sparse populations (for example, Kuwait, Saudi Arabia, the United Arab Emirates). Their citizens viewed most of the leaders who took control of these foreign creations "as autocratic heads of corrupt, authoritarian regimes . . . propped up by Western governments and multinational corporations" (p. 54).

When Arab armies from six states lost "so quickly, completely, and publicly" in a war with Israel in 1967, Arabs were forced to question the political and moral structure of their societies (Hourani 1991, p. 442). Had the leaders and the people abandoned Islamic principles or deviated too far from them? Could a return to a stricter Islamic way of life restore confidence to the Middle East and give it an identity independent of the West? Questions of social justice also arose. Oil wealth and modernization policies had led to rapid increases in population and urbanization and opened up a vast chasm between the oil-rich countries, such as Kuwait and Saudi Arabia, and the poor, densely populated countries, such as Egypt, Pakistan, and Bangladesh. Western capitalism, which was seen as one of the primary forces behind these trends, seemed blind to social justice, instead promoting unbridled consumption and widespread poverty. Likewise, the Marxist socialism

SSGT CECILIO RICARDO, USAF

An Afghan child holds up a leaflet warning against picking up unexploded ordinance. The warning extends to unexploded ordinance that has accumulated over the last 25 years of war. In light of history, is it any wonder that Afghans have rejected nationalism, socialism, and capitalism in favor of Islam?

Islamic revitalism Responses to the belief that existing political, economic, and social systems have failed—responses that include a disenchantment with, and even a rejection of, the West; soul-searching; a quest for greater authenticity; and a conviction that Islam offers a viable alternative to secular nationalism, socialism, and capitalism.

No Borders No Boundaries Countries with Significant Percentages of Muslims

The prophet Muhammad's first recitations of the Qur'an occurred in Arabia around AD 610. Islam's spread has made it one of the world's major religions. The map shows countries where a significant percentage of people practice Islam. Here, *significant percentage* is defined as "at least 2 percent of the population (or 1 in every 50 persons)."

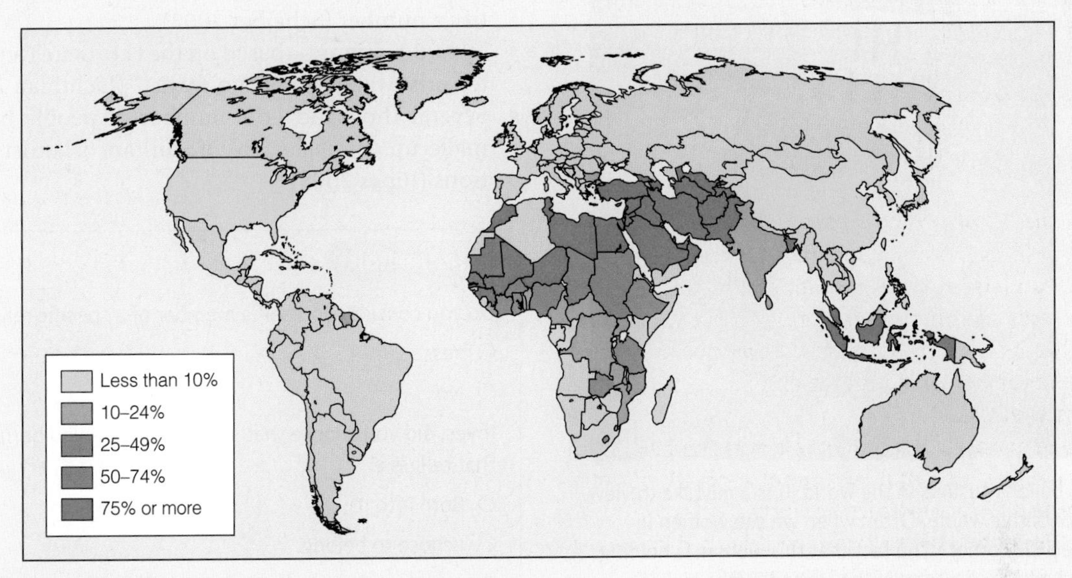

FIGURE 14.3

Source: Data from U.S. Central Intelligence Agency (2011)

(a godless alternative) of the Soviet Union had failed to produce social justice. It is no wonder that the Taliban and other Muslim groups in Afghanistan rejected nationalism, Western capitalism and Marxist socialism. After all, the disintegration of Afghanistan was a direct product of the cold war between the United States and the Soviet Union.

For many people, Islam offers an alternative vision for society. According to Esposito (1986), five beliefs guide Islamic activists (who follow many political persuasions, ranging from conservative to militant):

1. Islam is a comprehensive way of life relevant to politics, law, and society.
2. Muslim societies fail when they depart from Islamic ways and follow the secular and materialistic ways of the West.
3. An Islamic social and political revolution is necessary for renewal.
4. Islamic law must replace laws inspired or imposed by the West.
5. Science and technology must be used in ways that reflect Islamic values, to guard against the infiltration of Western values.

Muslim groups differ dramatically in their beliefs about how quickly and by what methods these principles should be implemented. Most Muslims, however, are willing to work within existing political arrangements; they condemn violence as a method of bringing about political and social change.

Jihad and Militant Islam

In thinking about the meaning of jihad, it is important to distinguish between religious and political jihad. Many Islamic scholars have pointed out that in the religious sense of the word, true *jihad* is the "constant struggle of Muslims to conquer their inner base instincts, to follow the path to God, and to do good in society" (Mitten 2002). But as Daniel Pipes (2003) points out in *Militant Islam Reaches America*, jihad as used by those who lead political organizations such as Egyptian Islamic Jihad, Islamic Jihad of Yemen, and International Islamic Front for Jihad against Jews and Christians means "armed struggle against non-Muslims" and against "Muslims who fail to live up to the requirements of their faith" (p. 264). Militant Islam is an "aggressive totalitarian ideology that ultimately discriminates barely, if at all, among those who stand in its path" (p. 249). In other words, non-Muslims as well as Muslims (who do not share the militants' outlook or who happen to be in the wrong place at the wrong time) can be targets of attack.

Courtesy of Bob Nabor

There are 1.6 billion Muslims in the world. It is a mistake to view them as a monolithic whole. Often when we see women in burkas, we see the "other." The fact that this woman is holding a sign saying hello to someone in the United States makes us wonder about her life and personality; the simple act of wondering releases this woman from the designation "other."

How many militant Islamist political *jihadists* exist in the world today in which there are 1.6 billion Muslims (Pew Research 2009)? Some estimates follow:

- 15,000—based on the number believed to have been trained in al-Qaida training camps
- 5,000 living in the United States—based on FBI figures created in response to pressure from Congress to identify a number (Scheiber 2003)
- 100,000 or more—based on the U.S. State Department's terrorist watch list or "no-fly list" (Lichtblau 2003)
- Several thousand—the number of people believed to make up the *inner core* of militant Islamist organizations (Pipes 2003)

Online Poll

Do you consider yourself a member of a specific religion?

○ Yes

○ No

If yes, did you choose that religion or were you born into that religion?

○ Born into it.

○ I chose to belong.

To see how other students responded to these questions, go to **www.cengagebrain.com.**

TECH SGT CECILIO M. RICARDO JR

Summary of CORE CONCEPTS

CORE CONCEPT 1 When sociologists study religion, they are guided by the scientific method and by the assumption that no religions are false.

Sociologists studying religion adhere to the scientific method, which requires them to study only observable and verifiable phenomena. When studying religions, sociologists must rid themselves of all preconceived notions of what religion should be.

CORE CONCEPT 2 Durkheim defined religion as a system of shared rituals and beliefs about the sacred that bind together a community of worshippers.

Religions can be classified into three categories depending on the type of phenomenon their followers consider sacred: sacramental, prophetic, or mystical. There are at least five broad types of religious organizations: ecclesiae, denominations, sects, established sects, and cults.

CORE CONCEPT 3 Civil religion is an institutionalized set of beliefs about a nation's past, present, and future and a corresponding set of rituals that take on a sacred quality and elicit feelings of patriotism.

Civil religion forges ties between religion and a nation's needs and political interests. Even in the face of internal divisions, national beliefs and rituals can inspire awe, respect, and reverence for country. In times of war, presidents draw upon a selected historical and mythological framework to justify military engagement, offer the public a hoped-for outcome, and instill patriotism. America's civil religion can be traced to a nineteenth-century political doctrine known as *manifest destiny*—the belief that the United States had a divine mission to serve as a democratic model to the rest of the world, to exert its good influence upon other nations, and to instill hope to the rest of the world.

CORE CONCEPT 4 The functionalist perspective maintains that religion serves vital social functions for individuals and for the group.

That some form of religion appears to have existed for as long as humans have lived encourages functionalists to maintain that religion must serve some vital social functions for individuals and the group. On the individual level, people embrace religion in the face of uncertainty, in intolerable circumstances, and to achieve a successful outcome. Religion can function in several ways to promote group unity and solidarity, including forging emotional bonds among believers; instilling a broader purpose that raises individuals above themselves; and working as a stabilizing force in times of severe social disturbance and abrupt change.

CORE CONCEPT 5 The variety of religious responses is endless, because people play a fundamental role in determining what is sacred and how they should act in its presence.

If we operate under the assumptions that all religions are true in their own fashion and that the variety of religious responses is virtually endless, we must realize the role people play in creating religion and in determining what is sacred and how to act in the presence of the sacred. Consequently, at some level, people worship what they (or their ancestors) have created.

CORE CONCEPT 6 Conflict theorists focus on ways in which people use religion to repress, constrain, and exploit others.

Conflict theorists focus on how religion turns people's attention away from injustices and on religion's repressive, constraining, and exploitative qualities. From this point of view, religion is used to rationalize existing inequities. Religion is not always a sign or tool of oppression; it has been used as a vehicle for protesting or working to change social and economic inequities. In particular, liberation theologians maintain that they have a responsibility to demand social justice for the marginalized peoples of the world, especially landless peasants and the urban poor, and to take an active role at the grassroots level to bring about political and economic justice.

CORE CONCEPT 7 Modern capitalism emerged and flourished in Europe and the United States because Calvinism supplied an ideologically supportive spirit or ethic.

Max Weber focused on understanding how norms generated by different religious traditions influenced adherents' economic orientations and motivations. Based on his comparisons, Weber maintained that this-worldly asceticism and predestination created a crisis prompting Calvinists to search for some concrete sign that they were destined for salvation. Accumulated wealth became that concrete sign. At the same time, this-worldly asceticism discouraged excessive consumption and encouraged people to accumulate wealth and make investments—important actions for the success of capitalism.

CORE CONCEPT 8 Secularization and fundamentalism fuel each other's growth in that secularization invites a fundamentalist response.

Secularization is a broad term used to describe the decline of religious influences over everyday life. In the face of uncertainty, people are less likely to turn to a supernatural power to intervene; rather, they rely on scientific explanation and technological interventions. Secularization invites a fundamentalist response. Fundamentalists believe sacred principles have been abandoned and they aim to revive them as the definitive and guiding blueprint for life. Fundamentalism is a more complex phenomenon than popular conceptions would lead us to believe. It is impossible to define a fundamentalist in terms of age, ethnicity, social class, political ideology, or sexual orientation, because this kind of belief appeals to a wide range of people.

Resources on the Internet

Login to CengageBrain.com to access the resources your instructor requires. For this book, you can access:

Access an integrated eBook, chapter-specific interactive learning tools, including flash cards, quizzes, videos, and more in your Sociology CourseMate.

Take a pretest for this chapter and receive a personalized study plan based on your results that will identify the topics you need to review and direct you to online resources to help you master those topics. Then take a posttest to help you determine the concepts you have mastered and what you will need to work on.

CourseReader

CourseReader for Sociology is an online reader providing access to readings, and audio and video selections to accompany your course materials.

Visit **www.cengagebrain.com** to access your account and purchase materials.

Key Terms

church 365
civil religion 368
cults 368
denomination 367
ecclesia 366
established sects 367
Islamic revitalism 380
liberation theology 377
modern capitalism 378
mystical religions 364
predestination 378
profane 365
prophetic religions 364
rituals 365
sacramental religions 364
sacred 363
sect 367
secularization 379
this-worldly asceticism 378

15 BIRTH, DEATH, AND MIGRATION

With Emphasis on EXTREME CASES

How many children do you think you would have if, in a given year, there is a one in 50 chance of dying in childbirth? These odds apply to Sierra Leone. You might be surprised to learn that women at the greatest risk of dying in child birth also have the most children; on average, as many as seven or eight. But what if the odds of dying in childbirth were virtually zero, as is the case in Ireland and Sweden? How might you feel about having children? Chances are you would have two or fewer children. Moreover, the chances that an Irish or Swedish woman between the ages of 40 and 44 is childless ranges from 15 and 30 percent (OECD 2010). The point is that knowing the chances of surviving childbirth offers broader insights about how people, especially women, see themselves and plan their futures.

Why Focus On EXTREME CASES?

Births, deaths, and migration are key experiences, not just for individuals but also for societies. Births represent the entry of new members into society; deaths represent their exit. Migration involves leaving one society for another. In this chapter, we focus on these three key experiences and pay special attention to countries where births, deaths, and migration occur at the highest and lowest rates. Specifically, we compare the countries that experience the highest and the lowest

- birth rates, including the teen birth rate;
- death rates including infant, maternal, and overall death rates; and
- migration rates, including the rates of migration into and out of the country.

We emphasize extreme cases because doing so allows us to frame the end points on the continuum of human experience. Generally, being at the extreme ends of an experience suggests vulnerability and sometimes special advantage. As one example, the teen birth rate in Niger is 199 of 1,000 teens and that rate is 1.2 in 1,000 teens in South Korea. On the surface, these birth rates might just seem like numbers with little human significance. But actually, knowing these rates allows us to think deeply about what personal lives are like and how a society is organized. For example, if we know the teen birth rate in Niger is 199 per 1,000 teens, we know that *each year* there are 199 births for every 1,000 teen females. To put it another way, each year, 20 percent or 1 in 5 teens have a baby. If we know that the teen birth rate in South Korea is 1 in 1,000, we know that each year there is 1 birth for every 1,000 teen females. What do these rates suggest about the lives of females who are teenagers in each country? If you were a teenage girl, how might you think about the future if you lived in Niger versus South Korea? The point is that knowing rates allows us to think more deeply about our own and others' lives.

Online Poll

How many children do you think you would have if you knew that 1 in 50 mothers die in childbirth each year?

- None
- One
- Two
- Three or more

To see how other students responded to these questions, go to **www.cengagebrain.com.**

The Study of Population

CORE CONCEPT 1 Demography, a subspecialty within sociology, focuses on births, deaths, and migration—major factors that determine population size and rate of growth.

Demography focuses on human populations and their characteristics, including size and rate of growth. Most organizations—private, public, and governmental—have an interest in knowing population characteristics, if only for planning purposes. For example, school officials need to know the size of the school-age population and whether, on the basis of births and in-migration, it is projected to decline or increase in coming years. These projections will affect decisions to expand or consolidate the number of schools. Health care planners need to know the size of the population age 65 and older and whether it is projected to decline or increase, as this age group has some of the greatest health care needs (see Working for Change).

The size and growth of a population depend on three key events—births, deaths, and migration. In the pages that follow, we will consider these population-related characteristics and how each is expressed as rates, giving special attention to extreme cases as endpoints on the continuum of human experiences (for example, the teen birth rate range from a low of 1.2 in 1,000 teens in South Korea to a high of 199 per 1,000 teens in Niger). Later in the chapter, we will learn the reason for very low and very high rates.

Births

Births add new people to a population. Each year, the world adds approximately 134 million people. For comparison, demographers often convert the number of births into a crude birth rate. The **crude birth rate** is the annual number of births per 1,000 people in a designated area. From a global perspective, the crude birth rate is 19.5 births for every 1,000 people in the world. The country with the highest crude birth rate is the African country of Niger, where each year there are approximately 51.4 births for every 1,000 people. The country with the lowest crude birth rate is Japan, with an annual birth rate of 7.3 births per 1,000 people. To calculate the birth rate, we divide the number of births in a year by the size of the population living in the geographic area of interest at the onset of that year and then multiply that figure by 1,000.

Sometimes demographers want to know age-specific rates for a specific age cohort within the population. Of particular interest is the teenage birth rate, the number of babies born each year to women who are in their teens. We have already learned that the country with the highest teen birth rate is Niger (199 babies for every 1,000 teens). So over the course of a year, 19.9 percent of teens give birth to a baby. We have also learned that South Korea has the lowest teen birth rate, which is 1.2 babies per every 1,000 teens.

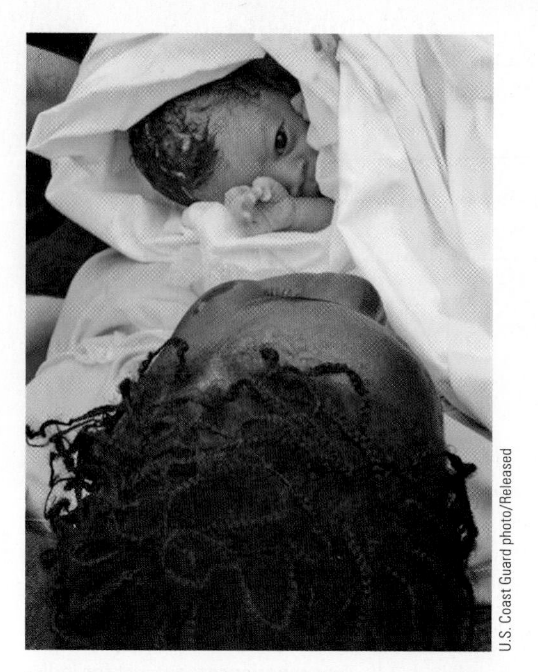

U.S. Coast Guard photo/Released

The average woman in the world bears 2.5 children over her lifetime. But the total fertility rate for a country ranges from 1.15 in China to 7.7 children in Niger.

In addition to the birth rate, demographers are interested in the **total fertility rate**, which states the average number of children that women bear over their lifetime. The average woman in the world bears 2.5 children over the course of her reproductive life. The country with the highest total fertility rate is Niger with 7.7 children; the country of China has the lowest rate (1.15 children).

Deaths

Deaths reduce the size of a population. Each year, the planet loses about 56.2 million people to death. This loss is often expressed as a rate. The **crude death rate** is the annual number of deaths per 1,000 people in a designated area. Like the crude birth rate, it is calculated by dividing the number of deaths in a year by a designated area's population size at the onset of that year and then multiplying that number by 1,000. The country with the highest death rate in the world is Angola with 23.4 deaths per 1,000 population, and the country with the lowest death rate is United Arab Emirates with a death rate of 2 deaths per every 1,000 people.

As with birth rates, we can calculate the death rates for specific segments of the population, such as for men, for women, or for specific age categories such as 1 year olds or younger. The death rate among those 1 year old or younger is called the **infant mortality rate**. Infant mortality is calculated by dividing the number of deaths among those 1 year old or younger by the total number of births in that year and then multiplying that result by 1,000. The

demography A subspecialty within sociology that focuses on the study of human populations and their characteristics, including size and rate of growth.

crude birth rate The annual number of births per 1,000 people in a designated geographic area.

total fertility rate The average number of children that women in a specific population bear over their lifetime.

crude death rate The annual number of deaths per 1,000 people in a designated geographic area.

infant mortality rate The annual number of deaths of infants 1 year old or younger for every 1,000 such infants born alive.

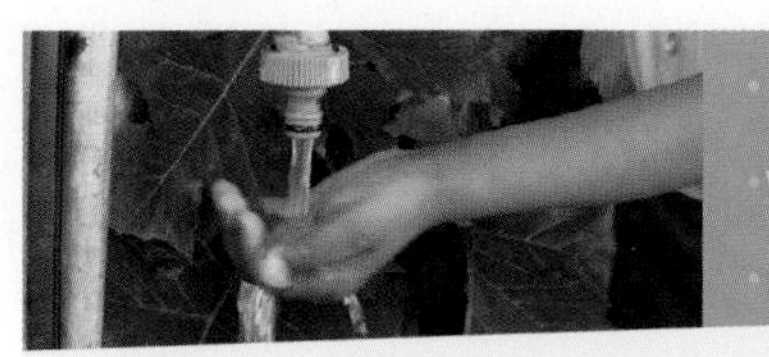

Working for Change U.S. Bureau of the Census

The U.S. Bureau of the Census serves as the leading source of quality data about the nation's people and economy. In collecting that data, the census bureau honors privacy, protects confidentiality, shares its expertise globally, and conducts its work openly. Among other things, the information gathered allows us to know how many people were born since the last census, moved from one location to another within the United States, and moved into the United States from a foreign country, as well as the age-sex composition of the population. The Bureau of the Census normally employs nearly 12,000 people, but it temporarily expands its workforce by about 800,000 when the census is taken every ten years. Some of its most important data products are:

- Population and Housing Census—every 10 years
- Economic Census—every 5 years
- Census of Governments—every 5 years
- American Community Survey—annually

The data collected has many uses, including to determine the distribution of congressional seats to states as mandated by the U.S. Constitution; to apportion seats in the U.S. House of Representatives; to define legislature districts, school district assignment areas, and other important functional areas of government to make decisions about services for the elderly; to define where to build new roads and schools; and where to locate job training centers. Census data affects how funding is allocated to communities' neighborhood improvements, public health, education, transportation, and much more.

Source: U.S. Bureau of the Census 2011.

infant mortality rate for the world is 41.6 deaths before age 1 for 1,000 babies born. The highest infant mortality in the world is Angola, with 175.9 deaths per 1,000 babies born; the lowest infant mortality is in Sweden where 2.7 babies die per 1,000 born before reaching age 1. The maternal mortality rate is also an important indicator of well-being. **Maternal mortality** is the death of a woman, while pregnant or within 42 days of a termination of pregnancy, from any cause related to or aggravated by pregnancy or the way it is managed (World Health Organization 2011). The country with the highest maternal mortality rate is Sierra Leone, with 199 deaths per 1,000 pregnancies. Sweden has the lowest rate, 5 deaths per 1,000 pregnancies.

Migration

Migration is the movement of people from one residence to another. Demographers use the term **in-migration** to denote the movement of people into a designated area and the term **out-migration** to denote the movement of people out of a designated area. That movement increases population size if the people are moving in, or reduces the population size if they are moving out. Sociologists calculate the **net migration**, the difference between the number moving into an area and the number moving out. This difference is typically converted into a rate by dividing that difference by the size of the relevant population, and then multiplying the result by 1,000. We can calculate the **migration rate** for towns, cities, counties, states, countries, or any other region of the world. The country with the highest net migration rate in the world is Zimbabwe; its rate is +24.8 per 1,000 residents, which means that 24.8 more people moved into the country than moved out for every 1,000 residents who lived there. The country with the lowest net migration is Jordan with a rate of −14.3, which means that 14.3 more people moved out of the country than moved in for every 1,000 residents.

Migration results from two factors. **Push factors** are the conditions that encourage people to move out of an area. Common push factors include religious or political persecution, discrimination, depletion of natural resources, lack of employment opportunities, and natural disasters (droughts, floods, earthquakes, and so on). A dramatic example of a push factor was the 2005 Hurricane Katrina, which pushed 60 percent of New Orleans's population out of the city, changing the city's size from 454,863 to

maternal mortality rate The death of a woman while pregnant or within 42 days of a termination of pregnancy from any cause related to or aggravated by pregnancy or the way it is managed (World Health Organization).

migration The movement of people from one residence to another.

in-migration The movement of people into a designated area.

out-migration The movement of people out of a designated area.

net migration The difference between the number moving into an area and the number moving out.

migration rate A rate based on the difference between the number of people entering and the number of people leaving a designated geographic area in a year. We divide that difference by the size of the relevant population and then multiply the result by 1,000.

push factors The conditions that encourage people to move out of a geographic area.

The Sociological Imagination Moving to the United States from Liberia

Photo by Lance Cpl. Dorian Gardner, USMC

More than 33,000 non-U.S. citizens are serving in the U.S. military, which means that each immigrated to the United States at some point in life. One among the 33,000 is a Liberian native named Nimley Tabue. Tabue's parents came from different tribes. He said his parents' tribal differences did not affect his family until a war between the tribes erupted in 1989. "My father refused to kill, so (rebels) tried to kill him," Tabue said.

Tabue remembers fleeing through the country for three days as a child. "We stopped by a river once to get some water," said Tabue, who was with his mother and siblings at the time. "I held my 4-month-old brother in my arms as he died." According to Tabue, his father, Aloysius Tabue, traveled to America searching for ways to improve his family's life, and he called home often. "I learned about the Marines from my father," Tabue said. "He would say, 'If you guys come over here, make sure you do something with your life. The Marines will give you something no other service can.'"

Because of the ongoing war around him, school became less of a priority, and Tabue was taken out of school following the second grade. He, along with his mother and sister, came to Chicago to live with his father. At 12 years old, Tabue jumped back into the school swing. But after four years without touching a book, school presented a new challenge. "I forgot how to do math, and my English was bad," Tabue said. "I had to go to school over the summer and take extra classes."

After years of extra classes, Tabue's name was added to the high school honor roll. Tabue had not planned on leaving Chicago, but he remembered what his father had always told him about the Corps. "He told me, 'This is where they separate the men from the boys,'" Tabue said. Adjusting to boot camp was harder than any English class. "The first day was horrible. I almost lost my temper when the drill instructor got in my face . . . . But I told myself it was just a mind game . . . . I had trouble speaking in third person (as required in boot camp). Instead of saying 'This recruit requests permission to use the head,' I would say, 'I would like to use the head.' Drill instructors didn't really like that." When the Crucible—the grueling 54-hour field exercise that is the culmination of boot camp—came, Tabue found his role in the platoon. "He stepped up," Nofziger said. "He wasn't a squad leader, but he acted as one." After Marine Corps recruit training, Tabue will become a mortarman in the Marine Corps Reserve. He said he'll be ready to fight.

From: "West African Immigrant Heeds Father's Words, Joins U.S. Marines" by Lance Cpl. Dorian Gardner, USMC Special to American Forces Press Service.

187,525 overnight (U.S. Department of Homeland Security 2006). If we consider the entire Gulf Coast population, the number of people pushed out of the area exceeds one million (Nossiter 2006).

Pull factors are the conditions that encourage people to move into an area. Common pull factors include employment opportunities, favorable climate, and tolerance for a particular lifestyle. Migration can be placed into two broad categories: international and internal (see The Sociological Imagination: "Moving to the United States from Liberia").

International Migration International migration involves the movement of people between countries. In reference to international migration, demographers use the term **emigration** to denote the act of *departing* from one country to take up residence elsewhere, and the term **immigration** to denote the act of entering one country after leaving another. Most governments restrict the numbers of people who can immigrate. Sometimes governments encourage the immigration of certain categories of people, such as nurses, to fill occupations characterized by a shortage of workers.

pull factors The conditions that encourage people to move into a geographic area.

emigration The act of departing from one country to take up residence elsewhere.

immigration The act of entering one country after leaving another.

Pull factors are those qualities that draw people into a geographic area to live. San Francisco has established a reputation of being a gay-friendly city. That friendliness is symbolized by the prevalence of colors associated with gay pride.

Internal Migration In contrast to international migration, **internal migration** involves movement of people within the boundaries of a single country—from one state, region, or city to another. One major type of internal migration is the rural-to-urban movement (urbanization) that accompanies industrialization.

The United States is a country characterized by high rates of internal migration. Consider that each year about 37.1 million Americans move (change residences). About 67 percent of that number moves from one residence to another within the same county. Approximately 17 percent move from one county to another within the same state. Another 12.6 percent (4.7 million people) move from one state to another (U.S. Bureau of the Census 2010).

Population Size and Growth

The population size of a geographic area constantly changes, depending on births, deaths, and migration flows. About 7 billion people live on planet Earth, and the world's population is distributed unevenly (see Table 15.1). Table 15.1 shows the ten most populous countries in the world. Two countries, China and India, top the list. Taken together, the two countries account for 36 percent of the world's population. The United States in the third most populous country in the world, with 313 million people.

Demographers calculate annual growth in population size according to the following formula: (number of births − number of deaths) + (in-migration − out-migration). Each year the planet increases its population size by approximately 77.8 million. That is, there is 77.8 million more births than deaths. Migration is not a factor in figuring *world* population growth because people cannot move off the planet unless we count the handful of people working in outer space who will eventually return to Earth.

To determine the rate of world population growth, simply divide the amount of change in population size over the course of a year by the population size at the beginning of the year. Using this formula, the annual growth rate for the planet is 1.1 percent. The country with the highest annual growth rate is Zimbabwe; its population size increased by 4.3 percent. The country with the lowest growth rate is Bulgaria; its population size declined by −.78 percent. Keep in mind that the growth rate is relative to the size of an existing population so often the country with the highest growth rate is not the country that adds the greatest number of people to its population over the course of a year. The country that adds the largest number of people per year is India—it adds about 15.9 million people per year. Given the size of India's population, that country's population growth rate is 1.3 percent.

Doubling time is the estimated number of years required for a country's population to double in size. India, with a population growth rate of 1.3 percent, will double its population of 1.2 billion in about 51 years. The United States, with a natural growth rate of 1.0 percent, will double its population in about 78 years. Figure 15.1 shows world population growth since A.D. 1. Note that the population has doubled five times in the last 2,000 years, and that the time between the doublings has decreased dramatically, even alarmingly.

Each year, India loses almost 9 million people through deaths for a net gain of 15 million people. India has 352 million children age 14 and under—that number is larger than the entire U.S. population. Each year, the country adds 24 million babies to its population (U.S. Bureau of the Census 2011).

internal migration The movement of people within the boundaries of a single country—from one state, region, or city to another.

doubling time The estimated number of years required for a country's population to double in size.

Global Comparisons The 10 Countries With Largest Populations

TABLE 15.1 The World's Most Populous Countries, 2011

Rank	Country or Area	Population	% of World's Population
1	China	1,336,718,015	19.1
2	India	1,189,172,906	16.9
3	United States	313,232,044	4.5
4	Indonesia	245,613,043	3.5
5	Brazil	203,429,773	2.9
6	Pakistan	187,342,721	2.7
7	Bangladesh	158,570,535	2.3
8	Nigeria	155,215,573	2.2
9	Russia	138,739,892	2.0
10	Japan	126,475,664	1.8

Source: Data from U.S. Bureau of the Census 2011.

TABLE 15.2 Population Size and Growth: The Role of Birth, Death and Migration.

This table shows the population size at two points in time; midyear 2010 and midyear 2011 for three countries: (1) India, the country that added the greatest number of people to its population between 2010 and 2011; (2) Zimbabwe, the country that increased its population size by the greatest percentage; and (3) Bulgaria, the country with the greatest percentage decrease in population size. How many births occurred in each country? How many deaths? How many people did each country gain or lose through migration?

		India	Zimbabwe	Bulgaria
Population (Midyear)	2010	1,173,190,000	11,563,000	7,038,000
Births		+24,937,000	+385,000	+66,000
Deaths		−8,895,000	−164,000	−102,000
Net Migration		−59,000	+300,000	−20,000
Population (Midyear)	2011	1,189,173,000	12,084,000	7,094,000
Growth Rate		1.3%	4.3%	−0.8%
Doubling Time		51 years	11.6 years	92.5 years country will disappear

Source: Data from U.S. Bureau of the Census 2011; U.S. Central Intelligence Agency 2011.

Age-Sex Composition

CORE CONCEPT 2 The age-sex composition of a population helps demographers predict birth, death, and migration rates.

A population's age and sex composition is commonly depicted as a **population pyramid**, a series of horizontal bar graphs, each representing a different five-year age cohort. A **cohort** is a group of people born around the same time—in this case, within a five-year time frame—who

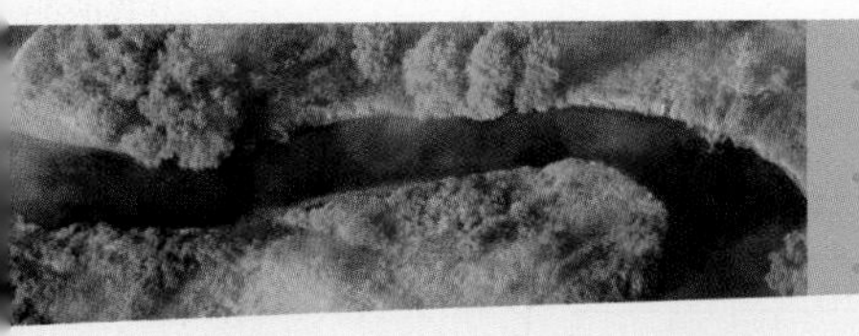

No Borders No Boundaries World Population Growth A.D. 1 to 2045

The graph shows that it took approximately 1,150 years for the world's population to double from 170 million in A.D. 1 to 340 million in 1150. Around 1930, the world's population reached 2 billion people, taking less than 100 years to double from 1 billion in 1850. By the 1960s, the world's population reached 3.04 billion, and it took just 30 years to double to 6.26 billion.

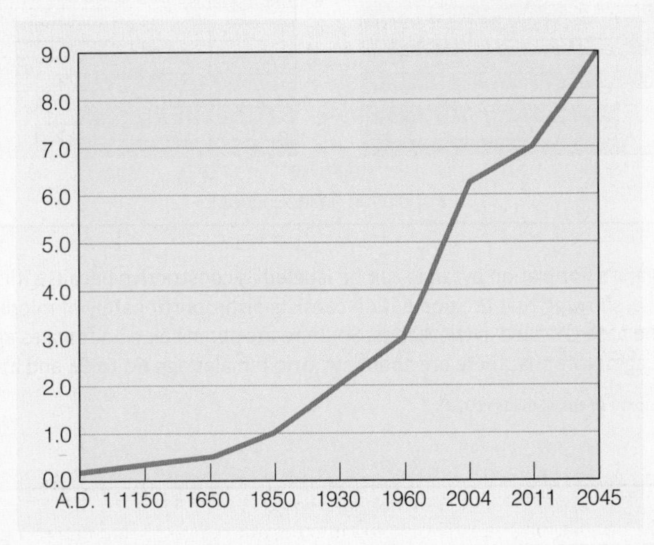

FIGURE 15.1

Source: Data from U.S. Bureau of the Census (2011)

share common experiences and perspectives by virtue of the time they were born. To create a population pyramid, we construct two bar graphs for each cohort—one for males and the other for females. We place the bars end to end, separating them by a line representing zero. Typically, the left side of the pyramid depicts the number or percentage of males that make up each cohort, and the right side depicts the number or percentage of females. We stack the bar graphs according to age—the age 0 to 4 cohort forming the base of the pyramid and the age 100+ cohort forming the apex. The population pyramid allows us to compare the sizes of the cohorts and to compare the numbers or percentages of males and females in each cohort.

The population pyramid offers a snapshot of the number of males and females in the various cohorts at a particular time. Generally, a country's population pyramid approximates one of three shapes: expansive, constrictive, or stationary. An **expansive pyramid** is triangular; it is broadest at the base, and each successive bar is smaller than the one below it. The relative sizes of the cohorts in expansive pyramids show that the population is increasing and consists disproportionately of young people. A **constrictive pyramid** is narrower at the base than in the middle. This shape shows that the population consists disproportionately of middle-aged and older people. A **stationary pyramid** is similar to a constrictive pyramid, except that all cohorts other than the oldest are roughly the same size (see Figure 15.2).

Knowing age-sex composition can help demographers predict a country's birth, death, and migration rates. Bulgaria's population pyramid shows that there are few people age 14 and under relative to the size of the age cohorts that could be their parents. This suggests that many women of reproductive ages 15 to 54 are not having children or a small number of children.

population pyramid A series of horizontal bar graphs, each representing a different five-year age cohort, that allows us to compare the sizes of the cohorts.

cohort A group of people born around the same time (such as a specified five-year period) who share common experiences and perspectives by virtue of the time they were born.

expansive pyramid A triangular population pyramid that is broadest at the base, with each successive cohort smaller than the one below it. This pyramid shows that the population consists disproportionately of young people.

constrictive pyramid A population pyramid that is narrower at the base than in the middle. It shows that the population consists disproportionately of middle-aged and older people.

stationary pyramid A population pyramid in which all cohorts (except the oldest) are roughly the same size.

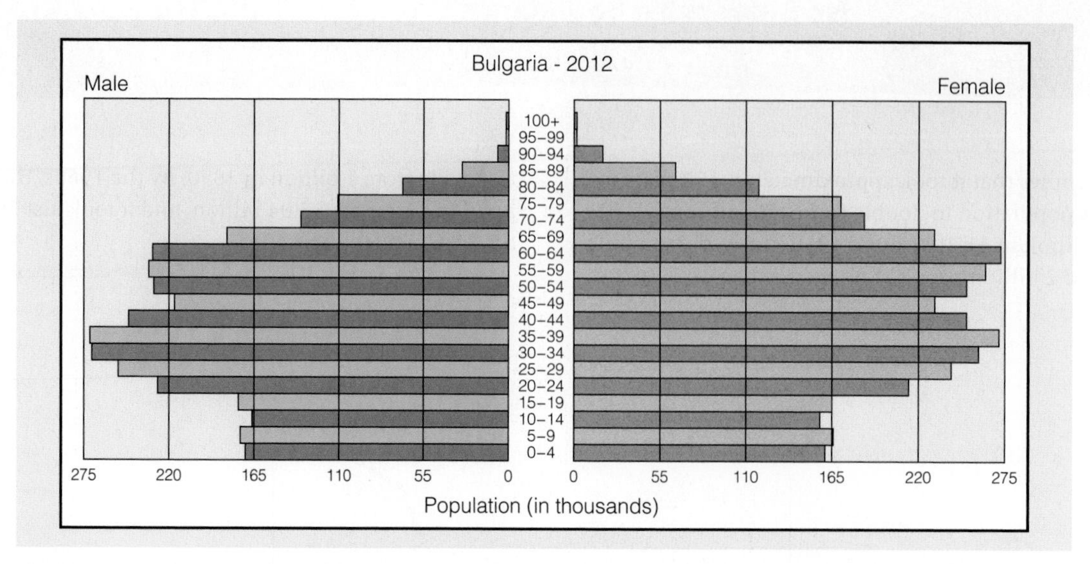

FIGURE 15.2a Bulgaria's population pyramid can be labeled as constrictive because it is narrower at the base than in the middle, showing that the population consists disproportionately of middle-aged and older people. Notice that the base is scaled in thousands. So there are about 165,000 females age 0 to 4 and about 171,000 males of that age. Note that there are about 275,000 females age 60 to 64 and about 231,000 males.

Source: Data from U.S. Bureau of the Census (2012)

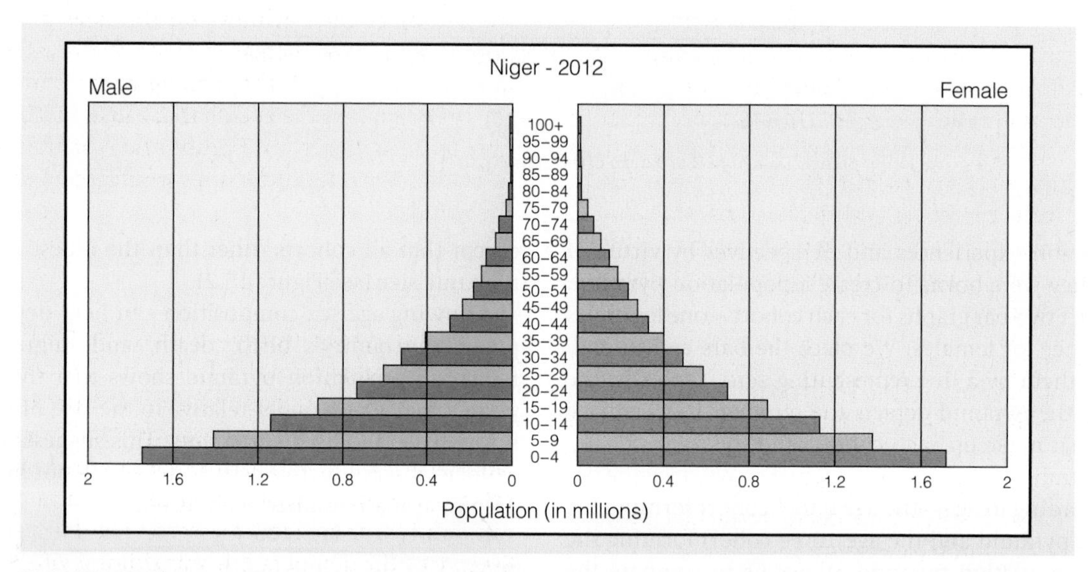

FIGURE 15.2b The population pyramid for Niger can be labeled as expansive because it is broadest at the base, and each successive bar is smaller than the one below it. The relative size of the bars indicates that Niger's population consists disproportionately of young people. Note that there are about 1.65 million females ages 0 to 4 and 1.7 million males of that age.

Source: Data from U.S. Bureau of the Census (2012)

When we know age-sex composition, we can calculate the **sex ratio**—the number of males for every 100 females (or another preferred constant, such as every 10, 100, or 10,000 males). The country of United Arab Emirates has the greatest imbalance in favor of males relative to females with 219 males for every 100 males. The country Russia has the greatest imbalance in favor of females: 86 males per 100 females (U.S. Central Intelligence Agency 2011).

The Theory of Demographic Transition

CORE CONCEPT 3 The demographic transition links the birth and death rates in western Europe and North America to the level of industrialization and economic development.

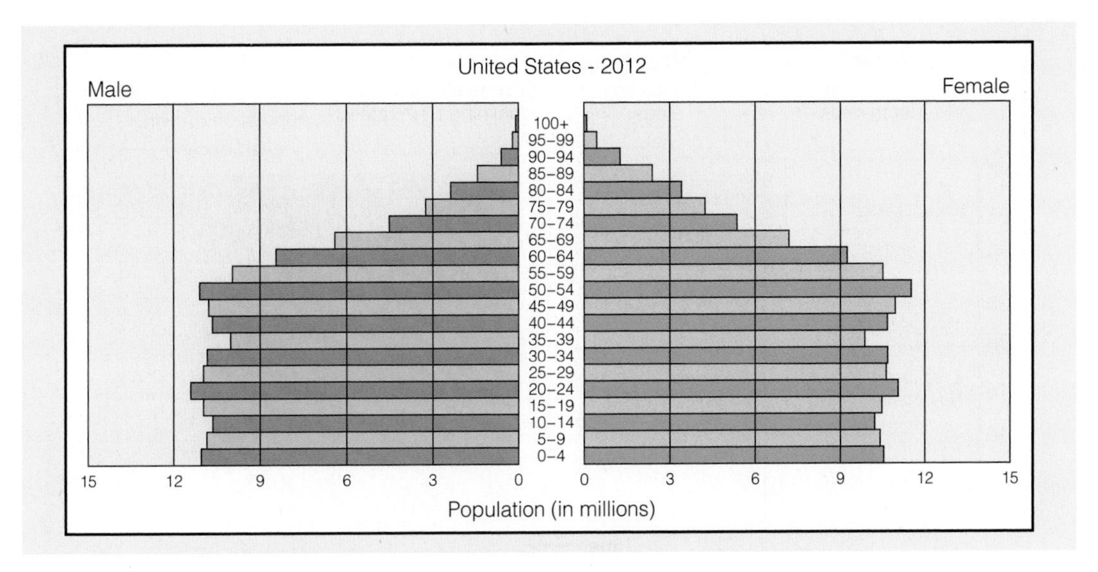

FIGURE 15.2c The population pyramid of the United States yields a near stationary pyramid, because, except for the older age categories, each cohort is roughly the same size. In the United States, there are about 10 million females age 0 to 4 and 10.5 million males of that age.

Source: Data from U.S. Census Bureau (2012)

In the 1920s and early 1930s, demographers observed birth and death rates in various countries. They soon noticed that both birth and death rates were high in Africa, Asia, and South America. In eastern and southern Europe, death rates were declining and birth rates remained high. In western Europe and North America, birth rates were declining and death rates were low. At that time, demographers observed that western Europe and North America had the following sequence of birth and death rates:

1. Birth and death rates remained high until the mid-eighteenth century, when death rates began to decline.
2. As the death rates decreased, the population grew rapidly, because more births than deaths occurred. The birth rates began to decline around 1800.
3. By 1920, both birth and death rates had dropped below 20 per 1,000 (see Figure 15.3).

Based on these observations, demographers put forth the theory of the demographic transition. They proclaimed that the characteristics of a country's birth and death rates are linked to its level of industrial or economic development, and they hypothesized that the less economically and industrially developed countries would follow the pattern of western Europe and North America.

Note that this four-stage model documents the general situation; it should not be construed as a detailed description of the experiences of any single country. Even so, we can say that for the most part the countries of the world have followed the essential pattern of the demographic transition, although they have differed in the timing of the declines and the rates at which their populations have increased since death rates began to fall. The theory of the demographic transition also sought to explain the events that caused birth and death rates to drop in western Europe and North America, and to predict when these declines would occur in the rest of the world.

Stage 1: High Birth and Death Rates

For most of human history—the first 2 to 5 million years—populations grew very slowly, if at all. The world population remained at less than 1 billion until around A.D. 1850, when it began to grow explosively. Demographers speculate that growth until that time was slow because **mortality crises**—violent fluctuations in the death rate, caused by war, famine, or epidemics—were a regular feature of life. Stage 1 of the demographic transition is often called the stage of high potential growth: If something happened to cause the death rate to decline—for example, improvements in agriculture, sanitation, or medical care—the population would increase dramatically. In this stage, life is short and brutal; the death rate almost always exceeds 50 per 1,000. When mortality crises occur, the death

sex ratio The number of females for every thousand males (or another preferred constant, such as 10, 100, or 10,000).

mortality crises Violent fluctuations in the death rate, caused by war, famine, or epidemics.

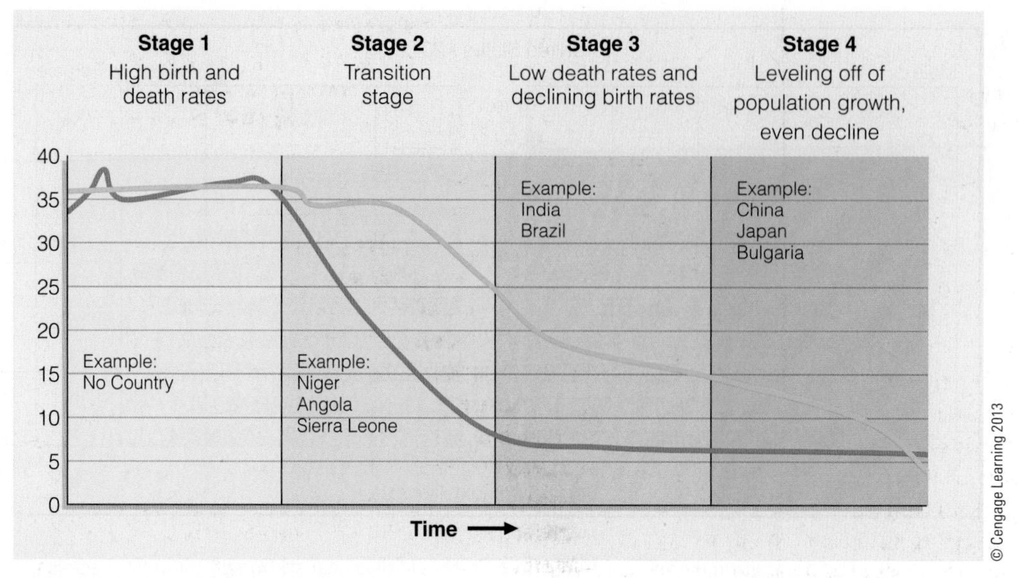

FIGURE 15.3 The Demographic Transition
The theory of the demographic transition is represented by a graph of historical changes in birth and death rates that reflect the path followed by western European countries and the United States.

rate seems to have no limit. Sometimes half the population is affected, as when the Black Death struck Europe, the Middle East, and Asia in the mid-fourteenth century (the plague recurred periodically for approximately 300 years). Within 20 years of its onset, the plague killed an estimated three-fourths of all people in the affected populations.

Another mortality crisis—but one that has not received as much attention as the Black Death—affected the indigenous populations of North America when Europeans arrived in the fifteenth century. A large proportion of the native population died because they had no resistance to diseases such as smallpox, measles, tuberculosis, and influenza, which the colonists brought with them. Historians continue to debate what proportion of the native population died because of this contact; estimates range from 50 to 90 percent.

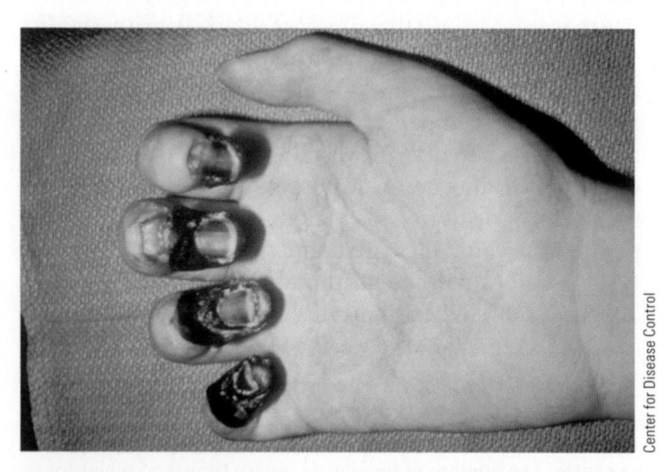

The plague known as the Black Death hit Europe, the Middle East, and Asia in the mid-fourteenth century, recurring periodically for approximately 300 years. The plague's name came from one of its symptoms: gangrene.

In stage 1, then, average life expectancy at birth remained short—perhaps 20 to 35 years—with the most vulnerable groups being women of reproductive age, infants, and children younger than age 5. It is believed that women gave birth to large numbers of children and that the crude birth rate was about 50 per 1,000—the highest rate recorded, and thus believed possible, for humans. Families remained small, however, because one of every three infants died before reaching age 1, and another died before reaching adulthood. If the birth rate had not remained high, the society would have become extinct. Demographer Abdel R. Omran (1971) estimates that in societies where life expectancy at birth is 30 years, each woman must have an average of seven live births to ensure that two children survive into adulthood. Theoretically, she must bear six sons to ensure that at least one son survives into adulthood. In western Europe before 1650, high mortality rates were associated closely with food shortages and famines. Even when people did not die directly from starvation, they died from diseases that preyed on their weakened physical state.

Thomas Malthus (1798), a British economist and an ordained Anglican minister, concluded that "the power of population is so superior to the power in the earth to produce subsistence for man, that premature death must in some shape or other visit the human race" (p. 140). According to Malthus, **positive checks** served to keep population size in line with the food supply. He defined *positive checks* as events that increase deaths, including epidemics of infectious and parasitic diseases, war, famine, and natural disasters. In

Department of Defense, U.S. Navy Photo

The March 11, 2011 earthquake in Japan registered a 9.0 magnitude. An estimated 30,000 people died, and more than 250,000 people lost their homes and were moved to evacuation shelters.

2010, 373 natural disasters worldwide killed about 300,000 and affected 208 million (Centre for Research on the Epidemiology of Disasters 2011). In terms of human life lost, two disasters stand out—the January 12 earthquake in Haiti in which 222,500 people died, and the Russian summer heat wave that resulted in 56,000 deaths. In addition to disasters, Malthus believed that the only moral ways to prevent populations from growing beyond what the food supply could support were delayed marriage and celibacy.

Stage 2: Transition

Around 1650, mortality crises became less frequent in western Europe; by 1750, the death rate there had begun to decline slowly. This decline was triggered by a complex array of factors associated with the onset of the Industrial Revolution. The two most important factors were (1) increases in the food supply, which improved the nutritional status of the population and increased its ability to resist diseases, and (2) public health and sanitation measures, including the use of cotton to make clothing and new ways of preparing food. The following excerpt elaborates on these trends:

> The development of winter fodder for cattle was important; fodder allowed the farmer to keep his cattle alive during the winter, thereby reducing the necessity of living on salted meats during half of the year. . . . [C]anning was discovered in the early nineteenth century. This method of food preservation laid the basis for new and improved diets throughout the industrialized world. Finally, the manufacture of cheap cotton cloth became a reality after mid-century. Before then, much of the clothes were seldom if ever washed, especially among the poor. A journeyman's or tradesman's wife might wear leather stays and a quilted petticoat until they virtually rotted away. The new cheap cotton garments could easily be washed, which increased cleanliness and fostered better health. (Stub 1982, p. 33)

Library of Congress Prints and Photographs Division

Preserving foods in airtight jars, cans, or pouches and then heating to destroy contaminating microorganisms improved the nutritional status of the population in industrialized societies, leading to lower death rates.

Contrary to popular belief, advances in medical technology had little influence on death rates until the turn of the twentieth century—well after improvements in nutrition and sanitation had caused dramatic decreases in deaths due to infectious diseases. Over a 100-year period, the death rate fell from 50 per 1,000 to less than 20 per 1,000, and life expectancy at birth increased to approximately 50 years of age. As the death rate declined, fertility remained high. Fertility may even have increased temporarily, because improvements in sanitation and nutrition enabled women to carry more babies to term. With the decrease in the death rate, the **demographic gap**—the difference between the birth rate and the death rate—widened, and the population grew substantially.

Accompanying the unprecedented growth in population was **urbanization**, an increase in the number of cities and

positive checks Events that increase deaths, including epidemics of infectious and parasitic diseases, war, famine, and natural disasters.

demographic gap The difference between a population's birth rate and death rate.

urbanization An increase in the number of cities in a designated geographic area and growth in the proportion of the area's population living in cities.

growth in the proportion of the population living in cities. (As recently as 1850, only 2 percent of the world's people lived in cities with populations of 100,000 or more.) Around 1880, fertility began to decline. The factors that caused birth rates to drop are unclear and continue to inspire debate among demographers. But one thing is clear: The decline was not caused by innovations in contraceptive technology, because the methods available in 1880 had been available throughout history. Instead, the decline in fertility seems to have been associated with several other factors.

First, the economic value of children declined in industrial and urban settings, as children no longer represented a source of cheap labor but rather became an economic liability to their parents. Second, with the decline in infant and childhood mortality, women no longer had to bear a large number of children to ensure that a few survived. Third, a change in the status of women gave them greater control over their reproductive life and made childbearing less central to their life.

TABLE 15.3 Percentage of Total U.S. Deaths by Age Cohort, 1900, 1950, 2005

At one time in the United States—1900—death was something people in every age group experienced, but especially those less than 1 year of age. Twenty percent of all deaths in 1900 involved infants less than 1 year old. As late as 1950, 20 percent of deaths (one in every five) involved people under age 45. By 2005, that percentage dropped to about 8 percent. Today, death is something we have come to associate primarily with people of older ages as those 55 and older account for 85 percent of all deaths.

Age	1900	1950	2005
less than 1	20.72	7.15	1.16
1 to 4	9.44	1.25	0.19
5 to 14	4.27	1.01	0.27
15 to 24	6.34	1.95	1.40
25 to 34	8.31	2.92	1.71
35 to 44	8.18	5.30	3.46
45 to 54	8.25	10.20	7.50
55 to 64	9.94	17.50	11.25
65 to 74	11.57	23.61	16.27
75 to 84	9.48	21.06	28.05
85 plus	3.19	8.02	28.72

Source: Data from U.S. Centers for Disease Control and Prevention 2009.

Stage 3: Low Death Rates and Declining Birth Rates

Around 1930, both birth and death rates fell to less than 20 per 1,000, and the rate of population growth slowed considerably. Life expectancy at birth surpassed 70 years—an unprecedented statistic. The remarkable successes in reducing infant, childhood, and maternal mortality rates permitted accidents, homicides, and suicide to become the leading causes of death among young people. The reduction of the risk of dying from infectious diseases ensures that people who would have died of infectious diseases in an earlier era can survive into middle age and beyond, when they face an elevated risk of dying from degenerative and environmental diseases (such as heart disease, cancer, and strokes). For the first time in history, people age 50 and older account for more than 70 percent of annual deaths. Before stage 3, infants, children, and young women accounted for the largest share of deaths (Olshansky and Ault 1986).

As death rates decline, disease prevention becomes an important issue. The goal is to live not merely a long life but a "quality life" (Olshansky and Ault 1986, Omran 1971). As a result, people become conscious of the link between health and lifestyle (sleep, nutrition, exercise, and drinking and smoking habits). In addition to low birth and death rates, stage 3 is distinguished by an unprecedented emphasis on consumption (made possible by advances in manufacturing and food production technologies).

Since the time the demographic transition was first proposed, a fourth stage has been added in which both birth rates and death rates are low. Birth rates drop to levels that fall below that needed to replace those who die. Although death rates are low, there is an increase in lifestyle diseases caused by lack of exercise, poor nutrition, and obesity. Birth rates fall below replacement when the average woman has fewer than two children over the course of her reproductive life. Italy and Japan are examples of countries in which this is the case.

When the theory of the demographic transition was put forth, a hypothesis was also put forth that the so-called developing countries would follow this model. In some ways, most of the developing countries have followed the broad overall pattern but with some differences that we will discuss.

Industrialization in Developing Countries: An Uneven Experience

CORE CONCEPT 4 Industrialization was not confined to western Europe and North America. It pulled people from across the planet into a worldwide division of labor and created long-lasting, uneven economic relationships between countries.

The Industrial Revolution was not confined to western Europe and the United States. In fact, during this revolution, people from even the most seemingly isolated and remote regions of the planet became part of a worldwide division

TABLE 15.4 Demographic Differences between Selected Labor-Intensive Poor Economies and Core Economies

Labor-intensive poor economies differ markedly from core economies on a number of important indicators, including doubling time, infant mortality, total fertility, and per capita income.

	Population Doubling Time (years)	Infant Mortality (per 1,000)	Total Fertility	Per Capita Income ($U.S.)
Labor-Intensive Poor Economies				
Afghanistan	20.7	149.2	5.39	1,000
Haiti	41.9	54.02	3.07	1,200
India	51	47.57	2.62	3,400
Core Economies				
United States	78	6.06	2.06	47,400
Japan	636	2.78	1.21	34,200
Germany	1,750	3.54	1.41	35,900

Source: Data from U.S. Central Intelligence Agency 2011.

of labor. Industrialization's effects were not uniform; they varied according to country and region of the world.

With regard to industrialization, the countries of the world are commonly placed into two broad categories, such as developed and developing. Comparable but equally misleading terms for this dichotomy include *industrialized/industrializing* and *first world/third world*. These terms are misleading because they suggest that a country is either industrialized or not industrialized. The dichotomy implies that a failure to industrialize is what makes a country poor, and it camouflages the fact that as Europe and North America plunged into industrialization, they took possession of Asia, Africa, and South America—establishing economies there that served the industrial needs of the colonizers, not the needs of the colonized. The point is that countries we label as "developing" or "industrializing" were actually part of the Industrial Revolution from the beginning.

The World Bank, the United Nations, and other international organizations use a number of indicators to distinguish between so-called developed and developing countries, including the following: doubling time, infant mortality, total fertility, per capita income, percentage of the population engaging in agriculture, and per capita energy consumption. Instead of the term *developing, industrializing*, or *third world*, it might be more accurate to think in terms of **labor-intensive poor economies**. Instead of the term *developed, industrialized*, or *first world*, we will use the term **core economies**. Table 15.4 shows how labor-intensive poor economies differ from core economies on a number of important indicators, such as per capita electricity consumption and doubling time.

The Demographic Transition in Labor-Intensive Poor Economies

CORE CONCEPT 5 Labor-intensive poor economies differ from core economies in several characteristics: In particular, they have experienced relatively high birth rates despite declines in their death rates, resulting in more rapid population growth and unprecedented levels of rural-to-urban migration (urbanization).

Birth and Death Rates in Labor-Intensive Economies

Sociologists Bernard Berelson (1978) and John Samuel (1997) have identified some important "thresholds" associated with declines in fertility:

1. Less than 50 percent of the labor force is employed in agriculture. (The economic value of children decreases in industrial and urban settings.)

labor-intensive poor economies Economies that have a lower level of industrial production and a lower standard of living than core economies. They differ markedly from core economies on indicators such as doubling time, infant mortality, total fertility, per capita income, and per capita energy consumption.

core economies Economies that have a higher level of industrial production and a higher standard of living than labor-intensive poor economies. They include the wealthiest, most highly diversified economies in the world.

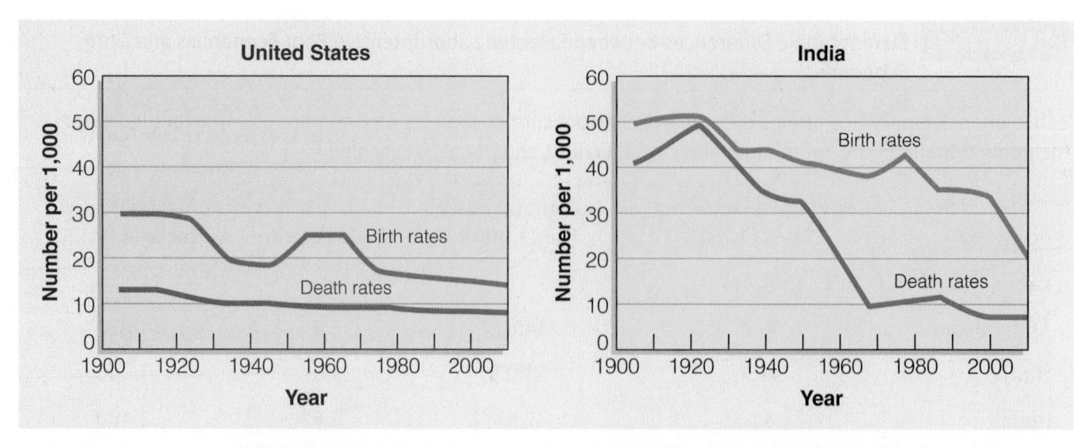

FIGURE 15.4 Birth and Death Rates in the United States and India, 1900–2010
Study the two graphs. What does it mean when the gap between birth rate and death increases? The answer explains why India's population is growing at a much faster pace than the U.S.

Source: Data from U.S. Bureau of the Census (2011)

2. At least 50 percent of people between the ages of 5 and 19 are enrolled in school. (Especially for women, education "widens horizons, sparks hope, changes status concepts, loosens tradition, and reduces infant mortality" (Samuel 1997).)
3. Life expectancy is at least 60 years. (With increased life expectancy, parents can expect their children to survive infancy and early childhood.)
4. Infant mortality is less than 65 per 1,000 live births. (When parents have confidence that their babies and children will survive, they limit the size of their families.)
5. Eighty percent of the females between the ages of 15 and 19 are unmarried. (Delayed marriage is important when it is accompanied by delayed sexual activity or protected premarital sex.)

Death rates in the labor-intensive poor economies, such as India, have declined much more rapidly than they did in the core economies, such as the United States. Demographers attribute the relatively rapid decline to cultural diffusion. That is, the labor-intensive poor economies imported Western technology—such as pesticides, fertilizers, immunizations, antibiotics, sanitation practices, and higher-yield crops—which caused an almost immediate decline in the death rates. Figure 15.4 shows that the death rate in India was so high at the beginning of the twentieth century that the gap between the birth rate and the death rate was relatively small. Around 1920, the death rate began to steadily decline because of medical advancements, especially mass inoculations. But the birth rates have remained high relative to death rate.

The swift decline in death rates and relatively slower decline in birth rates has caused the populations in India and other labor-intensive poor economies to grow very rapidly. Some demographers believe that such countries may be caught in a **demographic trap**—the point at which population growth overwhelms the environment's carrying capacity:

> Once populations expand to the point where their demands begin to exceed the sustainable yield of local forests, grasslands, croplands, or aquifers, they begin directly or indirectly to consume the resource base itself. Forests and grasslands disappear, soils erode, land productivity declines, water tables fall, or wells go dry. This in turn reduces food production and incomes, triggering a downward spiral. (Brown 1987, p. 28)

International agencies such as the World Food Programme (WFP) and World Health Organization reject the

Barbara Hougton

India's death rate has declined dramatically over the 50 to 60 years but its birth rate has lagged, declining at a much slower pace. Today, the birth rate is 20.97 per 1,000, and the average woman has 2.6 children over the course of her lifetime. Of course, average means that there are women who have more or fewer children than the average number for the women of the country.

demographic trap The point at which population growth overwhelms the environment's carrying capacity.

Missy Gish

Many people who are overweight do not have resources to buy nutritious food—food low in salt and sugar, for example. This bottle of apricot juice costs about $4.00. The bottle of Mountain Dew costs $1.50. For people who earn minimum wage, the apricot juice takes more than 30 minutes to earn before taxes. The Mountain Dew takes 12.5 minutes to earn.

idea that rapid population growth, by itself, overwhelms the environment's carrying capacity. In fact, enough food is produced each year to nourish the estimated 800 million people in the world who are chronically hungry or that go to bed hungry each night. There is also enough food produced to meet the needs of the 2 billion who suffer from food insecurity—that is, they do not have the financial resources to secure food that is consistently safe, sufficient, and nutritious. The simple fact is that these 2.8 million people simply do not have the income to buy the food they need. That is because the world economy is structured so that access to sufficient amounts of food remains highly uneven across the globe.

Urbanization

> **CORE CONCEPT 6 Urbanization is a transformative process by which people migrate from rural to urban areas and change the way they use land, interact, and make a living.**

Urbanization encompasses (1) the process by which a population becomes concentrated in urban areas and (2) the corresponding changes in land use, social interaction, economic activity, and landscape. What constitutes an urban area varies by country. The U.S. Bureau of the Census (2010) defines an urban area as a densely settled core with at least 1,000 people per square mile and adjacent territory containing nonresidential urban land uses. In addition, there are outlying densely settled territories that are considered part of the urban area. Greater San Francisco, for example, includes the downtown but extends its influence and reach to surrounding cities and suburbs.

Chris Caldeira

The U.S. Bureau of Census defines all areas not considered urban as rural. To be rural, the density or number of people per square mile must be fewer than 1,000.

The world has 483 **agglomerations**, urban areas with populations of 1 million or more. Of these, 80 are in China, 48 are in India, and 53 are within the United States. Within the agglomeration category is the **megacity,** cities in which at least 10 million people reside. According to this definition, 26 megacities exist in the world. Two lie in the United States: New York (22.2 million) and Los Angeles (17.9 million). The largest megacity in the world is Tokyo with 32.4 million people (Brinkhoff 2011).

Urbanization in Labor-Intensive Poor Economies versus Core Economies

Urbanization in labor-intensive poor economies differs in several major ways from urbanization in core economies. At comparable points in the demographic transition, the rate of urbanization in labor-intensive poor economies far exceeds that of the core economies. Consider that during the 25 years of its most rapid growth, New York City increased its population by 2.3 million. As one contrasting example, consider that during the 25 years of its most rapid growth, Bombay, India, added 11.2 million people (Brinkhoff 2011).

---

agglomerations Urban areas with populations of 1 million or more.

megacity An agglomeration of at least 8 million (UN definition) or 10 million (U.S. definition) people.

Why such a difference? For one thing, "new worlds" existed to siphon off the population growth of Europe (Light 1983). Millions of Europeans who were pushed off the land were able to migrate to sparsely populated places, such as North America, South America, South Africa, New Zealand, and Australia. If the people who fled Europe for other lands in the eighteenth and nineteenth centuries had been forced to make their living in the European cities, the conditions there would have been much worse than they actually were:

> Ireland provides the most extreme example. The potato famine of 1846–1849 deprived millions of peasants of their staple crop. Ireland's population was reduced by 30 percent in the period 1845–1851 as a joint result of starvation and emigration. The immigrants fled to industrial cities of Britain, but Britain did not absorb all the hungry Irish. North America and Australia also received Irish immigrants. Harsh as life was for these impoverished immigrants, the new continents nonetheless offered them a subsistence that Britain was unable to provide. (Light 1983, pp. 130–131)

In India, for example, the problem of urbanization is compounded by the fact that many people who migrate to the cities come from some of the most economically precarious sections of India. In fact, most rural-to-urban migrants are not pulled into the cities by employment opportunities; rather, they are forced to move there because they have no alternatives. When these migrants come to the cities, they face not only unemployment, but also a shortage of housing and a lack of services (electricity, running water, waste disposal). One distinguishing characteristic of cities in labor-intensive poor economies is the prevalence of slums and squatter settlements, which are much poorer and larger than even the worst slums in the core economies.

> It is a familiar sight in so-called underdeveloped countries to find somewhere, in the midst of great poverty, . . . a gleaming, streamlined new factory, created by foreign enterprise . . .. Immediately outside the gates you might find a shanty town of the most miserable kind teeming with thousands of people, most of whom are unemployed and do not seem to have a chance of ever finding regular employment of any kind. (Schumacher 1985, p. 490)

Sociologist Kingsley Davis uses the term **overurbanization** to describe a situation in which urban misery—poverty, unemployment, housing shortages, and insufficient infrastructure—is exacerbated by an influx of unskilled, illiterate, and poverty-stricken rural migrants who have been pushed into cities out of desperation. In this regard, the United Nations estimates that one billion people worldwide live in slums lacking essential services such as water and sanitation (Dugger 2007).

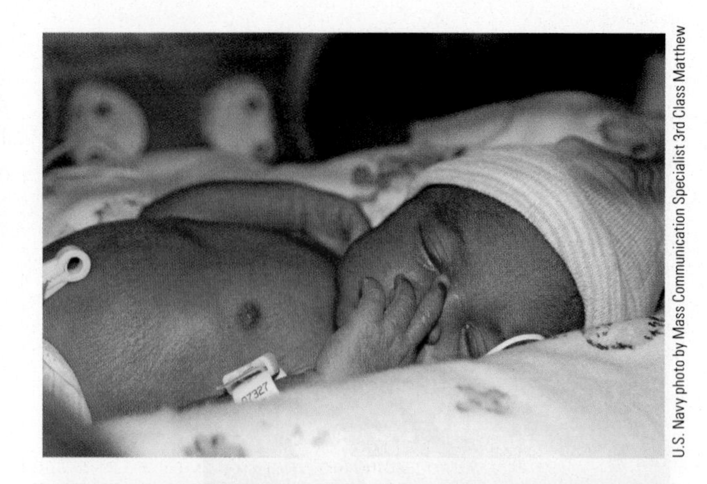

U.S. Navy photo by Mass Communication Specialist 3rd Class Matthew

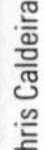

Chris Caldeira

Missy Gish

This chapter has focused on key experiences of human life, not just for individuals but also for societies. Those experiences are births, deaths, and migration. Specifically, we compared the countries that experience the highest and the lowest related rates. We emphasize extreme cases because they capture end points on the continuum of human experience. We learned how knowing a country's birth, death, and migration rates allows us to think deeply about the human experience and the way a society is organized.

Extreme Cases

To this point, we have reviewed key concepts and processes related to three key human experiences: birth, death, and migration. We have also identified countries with highest and lowest rates related to these human experiences (see Table 15.5). We now consider each country listed in Table 15.5.

United Arab Emirates

The most noticeable feature of the United Arab Emirates (UAE) population pyramid is that, with the exception of those 19 and under, males outnumber females in all age categories. This is because about 80 percent of the 5 million people who live in UAE are not citizens of the country. They are people who migrated to UAE in search of work. About 25 percent of UAE's population is from surrounding Arab countries and the Persian country, Iran. About 50 percent is from South Asian countries, most notably India, Pakistan, and Bangladesh. There are also significant numbers from the Philippines and Sri Lanka. The large influx of labor explains why the UAE is the highest positive net migration rate in the world.

TABLE 15.5 Key Demographic Indicators for Extreme Cases and the United States

This table includes countries we have named in this chapter as extreme cases; that is, they have a birth, death, and migration rate that is particularly high or low. In the pages that follow we consider some reasons why each country named in this table is an extreme case.

	Highest	Lowest	United States
Crude Birth Rate	Niger 51.4 per 1,000	Japan 7.3 per 1,000	14 per 1,000
Teen Birth Rate	Niger 199 per 1,000	South Korea 1.2 per 1,000	41 per 1,000
Fertility Rate Average # children per female	Niger 7.7	China 1.15	2.1
Crude Death Rate	Angola 23.4 per 1,000	United Arab Emirates 2 per 1,000	8 per 1,000
Infant Mortality Rate	Angola 175.9 per 1,000	Sweden 2.7 per 1,000	6.0 per 1,000
Maternal Mortality	Sierra Leone 2,000 per 100,000 live births	Sweden 5 per 100,000 live births	17 per 100,000 live births
Migration Rate	United Arab Emirates +22 per 1,000	Jordan −14.3 per 1,000	+4 per 1,000 residents
Population Growth Rate	Zimbabwe 4.3%	Bulgaria −0.78%	1.1%
Male to Female Sex Ratio	United Arab Emirates 219 males per 100 females	Russia 86 males per 100 females	97 males per 100 females
Life Expectancy (at birth)	Japan 82.3	Angola 38.7	78.4
% of population 14 and under	Niger 49.6%	Japan 13.1%	20.1%
% of population 65 and older	Japan 22.9%	United Arab Emirates .9 %	13.1%
Median Age	Japan 44.8 years	Niger 15.2 years	36.9 years
% working in agriculture	Niger 90%	Sweden 1.1%	1.6%

Source: Data from U.S. Bureau of the Census 2011, U.S. Central Intelligence Agency 2011.

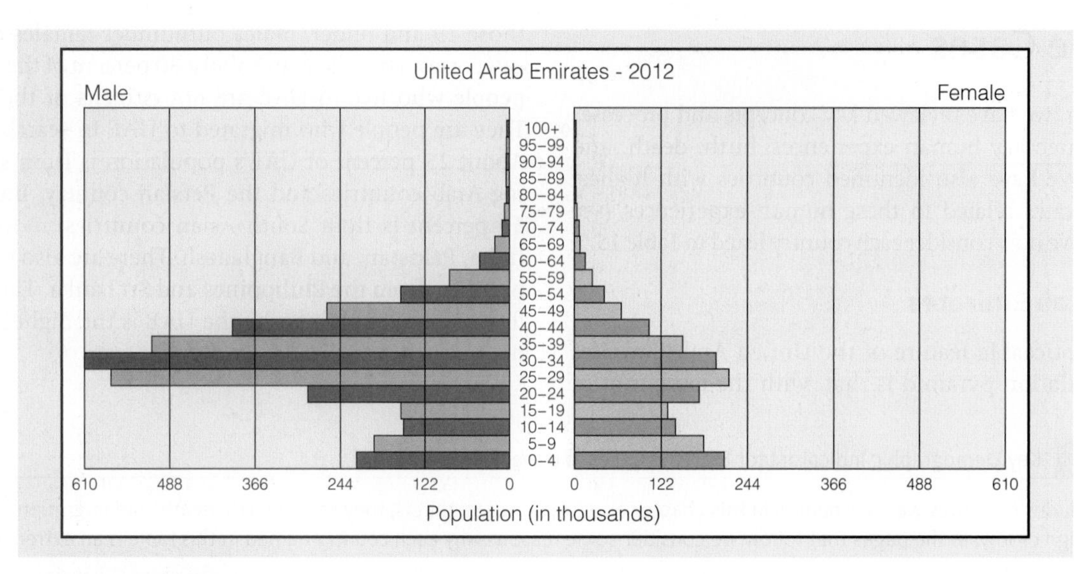

United Arab Emirates

Lowest percentage of population 65 and over: 0.9%
Lowest crude death rate: 2 per 1,000
Highest male to female sex ratio: 219 males per 100 females
Highest net migration: +22 per 1,000

Source: Data from U.S. Census Bureau (2012)

The UAE, with the third highest per capita income in the world, needs people to work in petroleum and natural gas sectors, which account for 80 percent of the country's wealth. In addition, UAE has a number of ambitious construction and tourism-oriented projects under way: The country is building the world's tallest building, a world-class international airport, the largest artificial islands in the world, Dubailand (which will be twice the size of Disney Land), the Dubai Sports City, and Dubai Mall, billed as the world's largest mall. The population pyramid suggests that when migrants' work life ends or when jobs end, they leave UAE to return home. The crude death rate and the percentage of the population 65 and older are low because the large migrant population remains young due to turnover and constant flux.

U.S. Bureau of the Census

The city of Dubai in the UAE has been labeled the richest city in the world. It has built many attractions for tourists. Migrants from outside the country work in the hotels, restaurants, and tourist destinations.

overurbanization A situation in which urban misery—poverty, unemployment, housing shortages, insufficient infrastructure—is exacerbated by an influx of unskilled, illiterate, and poverty-stricken rural migrants who have been pushed into cities out of desperation.

Bulgaria

Study the population pyramid for Bulgaria. Can you determine the year that Bulgaria's population began its decline? The age 0- to 19-year-old cohorts are dramatically smaller than most of the older cohorts. So the question becomes what happened 20 years ago, in the late 1980s and early 1990s? Bulgaria, being part of Eastern Europe, was under Soviet Union control from 1946 until 1989, the year the Berlin Wall fell and the Soviet Union collapsed. After Bulgaria held its first elections, it embarked on the long, hard transition of moving away from a communist government and centrally planned economy to a political democracy and market economy. In the process, Bulgarians experienced inflation, unemployment, corruption, and crime.

Recall that population decline occurs when the birth rate is lower than the death rate and when more people leave the country than enter it. The collapse of the Soviet Union allowed Bulgarians the opportunity to emigrate. In addition, hard economic times and stresses associated with dramatic economic and political change contributed to the low birth rate and total fertility. Finally, Bulgaria joined the European Union in 2007, which likely opened up further opportunities for people to emigrate.

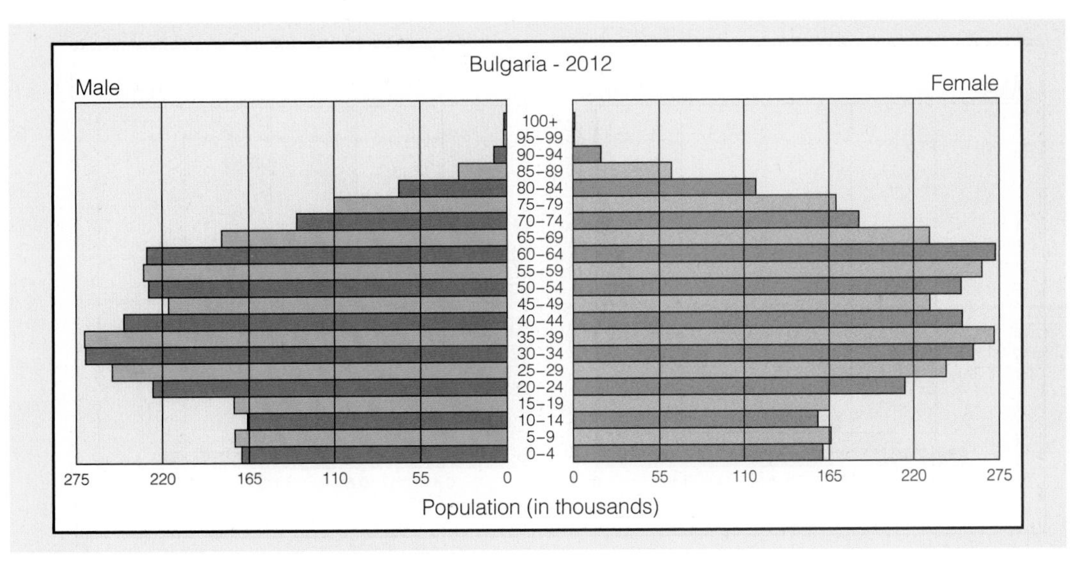

Bulgaria
Lowest growth rate: –.78

Source: Data from U.S. Census Bureau (2012)

Japan

What about the population pyramid tells you that Japan has the highest median age, the greatest percentage of people age 65 and older, the highest life expectancy at birth, the lowest percentage of people 14 and under, and the lowest crude birth rate? Perhaps one of the most striking features is that the number of 80- to 84-year-olds (4.3 million) is almost as large as the number of children aged 4 and under (4.9 million). In addition, each of the cohorts that make up the 35- to 74-year-old tiers are larger in size than any of the 19 and under cohorts. We should not be surprised that Japan's average life expectancy and median age are the highest in the world.

We can explain Japan's situation by noting that it is in stage 4 of the demographic transition. But we might also note that Japan is a stressful place for men, women, and children. Men must work long hours at companies where the expectation is that the jobs comes first and family life should not interfere. Women are expected to quit their jobs when they have children, and children must participate in the competitive examination system. These factors, along with the high cost of raising children, contribute to the low birth rate.

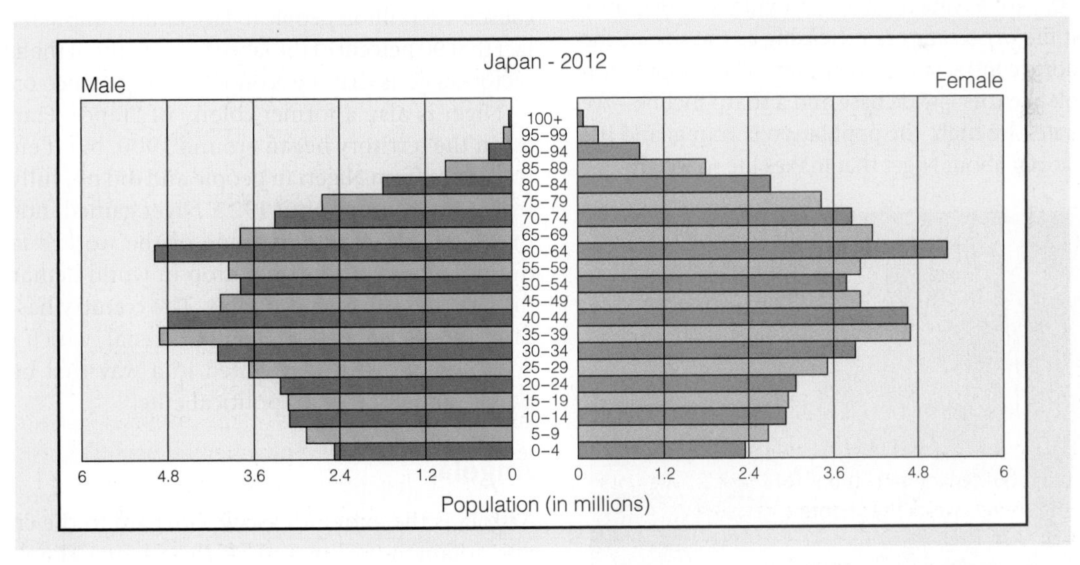

Japan
Highest median age: 44.8
Highest percentage of population age 65 and over: 22.9%
Highest live expectancy at birth: 82.3
Lowest percentage of population age 14 and under: 13.1%
Lowest crude birth rate: 7.3 per 1,000

Source: Data from U.S. Census Bureau (2012)

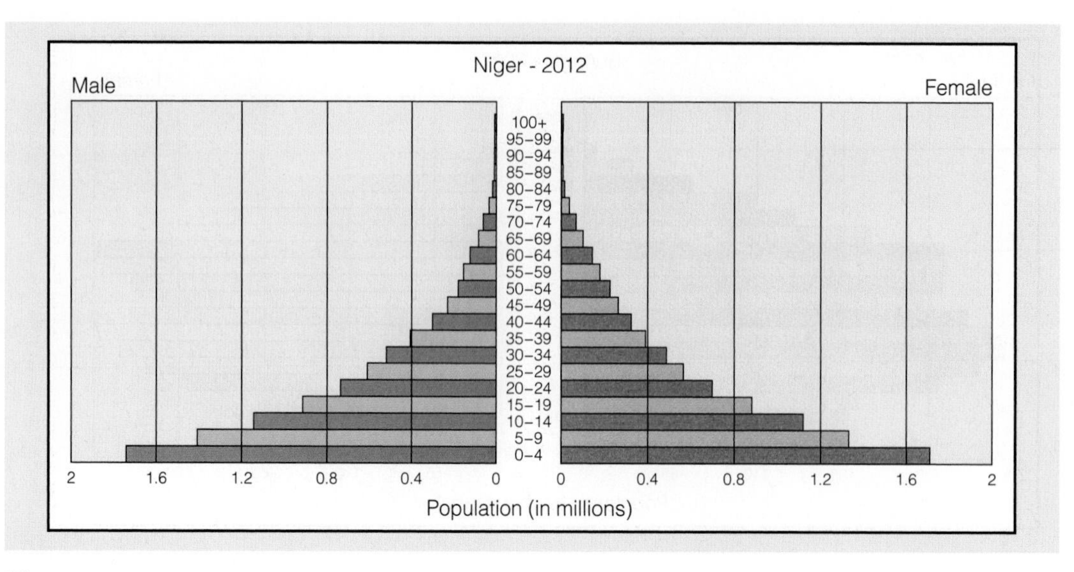

Niger

Lowest median age: 15.2
Highest percentage of population age 14 and under: 49.6%
Highest crude birth rate: 51.5 per 1,000
Highest percentage working in agriculture: 90%
Highest fertility rate: 7.7
Highest teen birth rate: 199 per 1,000

Source: Data from U.S. Census Bureau (2012)

Niger

Niger is the extreme case for six important demographic characteristics. How are these characteristics reflected in the population pyramid? The most distinguishing feature of this pyramid is the shear length of the bars for the 0 to 4 age groups and the fact that the number of people in each older cohort declines with advancing age. The pyramid shows little evidence that life expectancy is increasing as the size of the older age cohorts contracts with each tier. When population pyramids look like this—wide base and a sharp incline—we know death rates are high, the population is young, and life is tough. What is it about Niger that makes life so tough?

By Cpl. Enrique Saenz/US Marines

This photo of children who live in Niger gives some sense of the harsh environment of this landlocked country.

For one, Niger is a landlocked West African country, with over 80 percent of its territory within the Sahara desert. The remaining 20 percent of its territory is threatened by cyclical drought and desertification. The two most recent extreme weather events occurred in 2005, when drought and locust infestation created food shortages affecting 2.5 million people, and in 2010, when record heat waves affected crops and caused 1.5 million people to face famine and starvation. The fact that 90 percent of the labor force works in the agricultural sector suggests that the economy is subsistence-oriented.

Niger is also a former colony of France. France's interest in the territory began around 1900, but it encountered resistance from Nigerian people and did not fully gain control of the country until 1922. Niger gained independence in 1960. Niger contains some of the world's largest uranium deposits, although a drop in world demand for this mineral has hurt the economy. The country has a number of resources such as oil, gold, and coal, which could fuel economic growth if exploited in a way that benefits the people and not a small political elite.

Angola

Angola is the extreme case with regard to the crude death rate, infant mortality, and life expectancy. The shape of the pyramid with a wide base and steep "steps" from one age cohort to the next tells us that this is the case. Why is life so harsh? Angola was a Portuguese territory for about 400 years until 1975, when it gained independence. The country has fertile land and was considered the breadbasket of southern Africa. After gaining independence, the country

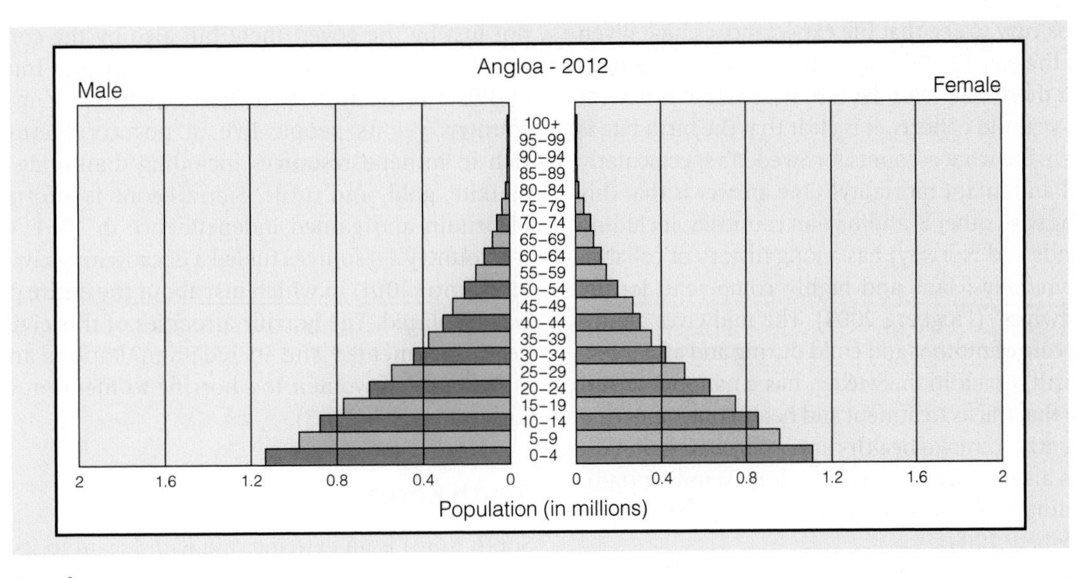

Angola

Highest crude death rate: 23.4 per 1,000
Highest infant morality: 175 per 1,000
Lowest life expectancy: 38.7 years

Source: Data from U.S. Census Bureau (2012)

experienced civil war for the next 27 years, until 2002, when a settlement was reached among the warring parties. The fertile land was destroyed and littered with landmines. Millions left the countryside for the cities or left the country altogether. In recent years, Angola has had some successes. As many as 4 million displaced people have returned to Angola. Agricultural production is increasing. It received a $5 billion loan from China to rebuild its infrastructure to be paid back in oil. Currently, diamonds and oil derived from exports account for 60 percent of Angola's economy. The problem is that, to date, only a small percentage of the population has benefited from the gains (BBC News 2011).

Sweden

Sweden stands out for having the lowest infant and maternal mortality rates in the world. It also has the lowest percentage of the population employed in the agricultural sector, one indicator that parents do not need to produce laborers to do farm work. The shape of the population pyramid for Sweden indicates that the country is in stage 4 of the demographic

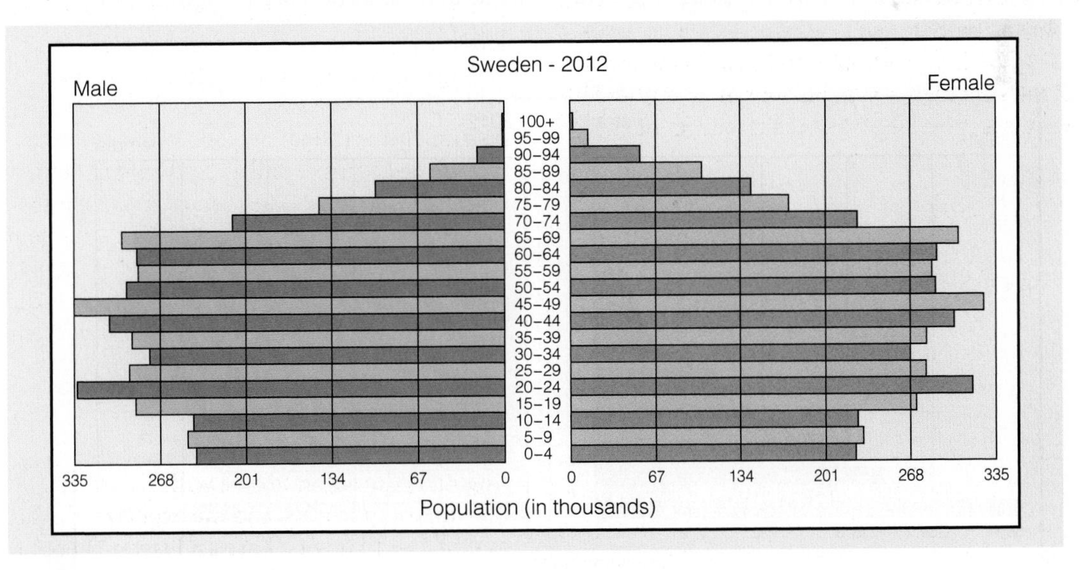

Sweden

Lowest infant mortality: 2.7 per 1,000
Lowest maternal mortality: 5 per 100,000
Lowest percentage or population working in agriculture: 1.1%

Source: Data from U.S. Census Bureau (2012)

transition. It is easy to see that life expectancy is high given the length of the bars for each age cohort. Given that length of the bars for those 14 and under is so much shorter that for the 29- to 49-year-old cohorts, it is clear that the birth rate is low for women. But what accounts for Sweden's spectacularly low maternal and infant mortality? One answer is that this country (along with other Scandinavian countries, including the Netherlands and Norway) has a long history of collaboration between "physicians and highly competent, locally available midwives" (Högberg 2004). The midwives are involved in the care of mother and child during and after pregnancy and birth. In addition, Sweden has a national health care database that tracks treatment and health outcomes that allow the country to make health care policy based on best practices. It is also a country that has a long-standing tradition of providing quality care to rural and low-income communities (Högberg 2004).

Sierra Leone

Sierra Leone has the highest maternal mortality rate in the world. Amnesty International has called this situation grave and has labeled it a human rights emergency. According to Amnesty International (2009), "thousands of women bleed to death after giving birth. Most die in their homes. Some die on the way to hospital, in taxis, on motorbikes, or on foot. In Sierra Leone, less than half of deliveries are attended by a skilled birth attendant and less than one in five are carried out in health facilities."

This situation exists because most women are too poor to afford health care. In Sierra Leone, women are held in low status and face discrimination. Girls are forced to marry, are denied access to education, and are victims of sexual violence. Women's health care needs are ignored, not just by the government but also by the communities and families in which they live (Amnesty International 2009). The irony is that Sierra Leone is a mineral-rich country, yet its people live in poverty. Sierra Leone is rich in mineral resources including diamonds, titanium, bauxite, gold, and rutile. Sierra Leone is a former colony of Britain and gained independence in 1961. Control of the country's resources fueled a decade-long civil war from 1991 until 2001, in which just about the entire population was displaced. The horrific atrocities of this civil war have been documented and include amputations and systematic abuses of women too horrific to mention here (Ben-Ari and Harsch 2005).

South Korea

South Korea is an extreme case with regard to its teen birth rate—the number of babies born each year for every 1,000 teens. Korea's rate is 1.2 babies born per 1,000 teens. To put this rate in perspective, consider that the rate is 41 per 1,000 in the United States. We can see by the shape of South Korea's population pyramid that its overall birth rate is extremely low. Simply look at the length of the 0 to 4 age cohort relative to age cohorts of women in their reproductive years. But why do teens, in particular, have such a low rate? There are several possible explanations: Korean society, influenced by Confucian beliefs, places a high value on chastity, and there is a stigma attached to being a single mother. On the other hand, the low teen birth rate is not the same as the pregnancy rate. It could be that Korean teens do get pregnant, but many get abortions. By one estimate, there are about 30 abortions each year for every 1,000 women between ages of 15 and 44. However, there is no way to determine the contribution teens make to the abortion rate (Sangwon 2010).

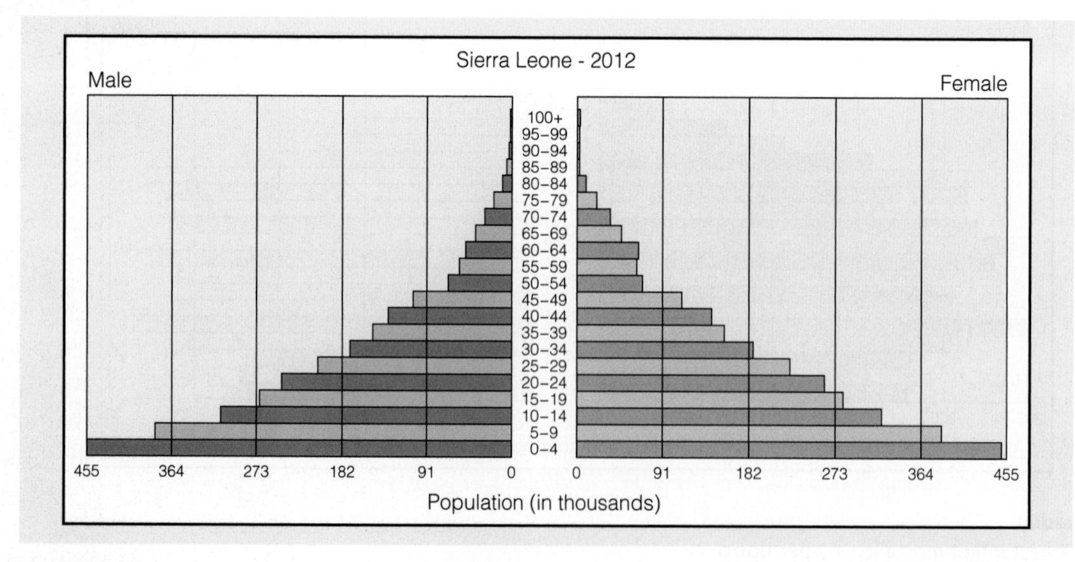

Sierra Leone
Highest maternal mortality: 2,000 per 100,000
Source: Data from U.S. Census Bureau (2012)

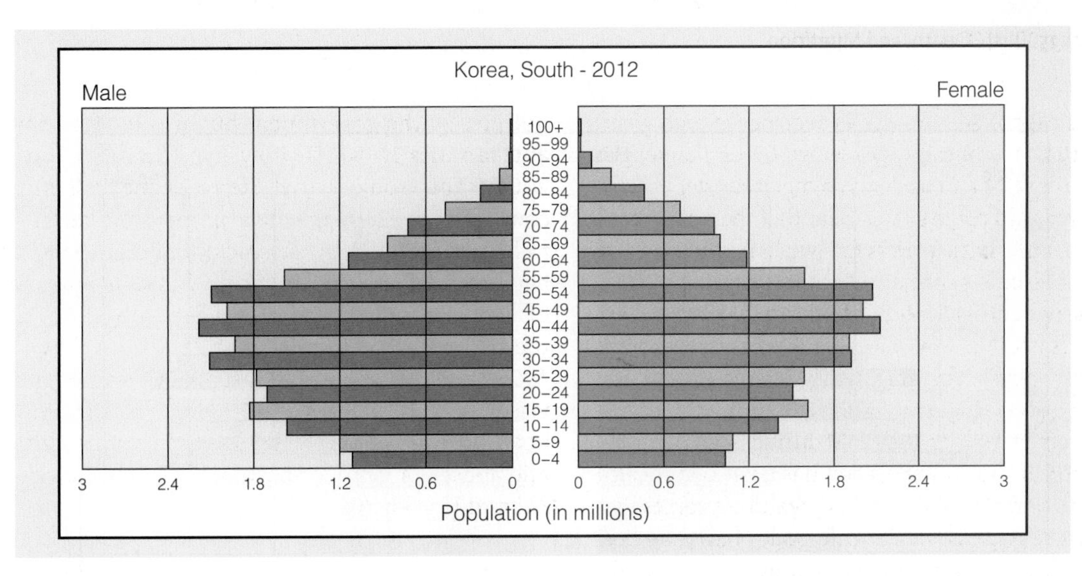

South Korea
Lowest teen birth rate: 1.2 per 1,000

Source: Data from U.S. Census Bureau (2012)

China

China has the largest population in the world but it also has the lowest fertility rate. On average, women have 1.1 children. In 1979, the government announced the one-child family planning policy. The population that is 31 years and younger was born after that policy went into effect. Except for the 20- to 24-year-old age cohort, the lengths of the bars since that announcement are shorter than the cohorts that preceded it. The 20 to 24 age cohort is likely so large because the size of their parents' cohorts—those who are now 37 to 49 years of age—was so large. Even if couples only had one child, so many couples had a child that it still created a large cohort. You might wonder why the 50 to 54 age cohort is small relative to the 45- to 49- and 55- to 59-year-old cohorts. In 1960 and 1961, China experienced natural disaster and famine. The death rate, including infant mortality, was high.

Russia

If you look closely at Russia's population pyramid on the next page, you can see that males outnumber females until age 25 to 29. At that point, the number of females always exceeds males, with greatest differences for age cohorts 50 to 84 (U.S. Bureau of the Census 2011). The life expectancy of the average Russian male is 60 years, compared to the average Russian female who lives to age 73. According to one estimate, an 18-year-old Russian male has a 50 percent chance of dying before reaching retirement age, compared to a 90 percent chance for an 18-year-old male living in the United States. To date, the best explanation relates to high levels of alcohol consumption and tobacco use among men in Russia, which increased dramatically after the breakup of the Soviet Union. The transition from a centrally planned economy to a market-oriented one has been especially difficult for men (Wong 2009).

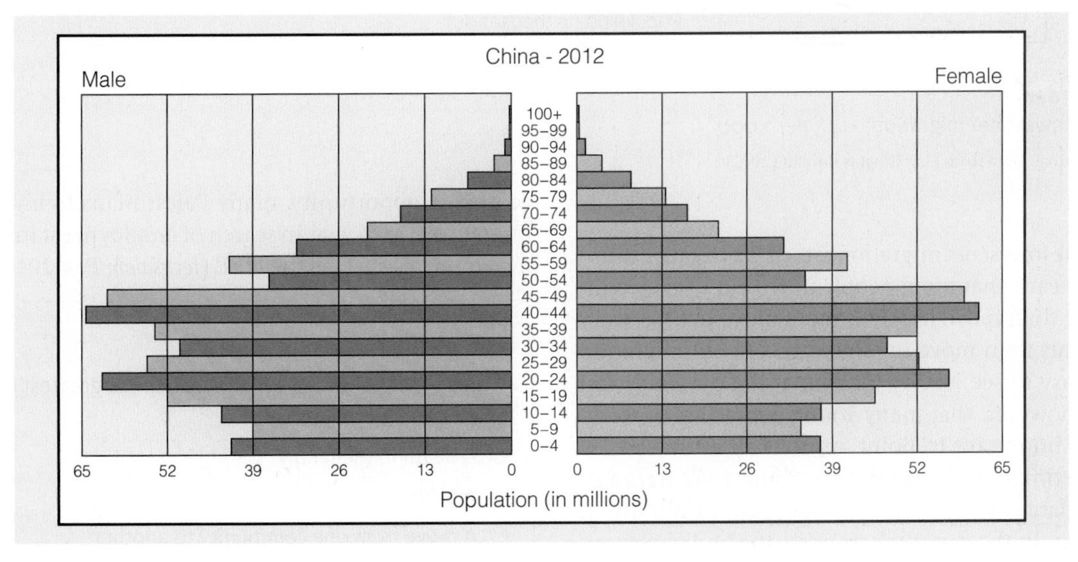

China
Lowest total fertility rate: 1.1 per woman

Source: Data from U.S. Census Bureau (2012)

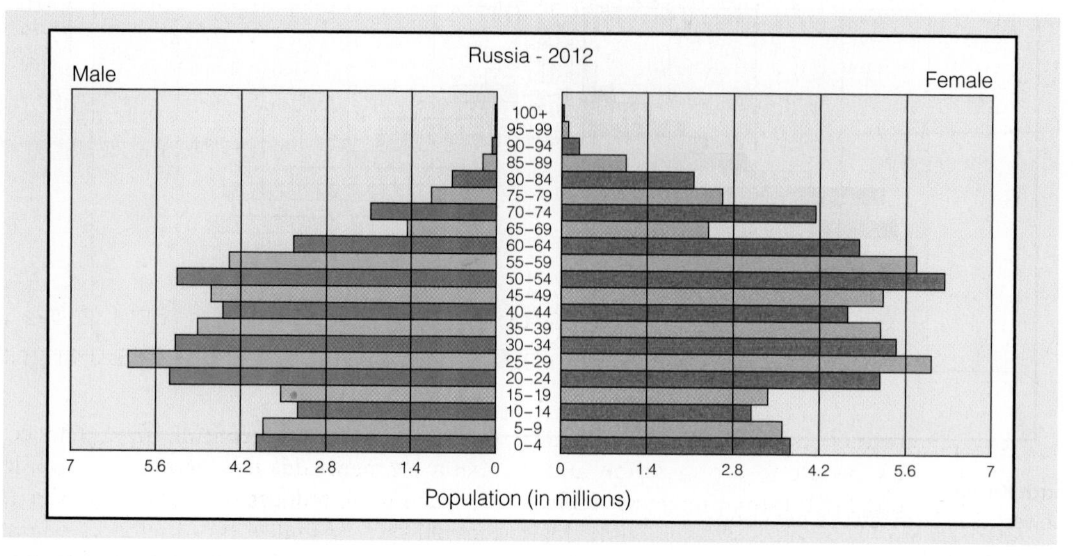

Russia

Lowest male to female sex ratio: 86.3 males per 1,000 females

Source: Data from U.S. Census Bureau (2012)

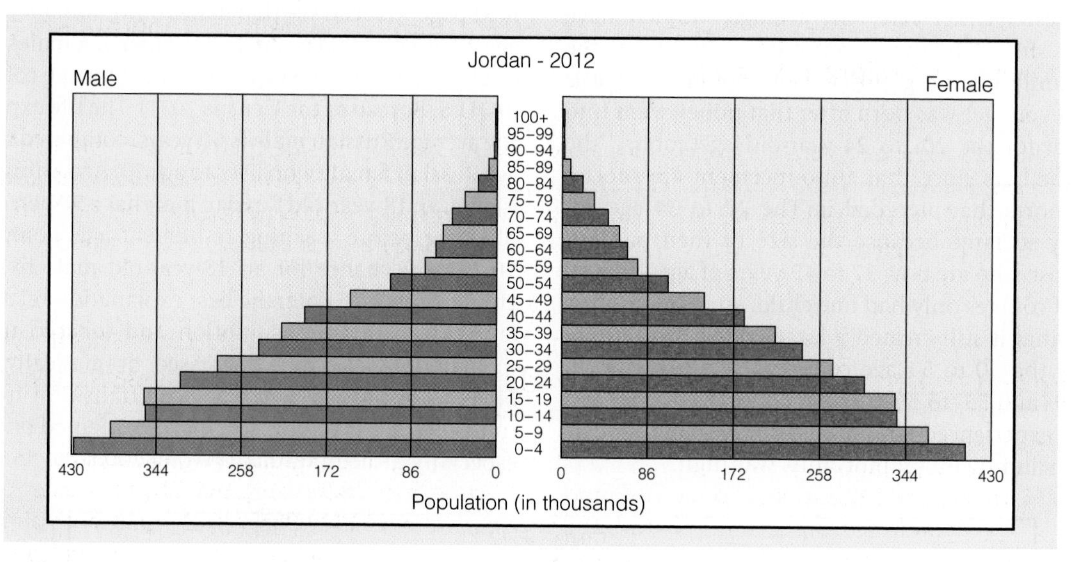

Jordan

Lowest net migration: –14.3 per 1,000

Source: Data from U.S. Census Bureau (2012)

Jordan

Jordan has the lowest net migration rate. It is a negative number, which means that more people move out of the country each year than move in—14.3 more move out for every 1,000 residents than move in. The effects of out-migration are not so easy to see just by looking at the pyramid. You will notice, however, that many young people are of working age. It is important to point out that an estimated 1.9 million Palestinian refugees from 1948 and 1967 wars and their descendants live in Jordan. About 350,000 still live in refugee camps. In the past 20 years or so, the Jordan government has refused to offer citizenship and it has revoked citizenship of some Palestinians in an effort to prevent Israel from pushing Palestinians out of the West Bank, Gaza, and Israel proper. Because of their precarious status and lack of economic opportunity, many Palestinians living in Jordan migrate out each year in search of employment in surrounding countries such as the UAE (*Jerusalem Post* 2010).

Online Poll

Which of the following events has had the greatest impact on you in the past year?

○ A death in the family

○ A birth

○ A move from one community to another

○ No such event has occurred

To see how other students responded to these questions, go to **www.cengagebrain.com.**

DANIEL SCOTT/Department of Defense

Summary of CORE CONCEPTS

CORE CONCEPT 1 Demography, a subspecialty within sociology, focuses on births, deaths, and migration—major factors that determine population size and rate of growth.

Population size is determined by births, deaths, and migration. Births add new people to a population and can be expressed as a crude birth rate or an age-specific birth rate. Deaths reduce the size of a population. As with the birth rate, we can calculate the death rate for entire populations or for specific segments. Migration is the movement of people from one residence to another. That movement adds new people to a population if they are moving in, reduces the population size if people are moving out, or makes no difference. Migration, which can be international or internal, results from push and pull factors.

CORE CONCEPT 2 The age-sex composition of a population helps demographers predict birth, death, and migration rates.

A population's age and sex composition is commonly depicted as a population pyramid, which offers a snapshot of the number of males and females in the various age cohorts at a particular time. Generally, a country's population pyramid approximates one of three shapes: expansive, constrictive, or stationary. Knowing age-sex composition helps demographers predict birth, death, and migration rates and calculate the sex ratio.

CORE CONCEPT 3 The demographic transition links the birth and death rates in western Europe and North America to the level of industrialization and economic development.

The theory of the demographic transition connects the fall in a population's birth and death rates to level of industrialization and economic development. Birth and death rates in western Europe and North America changed in the following sequence: (Stage 1) Birth and death rates remained high until the mid-eighteenth century, when death rates began to decline. (Stage 2) As the death rates decreased, the population grew rapidly, because more births than deaths occurred. The birth rates began to decline around 1800 (Stage 3). By 1920, both birth and death rates had dropped below 20 per 1,000. In Stage 4, the birth rate is smaller than the death rate. Based on these observations, demographers put forth the theory of the demographic transition.

CORE CONCEPT 4 Industrialization was not confined to western Europe and North America. It pulled people from across the planet into a worldwide division of labor and created long-lasting, uneven economic relationships between countries.

The Industrial Revolution was not unique to western Europe and North America. In fact, during this revolution, people from even the most seemingly isolated and remote regions of the planet became part of a worldwide division of labor. Industrialization's effects were not uniform; rather, they varied according to country and region of the world. Consequently, with regard to industrialization, we can place the countries of the world into two broad categories: core economies and labor-intensive poor economies.

CORE CONCEPT 5 Labor-intensive poor economies differ from core economies in several characteristics: In particular, they have experienced relatively high birth rates despite declines in their death rates, resulting in more rapid population growth and unprecedented levels of rural-to-urban migration (urbanization).

Birth rates have remained high relative to death rates and have taken longer to decline; the level of rural-to-urban migration has been unprecedented. Colonization and its legacy help explain why the model of the demographic transition does not exactly apply to India and most other former colonies. Labor-intensive poor economies have not reached a number of milestones associated with declines in birth rates. These milestones include the following: Less than 50 percent of the labor force is employed in agriculture; at least 50 percent of people between the ages of 5 and 19 are enrolled in school; life expectancy is at least 60 years; infant mortality is less than 65 per 1,000 live births; and 80 percent of the females between the ages of 15 and 19 are unmarried.

CORE CONCEPT 6 Urbanization is a transformative process by which people migrate from rural to urban areas and change the way they use land, interact, and make a living.

The world has 337 agglomerations, urban areas with populations of 1 million or more. Urbanization in labor-intensive poor economies far exceeds that in the core economies, if only because no "new worlds" exist to siphon off such countries' population growth as existed for Europe. In addition, rural migrants tend to come from the most economically precarious segments of the population. The result is overurbanization, a situation in which urban misery—poverty, unemployment, housing shortages, and insufficient infrastructure—is exacerbated by an influx of unskilled, illiterate, and poverty-stricken rural migrants, who have been pushed into cities out of desperation.

Resources on the Internet

Login to CengageBrain.com to access the resources your instructor requires. For this book, you can access:

Access an integrated eBook, chapter-specific interactive learning tools, including flash cards, quizzes, videos, and more in your Sociology CourseMate.

Take a pretest for this chapter and receive a personalized study plan based on your results that will identify the topics you need to review and direct you to online resources to help you master those topics. Then take a posttest to help you determine the concepts you have mastered and what you will need to work on.

CourseReader

CourseReader for Sociology is an online reader providing access to readings, and audio and video selections to accompany your course materials.

Visit **www.cengagebrain.com** to access your account and purchase materials.

Key Terms

agglomerations 401
cohort 393
constrictive pyramid 393
core economies 399
crude birth rate 388
crude death rate 388
demographic gap 397
demographic trap 400
demography 388
doubling time 391
emigration 390
expansive pyramid 393
internal migration 391
immigration 390
infant mortality rate 388
in-migration 389
labor-intensive poor economies 399
maternal mortality rate 389
megacity 401
migration 389
migration rate 389
mortality crises 395
net-migration 389
out-migration 389
overurbanization 404
population pyramid 393
positive checks 397
pull factors 390
push factors 389
sex ratio 395
stationary pyramid 393
total fertility rate 388
urbanization 397

SOCIAL CHANGE

16 With Emphasis on GREENLAND

A polar bear standing on an ice floe off Greenland is an image popularly used to represent climate change—a change affecting not just polar bears but the planet as a whole. Climate change is most evident in the melting glaciers in Greenland and other key locations in the world. When sociologists study any change, they ask at least three key questions: What has changed? What factors triggered that change? What are the consequences of that change to society and to the ways in which humans relate to each other and their environment?

Why Focus On GREENLAND?

In 2007, the United Nations (UN) issued a report that involved 1,250 authors and 2,500 scientific experts from 130 countries. That report, which took six years to write, announced that climate change can no longer be denied or doubted, and that human or social activity since 1750 has very likely caused the rise in the planet's temperatures. When (and since) this report was issued, it generated intense and bitter debate over whether climate change is real, and even if real, is it man-made or part of the planet's natural shifts in weather patterns?

Although climate change is a complex phenomenon, some of the most publicized and vivid images center on one element of the process—melting ice sheets—which the UN report names as very likely (90 to 99 percent probability) contributing to sea-level rise. Images of ice sheets collapsing and polar bears seemingly stranded on ice floes turn our attention to places like Greenland.

Greenland, a self-governing dependency of Denmark near the North Pole, possesses the largest reservoir of freshwater on the planet after Antarctica. If Greenland's ice sheet, which covers about 85 percent of its territory, melted entirely, it would release enough water to raise the global sea level by almost 23 feet. To date, the sea level has risen by .07 inches per year since 1961, for a total of 3.96 inches. Melting ice from Greenland and Antarctica has very likely contributed between .06 inches to .33 inches of that rise (representing 1.5 to 8.3 percent of total sea rise in the past 55 years) (United Nations 2007).

Apart from the debate over causes of climate change, we can all agree that since 1750, humans have burned fossil fuels to transport people and goods, to run appliances and light the night; and to regulate temperatures in houses, office buildings, and other indoor environments. We use the concepts and theories in this chapter to answer three important questions of this human dependence on fossil fuels: (1) What has changed since 1750, making human activity heavily dependent upon fossil fuels? (2) What factors trigger changes in social activity? (3) What are the consequences of those changes? It is not until we answer the third question that we will bring Greenland's situation in the mix—specifically how fossil fuel–dependent social activities and climate change have affected that country. Note that we use the term *climate change*, in keeping with the UN report, rather than *global warming*.

Online Poll

Do you think climate change is man-made or part of the natural cycle of weather?

○ Man-made

○ Natural cycle

○ Both man-made and part of natural cycle

To see how other students responded to these questions, go to **www.cengagebrain.com**.

Social Change

CORE CONCEPT 1 When sociologists study any social change, they take particular interest in identifying tipping points—situations in which previously rare events snowball into commonplace ones.

Sociologists define **social change** as any significant alteration, modification, or transformation in the organization and operation of social activity. When sociologists study change, they must first identify the social activity that has changed or is changing. The list of possible topics is virtually endless. It includes changes in the division of labor;

Jan Martin Will/Shutterstock.com

Missy Gish

How much further can this egg be pushed away from the tabletop such that it can no longer sustain its position and will fall to the floor? The point immediately before that moment is the tipping point.

in how people communicate; in the amounts of goods and services people produce, sell, or buy from others; in the size of the world population; and in the average life span (Martel 1986).

Sociologists are particularly interested in **tipping points**, situations in which a previously rare (or seemingly rare) event, response, or opinion, snowballs into something dramatically more common. The process by which the rare becomes commonplace is at first gradual; the change seems so small that few if any people notice it, and if they notice it, they dismiss it as not all that significant. But at some point, a critical mass is reached so that the next small increment of change "tips" the system in a dramatic way.

In sociology, the concept of tipping point was first applied to white flight—a situation in which residents classified as white decided en masse to move out of a neighborhood because there is "one too many" new residents classified as black. Researchers who studied white flight in the late 1950s and early 1960s found that white families did not move out when the first black family moved in; they stayed as long as the number of black families remained very small; whites moved out en masse at the point when they judged there to be "one too many" black households (Grodzins 1958).

Scientists warn that we may be approaching a tipping point with regard to the earth's temperature. Based on "satellite measurements of the planet's air from the surface up to 35,000 feet," the Earth's temperature has been above the average temperature of the 100 years of the twentieth century. The years 2005 and 2010 are tied as the hottest years on record. Since 1979, land surface temperatures have increased 0.25°C per decade, and ocean temperatures have increased 0.13°C per decade. Scientists seek to identify the point at which the next slight rise in land or ocean temperature will trigger an event that will trigger a far greater rise in temperature (Intergovernmental Panel on Climate Change [IPCC] 2007). For example, if Siberia's frozen peat bog was to thaw rapidly, it could release billions of tons

Chris Caldeira

When gas prices rise, they tend to rise in pennies at a time. Each penny added to the price by itself triggers little change to driving behavior. But at some point—the tipping point—the price becomes so high, large numbers of people begin to change their behavior.

social change Any significant alteration, modification, or transformation in the organization and operation of social life.

tipping points Situations in which a previously rare (or seemingly rare) event, response, or opinion becomes dramatically more common.

TABLE 16.1 World Wide Fossil Fuel Consumption since 1850 and U.S. Share of World Consumption

Energy Source	Year First Used for Industrial Purposes	Amount Used in 1850	Amount Used in 2010	% Consumed by U.S.
Coal (all)	1748	994 million tons	6.9 billion tons	19*
Oil	1859	5,475 barrels	32.2 billion barrels	19.0
Natural gas	1821	---	2,819 trillion cubic meters	20.8

Sources: Data from World Coal Association 2011, U.S. Central Intelligence Agency 2011.
*The United States produces more coal each year than it consumes.

of the greenhouse gas methane. That thawing could represent a tipping point, triggering a far greater increase in global surface temperature (Sample 2005). The warming global temperature does not mean snowstorms will cease to be. But warming is associated with probability of extreme weather events, such as record snowfall, winds, rain, and heat (Harris 2011).

Changes in Social Activity

CORE CONCEPT 2 When studying social change, sociologists ask three key questions: What has changed? What factors triggered that change? What are the consequences of the change?

Social change is an important sociological topic. In fact, sociology first emerged as a discipline attempting to understand an event that triggered dramatic and seemingly endless changes in every area of human life. That event was the Industrial Revolution. The discipline of sociology offers a number of broad concepts to help us answer the question: "What about human activity has changed since 1750 that has increased dependency on fossil fuels?" The changes in human activity are embodied in the following ongoing processes: industrialization and mechanization, globalization, rationalization, McDonaldization, urbanization, and the information explosion.

Industrialization and Mechanization

We learned in Chapter 1 that the Industrial Revolution is an ongoing process that started as early as 1300, but gained dramatic momentum between 1750 and 1850. The most critical factor driving the momentum was mechanization—the addition of external sources of power, such as coal, oil, and natural gas, to hand tools, appliances, and modes of transportation. Among other things, these new energy sources replaced wind- and human-powered sailboats with steamships and then freighters, and they replaced horse-drawn carriages with trains, cars, and trucks. The Industrial Revolution turned

Chris Caldeira

Almost every product we use involves oil. These commonplace products pictured are made of oil. In addition, fossil fuel–powered machines produced them and workers involved in their production drove cars to work. Finally, these products were transported from factory to stores by trucks, planes, and freighters that run on oil or diesel fuel (Ranken Energy Corporation 2011).

us into a **hydrocarbon society**—one in which the use of fossil fuels shapes virtually every aspect of our personal and social lives (see Table 16.1). The energy source that lights the night, cools our houses and offices, and powers appliances and tools is likely coal; the energy source that heats our houses, workplaces, and water is likely natural gas; and the energy source that enables trains, planes, cars, and buses to move people and goods short and long distances is most certainly oil. When we burn fossil fuels to make the energy to do all these things, we emit greenhouse gases (such as carbon dioxide and methane) into the air.

Globalization

In the most general sense, global interdependence is a situation in which social activity transcends national borders

hydrocarbon society One in which the use of fossil fuels shapes virtually every aspect of our personal and social lives.

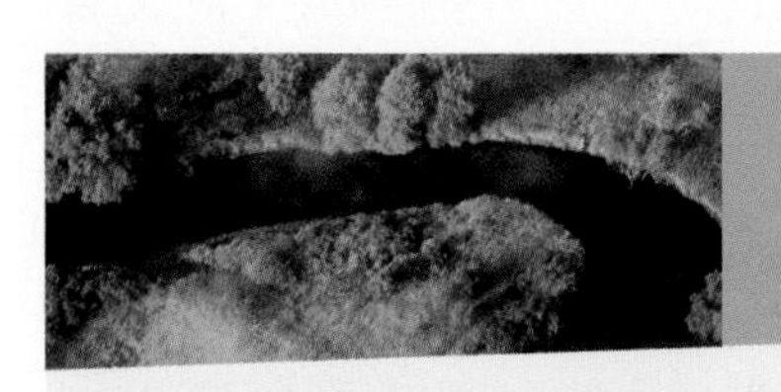

No Borders No Boundaries Facts about UPS, a Global Package Delivery System

One corporation whose 400,000 employees facilitate globalization is United Parcel Service (UPS), founded in 1907. That corporation depends on fossil-fueled vehicles to deliver or pick up 7.4 million packages per day to more than 200 countries.

	Employees	U.S.-based	330,600
		Foreign-based	70,000
	Daily Flights	International	815
		Domestic	942
	Delivery Fleet	Cars, vans, tractors, motorcycles	93,464*
		UPS Jet Aircraft	216
		Chartered Aircraft	311
	Countries served		200+
	Areas served in U.S.		Every address
	Packages Delivered per Day	International	2.3 million
		Domestic	13.8 million
	Packages Picked Up and Delivered per Day		7.4 million
	Hits on UPS.com per day		15 million

FIGURE 16.1

Source: Data from UPS (2011).

and in which social issues experienced locally are part of a larger global situation. Because the level of global interdependence is constantly changing, it is part of a dynamic process known as globalization—the ever-increasing flow of goods, services, money, people, information, and culture across political borders (Held et al. 1999). Sociologists debate the events that triggered globalization. Theoretically, one could trace its origins back 5 million years to East Africa (believed to be the cradle of human life) and to a time when humans began to spread out and eventually populate and dominate the planet. Other potential dates that mark the start of globalization include the invention of the printing press (1436) and the steam engine (1712). Regardless of the date, there is no question that globalization is a fossil fuel–driven phenomenon. Obviously, humans use trains, cars, buses, boats, planes, phones, and the Internet to deliver people, products, services, and information across national borders (see No Borders, No Boundaries: "Facts about UPS, a Global Package Delivery System"). And as globalization has increased, so has the demand for fossil fuels. One of the most profound measures of global interdependence is that 25,000 shipments of imported food products enter the United States each day. That translates into 20 million shipments per year. For example, "92 percent of all fresh and frozen seafood consumed is imported; 52 percent of the grapes; 75 percent of the apple juice; and 72 percent of the mushrooms" (Online Newshour 2007).

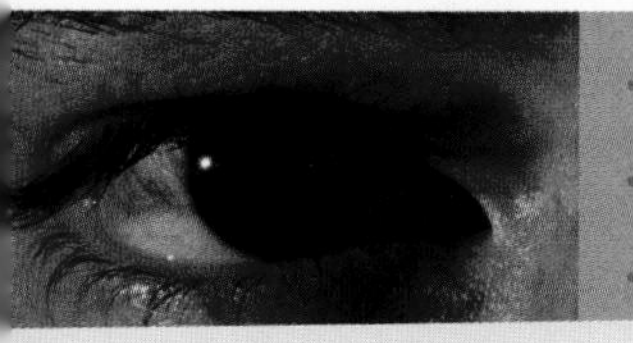

The Sociological Imagination Carbon Footprints

Do you think a Buddhist monk, who has dedicated his life to simple, nonmaterialist living, has a large carbon footprint? Carbon footprints are the imprints people make on the environment by virtue of their lifestyle. That impact is measured in terms of fossil fuels consumed or units of carbon dioxide emitted from burning that fuel. We can think of carbon footprints as primary or secondary. The primary carbon footprint is the total amount of carbon dioxide emitted as the result of someone's direct use of fossil fuels to heat and cool a home, run appliances, facilitate travel, and so on. The secondary footprint is the total amount of carbon dioxide emitted to manufacture a product or deliver a service to that person. Tim Gutowski (2008a), a professor of mechanical engineering at MIT, and 21 of his students studied 18 different lifestyles in the United States, including a person without a home (homeless), Buddhist monks, a patient in a coma, and a professional golfer. After extensive interviews, they estimated the energy each lifestyle requires. They estimated that no lifestyle uses less than 120 gigajoules of energy, even that of a Buddhist monk, whose lifestyle is devoted to simple living, modest dress, and a vegetarian diet. A gigajoule (GJ) is a metric measure of energy consumption. It is a particularly useful measure because it can be applied to different types of energy consumption, such as kilowatts of electricity, liters of heating oil or gasoline, and cubic feet of natural gas. One GJ is equivalent to the energy needed to cook over 2,500 hamburgers or to keep a 60-watt bulb lit for six months (Natural Resources Canada 2010).

Chris Caldeira

Gutowski and his students estimated that the Buddhist monk interviewed consumed 120 GJ of energy each year and the professional golfer interviewed consumed 8,000 GJ. The Buddhist monk consumes about one-third as much energy as the average American (350 GJ) but double the average amount of energy each person on the planet consumes (64 GJ). The Buddhist monk's 120 GJ emits 8.5 metric tons of carbon dioxide each year; the professional golfer's lifestyle (8,000 GJ) produces 566 metric tons of carbon emissions.

Gutowski's study (2008b) has important implications: The United States has a "very energy-intensive system" that in effect sets the lower limits on how much energy any American uses. If the Buddhist monk uses 120 GJ, then we might conclude that energy use is woven into the fabric of our society (Revkin 2005).

Rationalization

We learned in previous chapters that rationalization is a process whereby thought and action rooted in emotion (such as love, hatred, revenge, or joy), superstition, respect for mysterious forces, and tradition are replaced by instrumental-rational thought and action that involves employing the fastest and most cost-efficient means to achieve a desired result (for example, by any means necessary).

One important example of rationalization is the profit-making strategy known as **planned obsolescence**, which involves producing goods that are disposable after a single use, are designed to have a shorter life cycle than the industry is capable of producing, or go out of style quickly (Gregory 1947). The market offers an endless number of disposable (single-use) products, including paper cups, paper towels, diapers, cameras, razors, plastic utensils, and paper tablecloths. And the market offers many other products that do not seem to last as long as they once did. For example, major household appliances such as water heaters and refrigerators built since 2000 are projected to last 8 to 12 years, whereas those built in the 1970s and 1980s lasted 20 years or more (Repair2000.com 2007). Many people buy a new car even though their old car is still in excellent-to-good condition. Similarly, people tend to buy new clothes before they wear out the clothes they already have. Planned obsolescence is fossil fuel–driven in that we use those fuels to manufacture, deliver, operate, and eventually haul away obsolete products (see The Sociological Imagination: "Carbon Footprints").

planned obsolescence A profit-making strategy that involves producing goods that are disposable after a single use, have a shorter life cycle than the industry is capable of producing, or go out of style quickly even though the goods can still serve their purpose.

Library of Congress Prints and Photographs Division Washington, D.C.

Until 1971, the United States produced more oil than it consumed. This 1944 image shows workers managing large valves that regulated the flow of oil into tanker ships. The oil was to be shipped for use by the U.S. armed forces and allies. At that time, the United States was producing 1.6 billion barrels of oil per year. Today, the United States produces 3.5 billion barrels and consumes 6.9 billion.

The McDonaldization of Society

In Chapter 6, we studied the organizational trend McDonaldization, a process whereby the principles governing the fast food industry come to dominate other sectors of the American economy, society, and the world (Ritzer 1993). These principles are (1) efficiency, (2) quantification and calculation, (3) predictability, and (4) control. Efficiency means offering a product or service that can move consumers quickly from one state of being to another (say, from hungry to full, from fat to thin, from uneducated to educated, or from sleeplessness to sleep). Quantification and calculation involves providing numerical indicators by which customers can easily evaluate a product or service (for example: We deliver within 30 minutes! Lose 10 pounds in 10 days! Earn a college degree in 24 months! Limit menstrual periods to four times—or to no times—a year! Obtain eyeglasses in one hour!). Predictability ensures that a service or product will be the same no matter where or when it is purchased. Control means planning out in detail the process of producing and acquiring a service or product (for example, by filling soft drinks from dispensers that automatically shut off, or by having customers stand in line).

There is no question that McDonaldization is a fossil fuel–driven phenomenon. For example, pharmacies, banks, and car wash businesses have adopted "drive-through" services to facilitate their goal of moving customers from one state of being to another quickly. Of course, "drive-through" service is not all there is to McDonaldization. Whatever the service offered—a college degree in 18 months, a medical checkup integrated into a one-stop shopping establishment, matchmaking with success guaranteed in six weeks, a prepaid funeral, or the cheapest air flight—we can find one or more of the McDonaldization principles operating.

Urbanization

Another fossil fuel–driven phenomenon is urbanization—a transformative process in which people migrate from rural to urban areas and change the way they use land, interact, and make a living. In 1900, 13 percent of the world's population was considered urban; that percentage increased to 29 percent in 1950, and today stands at 50 percent (United Nations 2011). Urbanization has shifted a significant percentage of the population away from labor-intensive agricultural occupations into manufacturing, information, and service occupations—all of which depend heavily on fossil fuels not only to make, distribute, and deliver goods and services but to gather employees together at a workplace (World Resource Institute 1996).

From a global perspective, urban populations include not only city dwellers but also suburbanites and even residents of small towns. Spatial sociologists argue that highways and automobiles have created urban sprawl and have made it difficult to distinguish between city, suburbs, and non-urban environments. Urban sprawl spreads development

Chris Caldeira

Many businesses have borrowed the concept of "drive-through" service from fast food industry including this liquor store.

beyond cities by as much as 40 or 50 miles; puts considerable distance between homes, stores, churches, schools, and workplaces; and makes people dependent on automobiles (Sierra Club 2007). In addition, the automobile and highway have allowed people to live in more space than they need. In the past 30 years, the average house size has increased from 1,400 to 2,330 square feet (National Association of Homebuilders 2007).

The Information Explosion

Sociologist Orrin Klapp (1986) wrote about the **information explosion**, an unprecedented increase in the amount of stored and transmitted data and messages in all media (including electronic, print, radio, and television). One can argue that the information explosion began with the invention of the printing press. Today, the information explosion is driven by the Internet, a vast fossil fuel–powered computer network linking billions of computers around the world. The Internet has the potential to give users immediate access to every word, image, and sound that has ever been recorded (Berners-Lee 1996).

Although it is virtually impossible to catalog all the changes associated with the Internet, we can say that, thanks to the fossil fuels used to manufacture the hardware and to run the software that power, it has (1) sped up old ways of doing things, (2) given individuals access to the equivalent of a printing press, (3) allowed users to bypass the formalized hierarchy devoted to controlling the flow of information, (4) changed how students learn, and (5) allowed people around the world to exchange information on and communicate about any topic of interest.

Triggers of Social Change

When we think about a specific social change, such as fossil fuel dependence, we usually cannot pinpoint a single trigger of that change. That is because change tends to result from a complex series of interconnected events. An analogy may help clarify this point: Suppose that a wide receiver, after catching the football and running 50 yards, is tackled at the 5-yard line by a cornerback. One could argue that the cornerback *caused* the receiver to fall to the ground. Such an account, however, does not fully explain the cause. For one thing, the tackle was not the act of one person seizing and throwing his weight onto the person with the ball; it was more complex than that. The teammates of both the wide receiver and the tackler determined how that play developed and ended. Furthermore, the wide receiver was doing everything in his power to elude the tackler's grasp (Mandelbaum 1977).

The forces that result in change are complex. Regardless, we can identify some of the key triggers of change: innovation, revolutionary ideas, conflict, the pursuit of profit, and social movements.

Innovations

CORE CONCEPT 3 An innovation triggers changes in social activity. For an innovation to emerge, however, the cultural base must be large enough to support it.

Innovation is the invention or discovery of something, such as a new idea, process, practice, device, or tool. Innovations can be placed into two broad categories: basic or improving. The distinction between the two is not always clear-cut, however. **Basic innovations** are revolutionary, unprecedented, or groundbreaking inventions or discoveries that form the basis for a wide range of applications. Basic innovations include the cotton gin, steam engine, and first-generation PC (personal computer). The discoveries of the industrial uses for coal (1748), natural gas (1821), and oil (1853) certainly qualify as basic innovations.

Improving innovations, by comparison, are modifications of basic inventions that improve upon the originals—for example, making them smaller, faster, more user-friendly, more efficient, or more attractive. Each "upgrade" of the 1903 Wright Flyer (the first successful airplane, which Orville and Wilbur Wright designed, built, and flew) increased the airplane's capacity to fly farther, higher, faster, and with more passengers. Thirty years and many innovations after the 1903 Wright Flyer, Boeing unveiled the first modern passenger airliner, which could carry 10 passengers at the speed of 155 miles per hour. In 1958, the jet age arrived when Boeing unveiled the first U.S. passenger jet, capable of carrying 181 passengers at the speed of 550 miles per hour. In 1969, wide-body jets, capable of seating 450 passengers, made their debut (Airport Transport Association 2001). The significance of this series of improving innovations is evident when we consider that, in 2009, a total of 4.8 billion passengers traveled by airplane (Airports Council International 2010). The number of international passengers reached 935 million, up from

---

information explosion An unprecedented increase in the amount of stored and transmitted data and messages in all media (including electronic, print, radio, and television).

innovation The invention or discovery of something, such as a new idea, process, practice, device, or tool.

basic innovations Revolutionary, unprecedented, or groundbreaking inventions or discoveries that form the basis for a wide range of applications.

improving innovations Modifications of basic inventions that improve upon the originals—for example, making them smaller, faster, less complicated, more efficient, more attractive, or more profitable.

Department of Defense

Every year, fossil fuel–powered jets carry an estimated 4.8 billion people to destinations around the world.

an estimated 50,000 passengers in 1950 (World Tourism Organization 2011). This number is expected to increase to 1.6 billion by 2020.

Anthropologist Leslie White (1949) argued that once a basic or an improving innovation has emerged, it becomes part of the **cultural base**, the number of existing inventions. White defined an **invention** as a synthesis of existing inventions. For example, the first successful airplane was a synthesis of many preceding inventions, including the gasoline engine, the rudder, the glider, and the wheel.

White suggested that the number of inventions in the cultural base increases geometrically—1, 2, 4, 8, 16, 32, 64, and so on. (Geometric growth is equivalent to a state of runaway expansion.) He argued that for an invention to emerge, the cultural base must be large enough to support it. If the Wright brothers had lived in the fourteenth century, for example, they could never have invented the airplane, because the cultural base did not contain the ideas, materials, and innovations to support its invention. The process prompted White to ask a question: Are people in control of their inventions, or do inventions control people? For all practical purposes, he believed that inventions control us. White cited two arguments to support this conclusion.

First, he suggested that the old adage "Necessity is the mother of invention" is naive. In too many cases, the opposite idea—that invention is the mother of necessity—is true. That is, an invention becomes a necessity because we find uses for the invention after it comes into being: "We invent the automobile to get us between two points faster, and suddenly we find we have to build new roads. And that means we have to invent traffic regulations and put in stop lights [and build garages]. And then we have to create a whole new organization called the Highway Patrol—and all we thought we were doing was inventing cars" (Norman 1988, p. 483).

Second, White (1949) argued that when the cultural base is capable of supporting an invention, then the invention will come into being whether or not people want it. White supported this conclusion by pointing to **simultaneous-independent inventions**—situations in which more or less the same invention is produced by two or more people working independently of one another at about the same time (sometimes within a few days or months). He cited some 148 such inventions—including the telegraph, the electric motor, the steamboat, the car, and the airplane—as proof that someone will make the necessary synthesis if the cultural base is ready to support a particular invention. In other words, the light bulb and the airplane would have been developed regardless of whether Thomas Edison and the Wright brothers (the people we traditionally associate with these inventions) had ever been born. According to White's conception, inventors may be geniuses, but they must also be born in the right place and the right time—that is, in a society with a cultural base sufficiently developed to support their inventions. None of the inventors associated with the inventions just mentioned could have delivered their products if fossil fuels had not already been adapted for industrial use.

According to White's theory, if the parts are present, someone will eventually put them together. The implications are that people have little control over whether an invention comes into being and that they adapt to inventions after the fact. Sociologist William F. Ogburn (1968) calls the failure to adapt to a new invention *cultural lag*.

Cultural Lag In his theory of cultural lag, Ogburn (1968) distinguishes between material culture and nonmaterial culture. Material culture includes tangible things—including resources (such as oil, coal, natural gas, trees, and land), inventions (such as paper and the automobile), and systems (such as factories and package delivery)—that people have produced or, in the case of resources such as oil, have identified as having the properties to serve a particular purpose. Nonmaterial culture, by contrast, includes intangible things, such as beliefs, norms, values, roles, and language. Although Ogburn maintains that both material and nonmaterial culture are important agents of social change, his theory of cultural lag emphasizes the material component, which he suggests is the more important of the two. The case of the automobile illustrates how this piece of material culture changed the United States.

cultural base The number of existing innovations, which forms the basis for further inventions.

invention A synthesis of existing innovations.

simultaneous-independent inventions Situations in which more or less the same invention is produced by two or more people working independently of one another at about the same time.

> The availability of cheap energy and an inexpensive, mass-produced car soon transformed the American landscape. Suddenly there were roads everywhere—paid for, naturally enough, by a gas tax. Towns that had been too small for the railroads were now reachable by roads, and farmers could get to once unattainable markets. Country stores that stood at old rural crossroads and sold every conceivable kind of merchandise were soon replaced by specialized stores, catering to people who could drive off and shop where they wanted. Families that had necessarily been close and inwardly focused, in part because there was nowhere but home to go at night, weakened somewhat when family members could get in their cars and take off to do whatever they wanted to do. (Halberstam 1986, pp. 78–79)

Ogburn believes that one of the most urgent challenges facing people today is the need to adapt to material innovations in thoughtful and constructive ways. He uses the term **adaptive culture** to describe the nonmaterial component's role in adjusting to material innovations. One can argue that Americans adapted easily to the automobile because it supported deeply rooted norms—values and beliefs that applied to a nation composed mostly of immigrants, who by definition had separated from their native lands and traditions. The automobile simply extended a tradition of forsaking, voluntarily or involuntarily, "the region and habits of their parents" and striking out on their own (Halberstam 1986, p. 79). On the other hand, calls to cut back driving to lessen the nation's dependence on foreign sources of oil have made little to no impact on most Americans. They resist changing the norms that govern their driving habits, a value system that defines the car as the measure of personal freedom and independence, and a belief system that holds the car to be the most efficient method of transportation.

The case of the automobile suggests that adjustments are not always immediate. Sometimes they take decades; sometimes they never occur. Ogburn uses the term **cultural lag** to refer to a situation in which adaptive culture fails to adjust in necessary ways to a material innovation. Despite Ogburn's emphasis on material culture as the key force driving change, he is not a **technological determinist**—someone who believes that humans have no free will and are controlled entirely by their material innovations. Instead, he notes that people do not adjust to new material innovations in predictable and unthinking ways; rather, they choose to create them and only then choose how to use them. Ogburn argues that if people have the power to create material innovations, then they also have the power to destroy, ban, regulate, or modify those innovations. The challenge lies in convincing people that they need to address an innovation's potentially disruptive consequences before those consequences materialize.

Library of Congress

Homes built before 1904 did not have garages. Garages became a standard feature of homes probably decades after the invention of the automobile. In the meantime, people improvised.

Revolutionary Ideas

CORE CONCEPT 4 Social change occurs when someone breaks away from or challenges a paradigm. A scientific revolution occurs when enough people in the community break with the old paradigm and orient their research or thinking according to a new paradigm.

In *The Structure of Scientific Revolutions*, Thomas Kuhn (1975) maintains that people tend to perceive science as an evolutionary enterprise. That is, they imagine scientists as problem solvers building on their predecessors' achievements. Kuhn takes issue with this evolutionary view, arguing that some of the most significant scientific advances have been made when someone has broken away from or challenged a paradigm. He defines **paradigms** as the dominant and widely accepted theories and concepts in a particular field of study.

Paradigms gain their status not because they explain everything, but rather because they offer the "best way" of looking at the world at that time. On the one hand, paradigms are important thinking tools; they bind a group of people with common interests into a scientific or national community. Such a community cannot exist without agreed-on paradigms. On the other hand, paradigms can

adaptive culture The portion of nonmaterial culture (norms, values, and beliefs) that adjusts to material innovations.

cultural lag A situation in which adaptive culture fails to adjust in necessary ways to material innovation.

technological determinist Someone who believes that human beings have no free will and are controlled entirely by their material innovations.

paradigms The dominant and widely accepted theories and concepts in a particular field of study.

Chris Caldeira

Sustainable buildings represent a paradigm shift in the way we see and design buildings. Among other things, this building is completely sustainable, with a living roof and two silos to capture rainwater that can then be used for things like flushing toilets.

act as blinders, limiting the kinds of questions that people ask and the observations that they make.

The explanatory value—and hence the status—of a paradigm is threatened by any **anomaly**, an observation that the paradigm cannot explain. The existence of an anomaly by itself usually does not persuade people to abandon a particular paradigm. According to Kuhn (1975), before people discard an old paradigm, someone must articulate an alternative paradigm that accounts convincingly for the anomaly. He hypothesizes that the people most likely to put forth new paradigms are those who are least committed to the old paradigms—the young and those new to a field of study.

A scientific revolution occurs when enough people in the community break with the old paradigm and change their research or thinking to favor the new paradigm. Kuhn (1975) considers a new paradigm to be incompatible with the one it replaces, because it "changes some of the field's most elementary theoretical generalizations" (p. 85). The new paradigm causes converts to see the world in an entirely new light and to wonder how they could possibly have taken the old paradigm seriously. "When paradigms change, the world itself changes with them. Led by a new paradigm, scientists adopt new instruments and look in new places" (p. 111).

Perhaps the best example of a scientific revolution can be found in the work of Nicolaus Copernicus, author of *On the Revolutions of the Heavenly Spheres* (1543), which challenged a long-held belief that Earth (and by extension, humankind) was the stationary center of the solar system, with the sun, moon, and planets revolving around it. Copernicus maintained that the stationary center of the solar system was the sun and that Earth and the other planets revolved around that center. Copernicus's ideas did not take hold immediately. In 1633 (90 years later), powerful church inquisitors threatened to torture and kill Galileo, who had embraced Copernicus's theory, if he did not renounce it; upon renouncing it, Galileo was imprisoned for life. "Of all discoveries and opinions, none may have exerted a greater effect on the human spirit than the doctrine of Copernicus. The world had scarcely become known as round and complete in itself when it was asked to waive the tremendous privilege of being the center of the universe. Never, perhaps, was a greater demand made on mankind—for by this admission so many things vanished in mist and smoke!" (Goethe 2004).

anomaly An observation that a paradigm cannot explain.

Conflict

CORE CONCEPT 5 Conflict can create new norms, relationships, ways of thinking, and innovations.

Conflict occurs whenever a person or group takes action to increase its share of wealth, power, prestige, or some other valued resource and when those actions are resisted by those who benefit from the current distribution system. In other words, those in control of valued resources take action to protect what they have.

In general, any kind of change has the potential to trigger conflict between those who stand to benefit and those who stand to lose from that change. When the bicycle was invented in the 1840s, horse dealers organized against it because it threatened their livelihood. Physicians declared that people who rode bicycles risked getting "cyclist sore throat" and "bicycle stoop." Church groups protested that bicycles would swell the ranks of "reckless" women (because bicycles could not be ridden sidesaddle).

Conflict can be a constructive and invigorating force. The Internet, for example, emerged out of conflict between the then Soviet Union and the United States. That conflict was the cold war, when U.S. government leaders pulled together scientists from three sectors—military, industrial, and academic—to coordinate their research and expertise and thereby serve the war effort. Because these scientists worked in offices and laboratories across the United States, Department of Defense officials worried about the consequences of an attack on a military laboratory, defense contractor, or university site. The officials realized that they needed a computer network that would allow information stored at one site to be transferred to another site in the event of an attack, especially a nuclear attack. At the same time, the computer network had to be designed so that if one or more parts of the network failed or were knocked out by a bomb, the other parts could continue to operate.

Such a design meant that no central control could exist for the network. After all, if central control was destroyed, the entire network would crash. The Internet began in the late 1960s, as ARPANET (Advanced Research Projects Agency Network), linking four universities: the University of California–Los Angeles (UCLA), the University of

California–Santa Barbara (UCSB), the University of Utah, and Stanford University. Thus, the Internet was originally designed to (1) transfer information from one site to another quickly and efficiently in the event of war, and (2) create an information-sharing system without central control. Today, an estimated 2 billion people—28 percent of the world's population—are connected to the fossil fuel–powered Internet (Internet World Stats 2011).

The Pursuit of Profit

CORE CONCEPT 6 The profit-driven capitalist system drives change as it seeks to revolutionize production, create new products, and expand markets. In a capitalist system, profit is the most important measure of success.

Karl Marx believed that an economic system, capitalism, ultimately caused the explosion of technological innovation and the enormous and unprecedented increase in the amount of goods and services produced during the Industrial Revolution. The capitalist system acts as a vehicle of change, because the instruments of production must be revolutionized constantly. Marx believed that the capitalist thirst for profit "chases the bourgeoisie over the whole surface of the globe" (Marx 1881, p. 531). In fact, by the late nineteenth century, the capitalist world economy included virtually the whole inhabited earth. The search for profit is behind efforts to exploit the resources to be found in deserts, jungles, the seas, and the solar system (Wallerstein 1984, p. 165).

Capitalism helped create a global network of economic relationships that is fossil fuel–dependent. Corporations, by the products and services they offer, have helped to create voracious appetites for petroleum- and coal-based fuels (Goldman 2007). Now corporations that have helped to create fossil fuel dependence are asking us to consume our way out of this dependence—we are being asked to buy energy-efficient appliances, light bulbs, and solar panels, use recycled paper, and purchase hybrid vehicles. Consider that fossil fuels are essential to extracting resources, manufacturing and delivering products, and powering appliances (Goldman 2007).

Chris Caldeira

Thousands of products on the market claim to save the planet by going green. These products send the message that we can save the planet by consuming that product. Ethanol made from corn, for example, is touted as being a renewable fuel, but keep in mind fossil fuels are still needed to grow, transport, process, and deliver corn-based fuels. As well, diverting corn for fuel has helped to drive up the price of foods made from corn.

Social Movements

CORE CONCEPT 7 A social movement is formed when a substantial number of people organize to make a change, resist a change, or undo a change in some area of society.

Social movements depend on (1) an actual or imagined condition that enough people find objectionable; (2) a shared belief that something needs to be done about the condition; and (3) an organized effort to attract supporters, articulate the condition, and define a strategy for addressing the condition. Usually those involved in social movements work outside the system to advance their cause, because the existing system has failed to respond. To draw attention to their cause and accomplish objectives, supporters may strike, demonstrate, walk out, sit in, boycott, go on hunger strikes, riot, or terrorize.

Social movements can be placed into four broad categories, depending on the scope and direction of change being sought: regressive, reformist, revolutionary, and counterrevolutionary. Definitions and global examples of each category follow. Keep in mind that the distinctions between the four categories are not always clear-cut. As a result, you might find that some examples fit into more than one category.

Regressive or reactionary movements seek to turn back the hands of time to an earlier condition or state of being, often considered a "golden era." The "Buy Local" movements qualify as such. These movements push to create and support economies rooted in communities and accountable to community interest, rather than the interests of Wall Street and other outside financial investors. Buy Local movements also aim to change people's buying habits in the direction of locally produced products. Among other things, a locally driven economy would

social movement A situation in which a substantial number of people organize to make a change, resist a change, or undo a change in some area of society.

regressive or reactionary movements Social movements that seek to turn back the hands of time to an earlier condition or state of being, one sometimes considered a "golden era."

include restaurants that specialize in serving menu items made from locally grown food and locally raised animals and bartering systems where everyone's time is treated as of equal value so that a surgeon could exchange an hour operation for an hour of landscaping (Simmons 2010).

Reformist movements target a specific feature of society as in need of attention and change. The nonprofit organization Polar Bears International (2011) focuses on saving the polar bear population and its habitat from extinction. Its goals are to use education and research to conserve the world's polar bears, offer educational resources to the public, encourage constructive dialogue, and build an international organization dedicated to saving this population.

Revolutionary movements seek broad, sweeping, and radical structural changes to a society's basic social institutions or to the world order (Benford 1992). The Earth Liberation Front (ELF) is an underground eco-defense movement with no formal leadership or membership. Its members (who sometimes form cells) anonymously and autonomously engage in economic sabotage, including property destruction and guerrilla warfare, against those seen as exploiting and destroying the natural environment. ELF members have made news for setting fire to SUVs on dealership parking lots, a horse slaughterhouse, a scientific research center, a logging company, and a ski resort. Radical environmentalists claim to have committed 1,100 acts of arson and vandalism without killing a single person (Goldman 2007; see Working for Change: "Building Affordable Green Homes").

Counterrevolutionary movements seek to maintain a social order that reform and revolutionary movements are seeking to change. The Petition Project (2011) qualifies as such a movement, as it seeks to challenge reformist and revolutionary movements demanding reductions in greenhouse gas emissions. It recruits basic and applied scientists to sign a Global Warming Petition urging the U.S. government to reject the Kyoto Protocol, an international agreement to limit greenhouse gas emissions. The signers of the petition believe that *limiting* greenhouse gases will actually harm the environment: "There is no convincing scientific evidence that human release of carbon dioxide, methane, or other greenhouse gasses is causing or will, in the foreseeable future, cause catastrophic heating of the Earth's atmosphere and disruption of the Earth's climate. Moreover, there is substantial scientific evidence that increases in atmospheric carbon dioxide produce many beneficial effects upon the natural plant and animal environments of the Earth." The movement's website lists the names of 31,487 basic and applied American scientists who have signed the petition.

Sociologist Ralf Dahrendorf (1973) offers a broad theory that seeks to capture the life of a social movement. In trying to understand the circumstances under which people take action, Dahrendorf focused on structural sources of conflict. Dahrendorf argues that every society possesses formal authority structures (such as a state, a corporation, the military, the judicial system, and school systems). Generally, clear distinctions exist between those who exercise control over that structure and its system of rewards and punishments and those who must obey the commands or face the consequences (loss of job, jail, low grades, and so on). Thus, a distinction between "us" and "them" arises naturally from the unequal distribution of power built into authority structures. As long as there is an authority structure, conflict is inevitable and the potential for those without power to organize in opposition to those with power exists.

Dahrendorf's theory of social movements involves a three-stage process. In the first stage, those without power decide to organize. "It is immeasurably difficult to trace the path on which a person . . . encounters other people just like himself, and at a certain point . . . [says] 'Let us join hands, friends, so that they will not pick us off one by one'" (Dahrendorf 1973, p. 240). Often, a significant event—such as a natural disaster, a nuclear meltdown, a health scare, economic recession—makes seemingly powerless people aware that they share an interest in changing the system. At other times, people organize because they have nothing left to lose. "You don't need courage to speak out. . . . You just need not to care anymore—not to care about being punished or beaten. A point is reached where enough people don't care anymore about what would happen to them if they speak out" (Reich 1989, p. 20).

In the second stage of conflict, if those without authority have opportunities to communicate with one another, some freedom to meet together, the necessary resources, and a leader, then they organize. At the same time, those who hold authority often use the power of their positions to censor information, restrict resources, and undermine leaders' attempts to organize. **Resource mobilization** theorists maintain that having a core group of sophisticated strategists is key to getting a social movement off the ground. Effective strategists can harness a disaffected group's energies, attract money and supporters, capture

reformist movements Social movements that target a specific feature of society as needing change.

revolutionary movements Social movements that seek broad, sweeping, and radical structural changes to a society's basic social institutions or to the world order.

counterrevolutionary movements Social movements that seek to maintain a social order that reformist and revolutionary movements are seeking to change.

resource mobilization A situation in which a core group of sophisticated strategists works to harness a disaffected group's energies, attract money and supporters, capture the news media's attention, forge alliances with those in power, and develop an organizational structure.

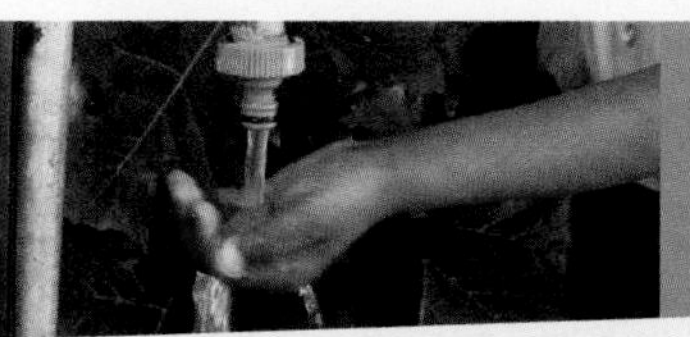

Working for Change Building Affordable Green Homes

All of us can probably remember the tragedy that occurred in 2005, when the violent whirlwind known as Hurricane Katrina came roaring into New Orleans and destroyed tens of thousands of homes and hundreds of neighborhoods. In total, Katrina affected approximately 303,274 people in some way. Prior to Katrina, the population of New Orleans was 484,674 (Information Collective 2011). Many Americans felt compelled to share their resources and donate supplies, money, and unpaid labor to help in recovery and reconstruction. One who wanted to help "make it right" was Brad Pitt. He offered a long-term plan for the Lower Ninth Ward, one of the hardest-hit communities in New Orleans where more than 4,000 homes were destroyed when the Industrial Canal levee failed to hold water. So the humble Hollywood star established the Make It Right Foundation and assembled a team of visionaries and staff to build 150 green, affordable, high-quality design homes.

The Make It Right (2011) team includes three core organizations: (1) William McDonough and Partners, a world leader in environmental architecture; (2) the Cherokee Gives Back Foundation, the nonprofit arm of Cherokee, a firm that specializes in remediation and sustainable redevelopment of environmentally impaired properties; and (3) Graft, an innovative architecture firm that Pitt has collaborated with on other projects around the world. There are also 27 staff members, including an executive director (Tom Darden), director of communications (Taylor Royle), director of development and government affairs (Steve Ragan), construction director (Jon Sader), chief operations officer (Veronica Taylor), and 22 supporting staff.

To launch the construction project, architects presented designs for energy-efficient green homes. Starting with the exterior, intelligent features are incorporated to save energy and reduce monthly water use. Metal roofs help to reduce air conditioning costs because they absorb less heat; and solar power panels decrease dependence on fossil fuels. The proper elevation of homes (on stilts) makes them flood resistant and creates space for parking under the home. Advanced framing techniques are used to increase the resilience of homes, allowing them to withstand 130+ mph winds. Fiber cement board sliding is used to prevent cracking, rotting, weather damage, termite infestation, and other problems. Landscaping is such that it requires minimal maintenance and is capable of handling drought conditions and temporary inundation of water. Rainwater harvesting allows homeowners to collect 600 gallons of water off the roof on an annual basis, giving homeowners water to wash their cars and water plants.

Just as impressive, the interiors of the homes adhere to the principles of cradle-to-cradle design. In the context of home building, this means the design should be not only energy-efficient but also strive to be waste-free. Examples of features incorporated in home design include

- tankless water heaters, which are 83 to 93 percent more efficient than traditional tank water heaters and can reduce the annual cost of heating water by 50 percent;
- energy-efficient appliances and light fixtures that use 30 percent less energy than conventional ones;
  - spray foam insulation placed under the roof, walls, and floors;
  - special energy-efficient windows and doors to help keep homes cool in summer and warm in winter; and
  - dual flush toilets that allow homeowners the option of using less water to flush liquid waste.

There are certain qualifications one must meet before moving into a Make It Right home. First and foremost, the person applying for a home must have lived in the Lower Ninth Ward when Hurricane Katrina hit (Make It Right can verify this once an applicant has filled out an online expression of interest form). Potential homeowners must prove they are able to afford the cost of maintenance, taxes, and insurance, and that they can devote one-third of their income to house payments. However, for those who don't qualify financially, Make It Right offers in-house counseling, financial literacy training, educational workshops, and other credit and budget counseling services. Once approved, those selected are able to choose between a single-family home costing an average of $150,000, or a duplex costing an average of $200,000 (depending on the size and design of the home). To date, the average resident contribution has been $75,000. The rest is covered by grants from Road Home, an outside mortgage financing firm, and Make It Right. To date, 73 people from the Lower Ninth Ward have returned to live in their new homes. Make It Right has received many awards but perhaps the most notable was the recognition by U.S. Green Building Council (USGBC) as being "the largest and greenest community of single-family homes in the world." Through hard work and clear visions, the Make It Right Foundation has demonstrated that social change is possible.

Source: Written by Dayna Schambach, Northern Kentucky University, class of 2011.

Chris Caldeira

Greenpeace is an international organization that uses nonviolent means to challenge the authority and government support of the big oil companies and draws public attention to the world's major environmental problems.

the news media's attention, forge alliances with those in power, and develop an organizational structure. Cell phones, text messaging, and the Internet have made organizing easier by allowing interested parties to connect in ways that "defy gravity and time" (Lee 2003).

In the third stage of conflict, those seeking change enter into direct conflict with those in power. The capacity of the ruling group to stay in power and the amount and kind of pressure exerted from below affects the speed and the depth of change. If protestors believe that their voices are being heard, the conflict is unlikely to turn violent or move in a revolutionary direction. If those who hold power decide that they cannot compromise and proceed to mobilize all of their resources to thwart protests, two results are possible: First, the protesters may believe that the sacrifices they will have to make to continue protests are too great, so they will withdraw from the fray. Alternatively, the protesters may decide to meet the "enemy" head-on, in which case the conflict may become bloody. If the power differential too greatly favors one side, the protestors or their opponents may resort to **terrorism**—the systematic use of anxiety-inspiring violent acts by clandestine or semiclandestine individuals, groups, or state-supported actors.

terrorism The systematic use of anxiety-inspiring violent acts by clandestine or semiclandestine individuals, groups, or state-supported actors for idiosyncratic, criminal, or political reasons.

Consequences of Change

CORE CONCEPT 8 Sociological concepts and theories can be applied to evaluating the consequences of any social change.

In the 16 chapters of this textbook, we have covered sociological concepts and theories that can be applied to evaluating the consequences of any social change. We close the book by selecting one key idea from each chapter to assess how climate change is affecting Greenland and how outsiders have pushed or pulled Greenland into the global arena. The chapter-specific ideas suggest questions that can help guide analysis of social change.

How is climate change being experienced in Greenland? More specifically, how is climate change shaping human activity and social interactions? (Chapter 1)

Sociologists are compelled to study human activity and social interactions as they are affected by larger social forces—whether those social forces be human created or human responses to the forces of nature. The following three examples highlight human activities brought about by climate change that involve Greenlanders (especially the island's indigenous population, the Inuit) and "outsiders":

- Two 19-year-old men hike through Greenland with Inuit hunting guides as part of a 22,000-mile North Pole–South Pole trek using human muscle- and wind-powered modes of transportation (foot, skis, bicycles, and sailboats) to draw attention to climate change (BBC News 2007).
- In 2009, 19,375 ecotourists came to see Greenland, observe the small island-size icebergs floating outside its harbor, and to observe the effect of climate change, specifically ice sheets emptying into open seas (Folger 2010).
- A Greenlander explains how winter fishing has become a problem as temperatures warm. "For 10 of the last 12 years, the bay has not frozen over in the winter." When the bay used to freeze, fisherman could dogsled out to go ice fishing. "I would spend a day and a night and bring back 200 or 500 pounds of halibut on my sled. Now winter fishing. . . is dangerous with a heavy load; the ice is too thin" (Folger 2010).

An estimated 20,000 ecotourists visit Greenland each year to observe icebergs and to witness the effects of climate change. Although 20,000 may seem like a small number, consider that Greenland's population is only 64,000.

How do sociologists frame a discussion about Greenland and climate change? (Chapter 2)

Each of the three major sociological theories—functionalist, conflict, and symbolic interaction—offers a central question to help guide thinking and a vocabulary for answering that question.

A functionalist asks: What are the anticipated (manifest) and unintended (latent) functions and dysfunctions of climate change on Greenland? One manifest function is an economic boom associated with a lengthened shipping season (once four months long and now six months long), which allows goods to move into and out of Greenland. A latent function is the emergence of working alliances between Inuit Greenlanders and tropical island peoples, who face cultural extinction from rising sea levels. Another latent function is a growing interest in Greenland, the Arctic, and Antarctica, so that popular films are set in or give prominent attention to these locations. Such films include *The Last Winter* (a horror film), *The Golden Compass, Pirates of the Caribbean, Arctic Tale, An Inconvenient Truth,* and *The March of the Penguins*. One manifest dysfunction is a growing ecotourism industry, in which the number of tourists visiting Greenland each year overwhelms the resident population of towns visited. Finally, a latent dysfunction connected to climate change is the loss of status among Inuit elders, who can no longer predict the weather (see the discussion of symbolic interaction following the next paragraph).

A conflict theorist asks: Who benefits from climate change, and at whose expense? Conflict theorists key in on the many industries that have expanded in or moved operations to Greenland because of the warming climate. These commercial interests include zinc, lead, and uranium mining companies; oil drilling and exploration companies; and water companies. From the conflict perspective, such companies and their customers will no doubt benefit at the expense of Greenland and its culture. Greenland Minerals and Energy Ltd., an Australian-based company, has discovered what it believes to be the largest deposits of rare earth metals. Rare earth metals are necessary to produce wind turbines, computer display screens, and hybrid car batteries. Currently, China controls 93 percent of the world's mined rare earth metals (Hiroko 2010, Folger 2010).

Finally, a symbolic interactionist asks: How do the involved parties experience, interpret, influence, and respond to what they and others are doing as interaction occurs? Symbolic interactionists are particularly interested in ways interaction among Greenlanders is changing because of climate change. For example, Inuit Greenlanders no longer turn to their elders for weather forecasts; because of climate change, the signs that once allowed the elders to accurately predict the weather no longer apply.

How is the culture of Greenland's Inuit and of other Arctic peoples changing because of climate change? (Chapter 3)

Sociologists define *culture* as the way of life of a people; more specifically, culture includes the human-created strategies for adjusting to the environment and to the creatures (including humans) that are part of that environment. Sociologists are interested in how climate change is affecting Greenland's Inuit and the other Arctic peoples who have adapted to an extreme weather environment that is now warming (see Intersection of Biography and Society: "Cultural Change in the Arctic"). One change affecting the Inuit is declines in, and even the gradual extinction of, marine species. This loss disrupts or destroys their hunting—and, by extension, their eating—habits. It

No roads connect towns and settlements in Greenland; boats and planes are the modes of transportation. In the winter, people travel using dogsleds and snowmobiles.

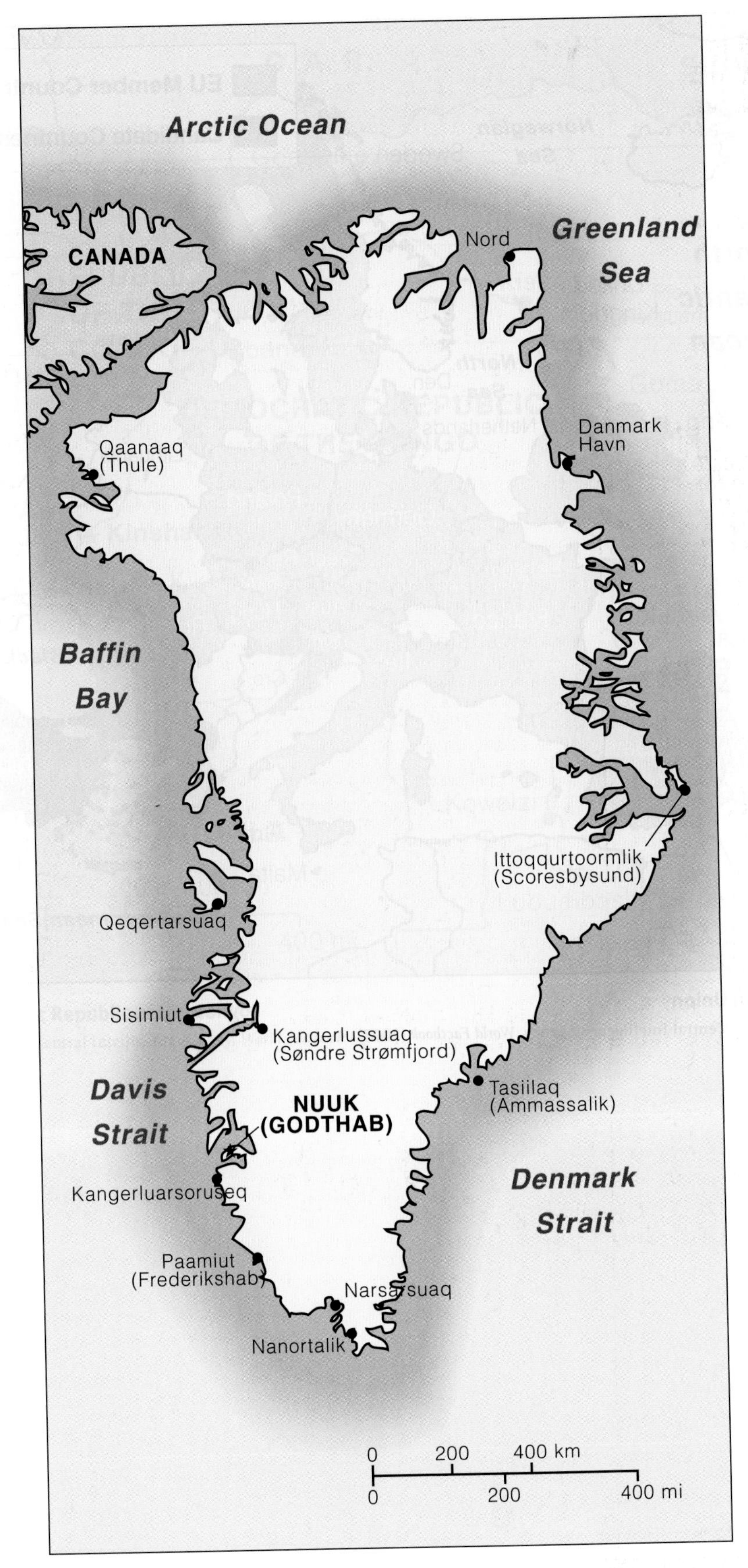

Greenland

Source: U.S. Central Intelligence Agency, *World Factbook* (2007)

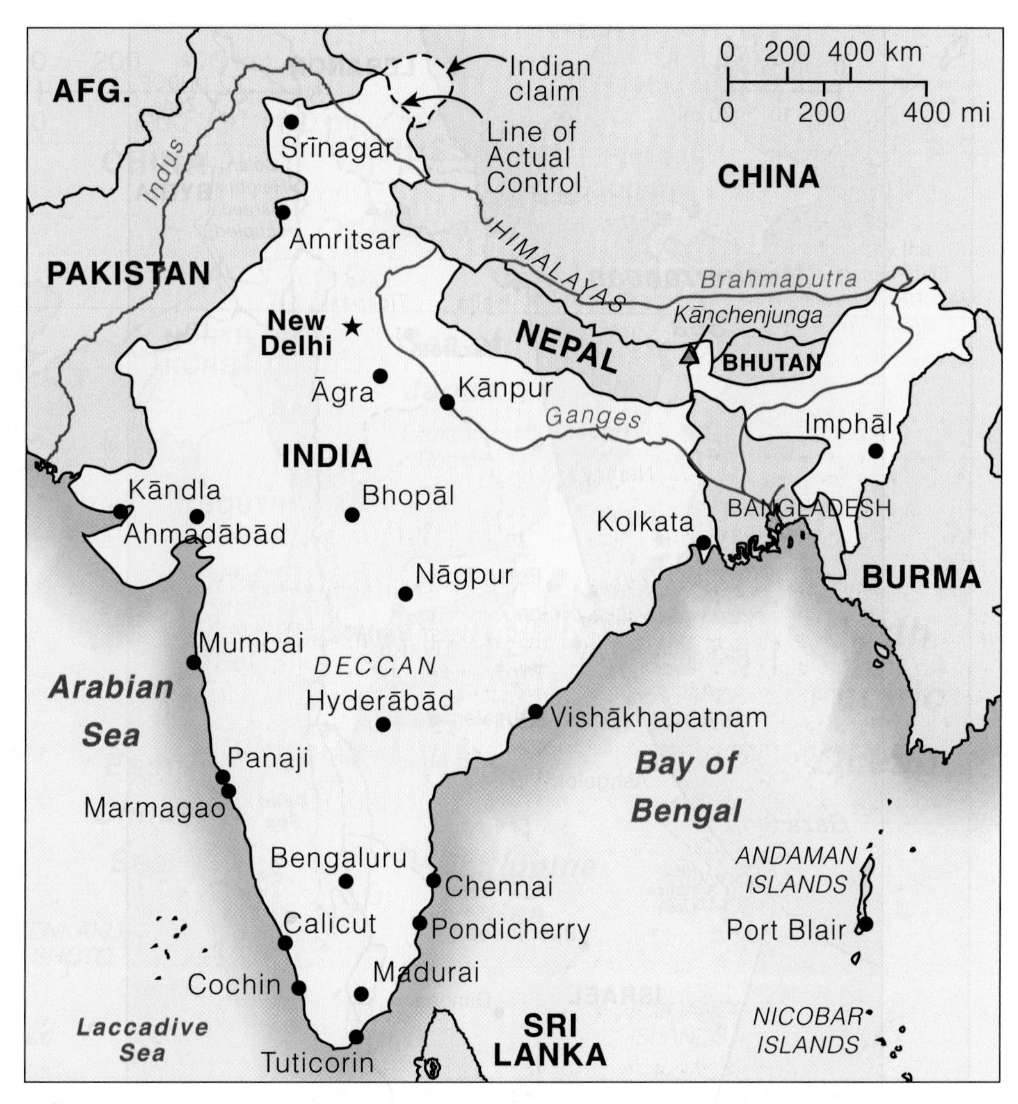

India

Source: U.S. Central Intelligence Agency, *World Factbook* (2007)

extreme wealth The most excessive form of wealth, in which a very small proportion of people in the world have money, material possessions, and other assets (minus liabilities) in such abundance that a small fraction of it (if spent appropriately) could provide adequate food, safe water, sanitation, and basic health care for the 1 billion poorest people on the planet.

facade of legitimacy An explanation that members of dominant groups give to justify the social arrangements that benefit them over others.

falsely accused People who have not broken the rules of a group but are treated as if they have.

family A social institution that binds people together through blood, marriage, law, and/or social norms. Family members are generally expected to care for and support each other.

fatalistic A state in which the ties attaching the individual to the group involve discipline so oppressive it offers no chance of release.

feeling rules Norms that specify appropriate ways to express internal sensations.

femininity The physical, behavioral, and mental and emotional traits believed to be characteristic of females.

feminism In its most basic sense, a perspective that advocates equality between men and women.

finance aristocracy Bankers and stockholders seemingly detached from the world of "work."

folkways Customary ways of handling the routine matters of everyday life.

formal curriculum The essential content of the various academic subjects—mathematical formulas, science experiments, key terms, and so on.

formal dimension The official aspect of an organization, including job descriptions and written rules, guidelines, and procedures established to achieve valued goals.

formal education A purposeful, planned effort aimed at imparting specific skills and information.

formal organization Coordinating mechanisms that bring together people, resources, and technology, and then channel human activity toward achieving a specific outcome.

formal sanctions Expressions of approval or disapproval backed by laws, rules, or policies that specify (usually in writing) the conditions under which people should be rewarded or punished and the procedures for allocating rewards and administering punishments.

fortified households Preindustrial arrangements characterized by no police force, militia, national guard, or other peacekeeping organization. Instead, the household acts as an armed unit, and the head of the household acts as its military commander.

front stage The area visible to the audience, where people feel compelled to present themselves in expected ways.

function The contribution a part of a society makes to the existing social order.

fundamentalism A belief in the timelessness of sacred writings and a belief that such writings apply to all kinds of environments.

games Structured, organized activities that usually involve more than one person and a number of constraints, such as established roles, rules, time, place, and outcome.

gender A social distinction based on culturally conceived and learned ideals about appropriate appearance, behavior, and mental and emotional characteristics for males and females.

gender polarization The organizing of social life around male-female ideals, so that people's sex influences every aspect of their life, including how they dress, the time they get up in the morning, what they do before they go to bed at night, the social roles they take on, the things they worry about, and even the ways they express emotion and experience sexual attraction.

gender-schematic A term describing decisions that are influenced by a society's polarized definitions of masculinity and femininity rather than by criteria such as self-fulfillment, interest, ability, and personal comfort.

generalizability The extent to which findings can be applied to the larger population from which a sample is drawn.

generalized other A system of expected behaviors, meanings, and viewpoints that transcend those of the people participating.

glass ceiling A barrier that prevents women from rising past a certain level in an organization, especially when women work in male-dominated workplaces and occupations.

glass escalator A term that applies to the invisible upward movement that puts men in positions of power, even within female-dominated occupations.

global interdependence A situation in which social activity transcends national borders and in which one country's problems—such as unemployment, drug abuse, water shortages, natural disasters, and the search for national security in the face of terrorism—are part of a larger global situation.

globalization The ever-increasing flow of goods, services, money, people, information, and culture across political borders.

goods Any products that are extracted from the earth, manufactured, or grown, such as food, clothing, petroleum, natural gas, automobiles, coal, computers, and so on.

government An organizational structure that directs and coordinates human activities in the name of a country or some other territory, such as a city, county, or state.

gross domestic product (GDP) The monetary value of the goods and services that a nation's workforce produces in a year or some other time period.

group Two or more people who share a distinct identity, feel a sense of belonging, and interact directly or indirectly with one another.

habitus Objective reality internalized. That internalized reality becomes the mental filter through which people understand the social world and their place in it.

Hawthorne effect A phenomenon in which research subjects alter their behavior when they learn they are being observed.

hegemony A process by which a power maintains its dominance over foreign entities.

hidden curriculum Important messages conveyed to students unrelated to subject content per se.

hidden ethnicity A sense of self that is based on little to no awareness of an ethnic identity because its culture is considered normative, or mainstream.

households All related and unrelated people who share the same dwelling.

hydrocarbon society One in which the use of fossil fuels shapes virtually every aspect of our personal and social lives.

hypothesis A trial explanation put forward as the focus of research; it predicts how independent and dependent variables are related and how a dependent variable will change when an independent variable changes.

I The active and creative aspect of the self that questions the expectations and rules for behavior.

ideal A standard against which real cases can be compared.

ideal type A deliberate simplification or caricature that exaggerates defining characteristics, thus establishing a standard against which real cases can be compared.

illegitimate opportunity structures Social settings and arrangements that offer people the opportunity to commit particular types of crime.

illiteracy The inability to understand and use a symbol system, whether it is based on sounds, letters, numbers, pictographs, or some other type of symbol.

immigration The act of entering one country after leaving another.

impairment A physical or mental condition that interferes with someone's ability to perform an activity that the average person can perform without technical or human assistance.

imperialistic power A political entity that exerts control and influence over foreign entities either through military force or through policies and economic pressures.

impression management The process by which people in social situations manage the setting, their dress, their words, and their gestures to correspond to the impression they are trying to make or the image they are trying to project.

improving innovations Modifications of basic inventions that improve upon the originals—for example, making them smaller, faster, less complicated, more efficient, more attractive, or more profitable.

income The money a person earns, usually on an annual basis through salary or wages.

independent variable The variable that explains or predicts the dependent variable.

individual discrimination Any individual or overt action aimed at someone in an out-group that depreciates, denies opportunities, or does violence to life or property.

industrial food system One that produces high-calorie, nutrient-low, processed food that is more available, affordable, and aggressively marketed than nutritious food. It is a food system in which the goal is to maximize profit, achieved by speeding up the production process, increasing the amount produced, cutting labor costs, and finding the lowest-cost ingredients. Industrial food depends heavily on pharmaceuticals, chemicals, and fossil fuels to produce, manufacture, and transport food products.

infant mortality rate The annual number of deaths of infants 1 year old or younger for every 1,000 such infants born alive.

informal dimension The unofficial aspect of an organization, including behaviors that depart from the formal dimension, such as employee-generated norms that evade, bypass, or ignore official rules, guidelines, and procedures.

informal education Education that occurs in a spontaneous, unplanned way.

informal sanctions Spontaneous, unofficial expressions of approval or disapproval that are not backed by the force of law.

information explosion An unprecedented increase in the amount of stored and transmitted data and messages in all media (including electronic, print, radio, and television).

in-group A group to which a person belongs, identifies, admires, and/or feels loyalty.

in-migration The movement of people into a designated area.

innovation (as a response to structural strain) The acceptance of cultural goals but the rejection of the legitimate means to achieve them.

innovation The invention or discovery of something, such as a new idea, process, practice, device, or tool.

institutionalized discrimination The established, customary way of doing things in society—the unchallenged rules, policies, and day-to-day practices established by advantaged groups that impede or limit the opportunities and achievements of those in disadvantaged groups. It is "systematic discrimination through the regular operations of societal institutions" (Davis 1978, p. 30).

institutions Relatively stable and predictable social arrangements created and sustained by people that have emerged over time with the purpose of coordinating human activities to meet some need, such as food, shelter, or clothing. Institutions consist of statuses, roles, and groups.

insurgents Groups who participate in armed rebellion against some established authority, government, or administration with the hope that those in power will retreat or pull out.

internal migration The movement of people within the boundaries of a single country—from one state, region, or city to another.

internalization The process in which people take as their own and accept as binding the norms, values, beliefs, and language that their socializers are attempting to pass on.

intersectionality The interconnections among socially constructed categories of sex, gender, race, class, sexual orientation, religious affiliation, age (generation), nationality, disability, and other statuses. These statuses combine in complex ways to influence advantages and disadvantages.

intersexed A broad term used by the medical profession to classify people with some mixture of male and female biological characteristics.

interviews Face-to-face or telephone conversations between an interviewer and a respondent, in which the interviewer asks questions and records the respondent's answers.

invention A synthesis of existing innovations.

involuntary ethnicity When a government or other dominant group creates an umbrella ethnic category and assigns people from many different cultures and countries to it.

involuntary minorities Ethnic or racial groups that were forced to become part of a country by slavery, conquest, or colonization.

iron cage of rationality The set of irrationalities that rational systems generate.

Islamic revitalism Responses to the belief that existing political, economic, and social systems have failed—responses that include a disenchantment with, and even a rejection of, the West; soul-searching; a quest for greater authenticity; and a conviction that Islam offers a viable alternative to secular nationalism, socialism, and capitalism.

issue A matter that can be explained only by factors outside an individual's control and immediate environment.

labor-intensive poor economies Economies that have a lower level of industrial production and a lower standard of living than core economies. They differ markedly from core economies on indicators such as doubling time, infant mortality, total fertility, per capita income, and per capita energy consumption.

language A symbol system involving the use of sounds, gestures (signing), and/or characters (such as letters or pictures) to convey meaning.

latent dysfunctions Unintended, unanticipated disruptions to an existing social order.

latent functions Unintended or unanticipated effects that a part has on the existing order.

laws Rules or policies that specify (usually in writing) the conditions under which people should be rewarded or punished and the procedures for allocating rewards and administering punishments.

legal-rational authority A type of authority that rests on a system of impersonal rules that formally specifies the qualifications for occupying a powerful position.

liberation theology A religious movement based on the idea that organized religions have a responsibility to demand social justice for the marginalized peoples of the world, especially landless peasants and the urban poor, and to take an active role at the grassroots level to bring about political and economic justice.

life chances The probability that an individual's life will follow a certain path and will turn out a certain way.

life chances A critical set of potential social advantages, including the chance to survive the first year of life, to live independently in old age, and everything in between.

linguistic relativity hypothesis The idea that "no two languages are ever sufficiently similar to be considered as representing the same social reality. The worlds in which different societies live are distinct worlds, not merely the same world with different labels attached" (Sapir 1949, p. 162).

looking-glass self A process in which a sense of self develops, enabling one to see oneself reflected in others' real or imagined reactions to one's appearance and behaviors.

low-technology tribal societies Hunting-and-gathering societies with technologies that do not permit the creation of surplus wealth.

manifest dysfunctions A part's anticipated disruptions to an existing social order.

manifest functions Intended or anticipated effects that a part has on the existing social order.

masculinity The physical, behavioral, and mental and emotional traits believed to be characteristic of males.

mass media Forms of communication designed to reach large audiences without face-to-face contact between those conveying and those receiving the messages.

master status One status in a status set that overshadows the others such that it shapes every aspect of life and dominates social interactions.

master status of deviant An identification that proves to be more important than most other statuses that person holds, such that he or she is identified first and foremost as a deviant.

material culture All the natural and human-created objects to which people have attached meaning.

maternal mortality rate The death of a woman while pregnant or within 42 days of a termination of pregnancy from any cause related to or aggravated by pregnancy or the way it is managed (World Health Organization).

McDonaldization A process whereby the principles governing the fast-food industry come to dominate other sectors of the American economy, society, and the world.

me The social self—the part of the self that is the product of interaction with others and that has internalized the rules and expectations.

mechanical solidarity Social order and cohesion based on a common conscience, or uniform thinking and behavior.

mechanization The process of replacing human and animal muscle as a source of power with external sources derived from burning wood, coal, oil, and natural gas.

mega city An agglomeration of at least 8 million (UN definition) or 10 million (U.S. definition) people.

melting pot assimilation Cultural blending in which groups accept many new behaviors and values from one another. The exchange produces a new cultural system, which is a blend of the previously separate systems.

methods of data collection The procedures a researcher follows to gather relevant data.

migration The movement of people from one residence to another.

migration rate A rate based on the difference between the number of people entering and the number of people leaving a designated geographic area in a year. We divide that difference by the size of the relevant population and then multiply the result by 1,000.

militaristic power One that believes military strength, and the willingness to use it, is the source of national—and even global—security.

minority groups Subgroups within a society that can be distinguished from members of the dominant group by visible identifying characteristics, including physical and cultural attributes. These subgroups are systematically excluded, whether consciously or unconsciously, from full participation in society and denied equal access to positions of power, privilege, and wealth.

mixed contacts Interactions between stigmatized persons and so-called "normals."

modern capitalism An economic system that involves careful calculation of costs of production relative to profits, borrowing and lending money, accumulating all forms of capital, and drawing labor from an unrestricted global labor pool.

modernization A process of economic, social, and cultural transformation in which a country "evolves" from preindustrial or underdeveloped status to a modern society in the image of the most developed countries.

mores Norms that people define as critical to the well-being of a group. Violation of mores can result in severe forms of punishment.

mortality crises Violent fluctuations in the death rate, caused by war, famine, or epidemics.

multinational corporations Enterprises that own, control, or license production or service facilities in countries other than the one where the corporations are headquartered.

mystical religions Religions in which the sacred is sought in states of being that, at their peak, can exclude all awareness of one's existence, sensations, thoughts, and surroundings.

nature Human genetic makeup or biological inheritance.

negative sanction An expression of disapproval for noncompliance.

negatively privileged property class Weber's category for people completely lacking in skills, property, or employment or who depend on seasonal or sporadic employment; they constitute the very bottom of the class system.

negotiated order The sum of existing expectations and newly negotiated ones.

neocolonialism A new form of colonialism where more powerful foreign governments and foreign-owned businesses continue to exploit the resources and labor of the postcolonial peoples.

net migration The difference between the number moving into an area and the number moving out.

nonmaterial culture Intangible human creations, which we cannot identify directly through the senses.

nonparticipant observation A research technique in which the researcher observes study participants without interacting with them.

nonprejudiced discriminators (fair-weather liberals) People who believe in equal opportunity but discriminate because doing so gives them an advantage or because they simply fail to consider the discriminatory consequences of their actions.

nonprejudiced nondiscriminators (all-weather liberals) People who accept the creed of equal opportunity, and their conduct conforms to that creed. They not only believe in equal opportunity but also take action against discrimination.

norms Written and unwritten rules that specify behaviors appropriate and inappropriate to a particular social situation.

nurture The social environment, or the interaction experiences that make up every individual's life.

objective deprivation The condition of the people who are the worst off or most disadvantaged—those with the lowest incomes, the least education, the lowest social status, the fewest opportunities, and so on.

objectivity A stance in which researchers' personal, or subjective, views do not influence their observations or the outcomes of their research.

observation A research technique in which the researcher watches, listens to, and records behavior and conversations as they happen.

oligarchy Rule by the few, or the concentration of decision-making power in the hands of a few people, who hold the top positions in a hierarchy.

operational definitions Clear, precise definitions and instructions about how to observe and/or measure the variables under study.

organic solidarity Social order or system of social ties based on interdependence and cooperation among people performing a wide range of diverse and specialized tasks.

Out-group A any group to which a person does not belong.

out-migration The movement of people out of a designated area.

overurbanization A situation in which urban misery—poverty, unemployment, housing shortages, insufficient infrastructure—is exacerbated by an influx of unskilled, illiterate, and poverty-stricken rural migrants who have been pushed into cities out of desperation.

paradigms The dominant and widely accepted theories and concepts in a particular field of study.

participant observation A research technique in which the researcher observes study participants while directly interacting with them.

penalties Constraints on a person's opportunities and choices, as well as the price paid for engaging in certain activities, appearances, or choices deemed inappropriate of someone in a particular category.

planned obsolescence A profit-making strategy that involves producing goods that are disposable after a single use, have a shorter life cycle than the industry is capable of producing, or go out of style quickly even though the goods can still serve their purpose.

play A voluntary and often spontaneous activity with few or no formal rules that is not subject to constraints of time or place.

pluralist model A model that views politics as an arena of compromise, alliances, and negotiation among many competing special interest groups, and it views power as something dispersed among those groups.

pluralist models of power A view that sees politics as a plurality of special interest groups competing, compromising, forming alliances, and negotiating with each other for power.

political action committees (PACs) Committees that raise money to be donated to the political candidates most likely to support their special interests.

political parties According to Weber, "organizations oriented toward the planned acquisition of social power [and] toward influencing social action no matter what its content may be."

political system A socially created institution that regulates the use of and access to power that is essential to articulating and realizing individual, local, regional, national, international, or global interests and agendas.

population pyramid A series of horizontal bar graphs, each representing a different five-year age cohort, that allows us to compare the sizes of the cohorts.

populations The total number of individuals, traces, documents, territories, households, or groups that could be studied.

positive checks Events that increase deaths, including epidemics of infectious and parasitic diseases, war, famine, and natural disasters.

positive sanction An expression of approval and a reward for compliance.

positively privileged property class Weber's category for the people at the very top of the class system.

positivism A theory stating that valid knowledge about the world can be derived only from *sense experience* or knowing the world through the senses of sight, touch, taste, smell, and hearing, and from empirical associations.

postindustrial society A society that is dominated by intellectual technologies of telecommunications and computers, not just "large computers but computers on a chip." These intellectual technologies have had a revolutionary effect on virtually every aspect of social life.

power The probability that an individual can achieve his or her will even against another individual's opposition.

power elite Those few people who occupy such lofty positions in the social structure of leading institutions that their decisions have consequences affecting millions, even billions, of people worldwide.

predestination The belief that God has foreordained all things, including the salvation or damnation of individual souls.

predictability The expectation that a service or product will be the same no matter where or when it is purchased.

prejudice A rigid and usually unfavorable judgment about an out-group that does not change in the face of contradictory evidence and that applies to anyone who shares the distinguishing characteristics of the out-group.

prejudiced discriminators (active bigots) People who reject the notion of equal opportunity and profess a right, even a duty, to discriminate. They express with deep conviction that anyone from the in-group (including the village idiot) is superior to any members of the out-group (Merton 1976).

prejudiced nondiscriminators (timid bigots) People who reject the creed of equal opportunity but refrain from discrimination, primarily because they fear possible sanctions or being labeled as racists.

primary deviants Those people whose rule breaking is viewed as understandable, incidental, or insignificant in light of some socially approved status they hold.

primary group A social group that has face-to-face contact and strong emotional ties among its members.

primary sector (of the economy) Economic activities that generate or extract raw materials from the natural environment.

prison-industrial complex The corporations and agencies with an economic stake in building and supplying correctional facilities and in providing services.

productive work Work that involves "the production of the means of existence, of food, clothing, and shelter and the tools necessary for that production" (Engels 1884, pp. 71–72).

profane A term describing everything that is not sacred, including things opposed to the sacred and things that stand apart from the sacred, albeit not in opposition to it.

professionalization A trend in which organizations hire experts with formal training in a particular subject or activity—training needed to achieve organizational goals.

proletariat Those individuals who must sell their labor to the bourgeoisie.

prophetic religions Religions in which the sacred revolves around items that symbolize significant historical events or around the lives, teachings, and writings of great people.

pull factors The conditions that encourage people to move into a geographic area.

pure deviants People who have broken the rules of a group and are caught, punished, and labeled as outsiders.

push factors The conditions that encourage people to move out of a geographic area.

quantification and calculation Numerical indicators that enable customers to evaluate a product or service easily.

racial common sense shared ideas believed to be so obvious or natural they need not be questioned.

racism The belief that genetic or biological differences explain and even justify inequalities that exist between advantaged and disadvantaged racial and ethnic groups.

random sample A type of sample in which every case in the population has an equal chance of being selected.

rationalization A process in which thought and action rooted in custom, emotion, or respect for mysterious forces is replaced by instrumental-rational thought and action.

rebellion The full or partial rejection of both cultural goals and the means of attaining them and the introduction of a new set of goals and means.

reentry shock Culture shock in reverse; it is experienced upon returning home after living in another culture.

reformist movements Social movements that target a specific feature of society as needing change.

regressive or reactionary movements Social movements that seek to turn back the hands of time to an earlier condition or state of being, one sometimes considered a "golden era."

reify Treating racial labels and categories as if they are real and meaningful, forgetting that they are made up or human constructed.

relative deprivation A social condition that is measured not by objective standards, but rather by comparing one group's situation with the situations of groups who are more advantaged.

relative poverty Measured not by some objective standard, but rather by comparing a person's situation against that of others who are more advantaged in some way.

reliability The extent to which an operational definition gives consistent results.

representative sample A type of sample in which those selected for study have the same distribution of characteristics as the population from which it is selected.

reproductive work Work that involves bearing children, caregiving, managing households, and educating children.

research A data-gathering and data-explaining enterprise governed by strict rules.

research design A plan for gathering data that specifies who or what will be studied and the methods of data collection.

research methods Techniques that sociologists and other investigators use to formulate or answer meaningful research questions and to collect, analyze, and interpret data in ways that allow other researchers to verify the results.

resocialization The process that involves breaking with behaviors and ways of thinking that are unsuited to existing or changing circumstances, and replacing them with new, more appropriate ways of behaving and thinking.

resource mobilization A situation in which a core group of sophisticated strategists works to harness a disaffected group's energies, attract money and supporters, capture the news media's attention, forge alliances with those in power, and develop an organizational structure.

retreatism The rejection of both culturally valued goals and the means of achieving them.

reverse ethnocentrism A type of ethnocentrism in which the home culture is regarded as inferior to a foreign culture.

revolutionary movements Social movements that seek broad, sweeping, and radical structural changes to a society's basic social institutions or to the world order.

right A behavior that a person assuming a role can demand or expect from another.

ritualism The rejection of cultural goals but a rigid adherence to the legitimate means of achieving them.

rituals Rules that govern how people must behave in the presence of the sacred to achieve an acceptable state of being.

role The behavior expected of a status in relation to another status.

role conflict A predicament in which the roles associated with two or more distinct statuses that a person holds conflict in some way.

role expectations Behaviors expected of someone enacting a role in relationship to a particular status.

role performance The actual behavior of the person occupying a role.

role set The array of roles associated with a given social status.

role strain A predicament in which there are contradictory or conflicting role expectations associated with a single status.

role-taking The process of stepping into another person's shoes by which to imaginatively view and assess our (and others) behavior, appearance, and thoughts.

sacramental religions Religions in which the sacred is sought in places, objects, and actions believed to house a god or spirit.

sacred A domain of experience that includes everything regarded as extraordinary and that inspires in believers deep and absorbing sentiments of awe, respect, mystery, and reverence.

samples Portions of the cases from a larger population.

sampling frame A complete list of every case in a population.

sanctions Reactions of approval or disapproval to others' behavior or appearance.

scapegoat A person or group blamed for conditions that (a) cannot be controlled, (b) threaten a community's sense of well-being, or (c) shake the foundations of an important institution.

schooling A program of formal, systematic instruction that takes place primarily in classrooms but also includes extracurricular activities and out-of-classroom assignments.

scientific method An approach to data collection in which knowledge is gained through observation and its truth is confirmed through verification.

scientific racism The use of faulty science to support systems of racial rankings and theories of social and cultural progress that placed whites in the most advanced ranks and stage of human evolution.

secondary deviants Those whose rule breaking is treated as something so significant that it cannot be overlooked or explained away.

secondary groups Two or more people who interact for a specific purpose. Secondary group relationships are confined to a particular setting and specific tasks.

secondary sector (of the economy) Economic activities that transform raw materials into manufactured goods.

secondary sex characteristics Physical traits not essential to reproduction (such as breast development, quality of voice, distribution of facial and body hair, and skeletal form) that result from the action of so-called male (androgen) and female (estrogen) hormones.

secondary sources (archival data) Data that have been collected by other researchers for some other purpose.

secret deviants People who have broken the rules of a group but whose violation goes unnoticed or, if it is noticed, prompts those who notice to look the other way rather than reporting the violation.

sect A small community of believers led by a lay ministry, with no formal hierarchy or official governing body to oversee its various religious gatherings and activities. Sects are typically composed of people who broke away from a denomination because they came to view it as corrupt.

secularization A process by which religious influences on thought and behavior are reduced.

secure parental employment A situation in which at least one parent or guardian is employed full-time (35 or more hours per week for at least 50 weeks in the past year).

segregation The physical and/or social separation of people by race or ethnicity.

selective forgetting A process by which people forget, dismiss, or fail to pass on a connection to one or more ethnicities.

selective perception The process in which prejudiced people notice only those things that support the stereotypes they hold about an out-group.

self-administered questionnaire A set of questions given to respondents who read the instructions and fill in the answers themselves.

self-fulfilling prophecy A concept that begins with a false definition of a situation. Despite its falsity, people assume it to be accurate and behave accordingly. The misguided behavior produces responses that confirm the false definition.

services Activities performed for others that result in no tangible product, such as entertainment, transportation, financial advice, medical care, spiritual counseling, and education.

sex ratio The number of females for every thousand males (or another preferred constant, such as 10, 100, or 10,000).

sexism The belief that one sex—and by extension, one gender—is innately superior to another, justifying unequal treatment of the sexes.

significant others People or characters who are important in an individual's life, in that they greatly influence that person's self-evaluation or motivate him or her to behave in a particular manner.

significant symbol Gestures that convey the same meaning to the persons transmitting them and receiving them.

simultaneous-independent inventions Situations in which more or less the same invention is produced by two or more people working independently of one another at about the same time.

situational causes Forces outside an individual's immediate control—such as weather, chance, and others' incompetence.

U.S. Department of Health and Human Resources. 2011. "Nursing Aides, Home Health Aides, and Related Health Care Occupations." http://bhpr.hrsa.gov/healthworkforce/reports/nursing/nurseaides/default.htm.

U.S. Department of State. 2011. Budget: Department of State and Other International Programs. www.gpoaccess.gov/usbudget/fy11/pdf/budget/state.pdf.

U.S. Office of Management and Budget. 2011. www.whitehouse.gov/omb.

Wacquant, Loic J. D. 1989. "The Ghetto, the State, and the New Capitalist Economy." *Dissent* (Fall): 508–520.

Weber, Max. (1947) 1985. "Social Stratification and Class Structure." Pages 573–576 in *Theories of Society: Foundations of Modern Sociological Theory*, ed. T. Parsons, E. Shils, K. D. Naegele, and J. R. Pitts. New York: Free Press.

______. 1948. "Class, Status, and Party," Pp. 185–195 in H. Gerth and C. W. Mills, *Essays from Max Weber.* New York: Routledge and Kegan Paul.

______. 1982. "Status Groups and Classes." Pages 69–73 in *Classical and Contemporary Debates*, ed. A. Giddens and D. Held. Los Angeles: University of California.

Wilson, William Julius. 1983. "The Urban Underclass: Inner-City Dislocations." *Society* 21: 80–86.

______. 1987. *The Truly Disadvantaged.* Chicago: University of Chicago Press.

______. 1991. "Studying Inner-City Social Dislocations: The Challenge of Public Agenda Research" (1990 presidential address). *American Sociological Review* (February): 1–14.

______. 1994. "Another Look at the Truly Disadvantaged." *Political Science Quarterly* 106 (4): 639–656.

World Bank. 2009. "Understanding Poverty." http://web.worldbank.org.

______. 2011. *World Development Indictors.* http://data.worldbank.org/.

Wright, Erik Olin. 2004. "Social Class." In *Encyclopedia of Sociological Theory*, ed. by G. Ritzer. Thousand Oaks, CA: Sage.

Yeutter, Clayton. 1992. "When Fairness Isn't Fair." *New York Times* (March 24): A13.

Chapter 9

About.com. 2004. "Immigration Before and After: Putting It in Perspective." immigration.about.com.

Alba, Richard D. 1992. "Ethnicity." Pages 575–584 in *Encyclopedia of Sociology*, vol. 2, ed. by E. F. Borgatta and M. L. Borgatta. New York: Macmillan.

Amnesty International. 2008. "Brazil Upholds Indigenous Rights in Key Case." (December 15). http://us.oneworld.net/article/359082-brazil-upholds-indigenous-rights-key-case.

Arab Institute. 2003. "First Census Report on Arab Ancestry Marks Rising Civic Profile of Arab Americans." www.aaiusa.org.

Bailey, Stanley. 2008. "Unmixing for Race Making in Brazil." *American Journal of Sociology* 114(3):577–614.

_____. 2009. *Legacies of Race: Identities, Attitudes and Politics in Brazil.* Stanford, CA: Stanford University Press.

Burdick, John. 1992. "Brazil's Black Consciousness Movement." *Report on the Americas* 25(4):23–27.

Cho, Margaret. 2002. Interview with Michele Norris, "Comedy and Race in America: Three Comedians Who Get Serious Laughs from Thorny Issues." *All Things Considered*, December 9–11. www.npr.org/programs/atc/features/2002/dec/comedians.

Cornell, Stephen. 1990. "Land, Labour and Group Formation: Blacks and Indians in the United States." *Ethnic and Racial Studies* 13(3):368–388.

Cornell, Stephen, and Douglas Hartmann. 2007. *Ethnicity and Race: Making Identities in a Changing World,* 2nd ed. Thousands Oaks, CT: Pine Forge.

Crapanzano, Vincent. 1985. *Waiting: The Whites of South Africa.* New York: Random House.

Dacey, Jessica. 2010. "Obama's Roots Traced to Swiss Villager." SwissInfo.ch (July 3). www.swissinfo.ch/eng/swiss_news/Obama_s_roots_traced_to_Swiss_villager.html?cid=15349746.

Davi, Frei. 1992. Quoted in "Brazil's Black Consciousness Movement" by John Burdick. *Report on the Americas* 25(4):23–27.

Davis, F. James. 1978. *Minority-Dominant Relations: A Sociological Analysis.* Arlington Heights, IL: AHM.

Doane, Ashley. 1997. "Dominant Group Ethnic Identity in the United States: The Role of 'Hidden' Ethnicity in Intergroup Relations." *The Sociological Quarterly* 38(3):375–397.

Gasnier, Anne. 2010. "Brazil Passes Racial Equality Law but Fails to Endorse Affirmative Action." *Guardian Weekly* (June 29). www.guardian.co.uk/world/2010/jun/29/brazil-race.

Goffman, Erving. 1963. *Stigma: Notes on the Management of Spoiled Identity.* Upper Saddle River, NJ: Prentice Hall.

Gordon, Milton M. 1978. *Human Nature, Class, and Ethnicity.* New York: Oxford University Press.

Gore, Rick. 2000. "What it Takes to Build the Unbeatable Body." *National Geographic* 198(3):2–32.

Gradín, Carlos. 2007. "Why Is Poverty So High among Afro-Brazilians? A Decomposition Analysis of the Racial Poverty Gap." Institute for the Study of Labor: Forschungsinstitut zur Zukunft der Arbeit (May 2007). http://papers.ssrn.com/sol3/papers.cfm?abstract_id=995479.

Graham, Lawrence Otis. 2001. "Black Men with a Nose Job." Pages 33–38 in *The Social Construction of Race and Ethnicity in the United States*, ed. by J. Ferrante and P. Brown. Upper Saddle River, NJ: Prentice-Hall.

Haney Lopez, Ian F. 1994. "The Social Construction of Race: Some Observations on Illusion, Fabrication, and Choice." *Harvard Civil Rights—Civil Liberties Law Review* 29:39–53.

Hertz, Todd. 2003. "Are Most Arab Americans Christian?" *Christianity Today* (March 25). www.christianitytoday.com.

Htun, Mala. 2005. Racial Quotas for a "Racial Democracy." *Report on the Americas* (February):20–25.

Hudson, Rex A. 1997. *Country Studies: Brazil.* Washington: Library of Congress.

Ikawa, Daniel, and Laura Mattar. 2009. "Racial Discrimination in Access to Health: The Brazilian Experience." *Kansas Law Review* 57:949–970. www.law.ku.edu/publications/lawreview/pdf/9.0-Ikawa_Final.pdf.

Instituto Brasileiro de Geografia e Estatística (IBGE). 2009. Population. www.ibge.gov.br/english.

_____. 2010. www.ibge.gov.br/english/#sub_populacao.

Jeter, Jon. 2003. "Affirmative Action Debate Forces Brazil to Take Look in the Mirror." *Washington Post* (June 16):A01.

Kimbrough v. United States. 2007. Certiorari to the United States Court of Appeals for the Fourth Circuit, Supreme Court of the United States NO. 06-6330. Argued October 2, 2007; Decided December 10, 2007.

Kochhar, Rakesh. 2006. *Latino Labor Report 2006.* http://pewhispanic.org/reports/report.php?ReportID=70.

Madison, Richard. 2002. "Changes in Immigration Law and Procedures." www.lawcom.com/chngs.shtml.

Marger, Martin. 2012. *Race and Ethnic Relations: American and Global Perspectives.* Belmont: Wadsworth.

Matijasevich A., et al. 2008. "Widening Ethnic Disparities in Infant Mortality in Southern Brazil: Comparison of 3 Birth Cohorts." *American Journal of Public Health.* 98(4):692–68.

McIntosh, Peggy. 1992. "White Privilege and Male Privilege: A Personal Account of Coming to See Correspondences through Work in Women's Studies." Pages 70–81 in *Race, Class, and Gender: An Anthology*, ed. by M. L. Andersen and P. H. Collins. Belmont, CA: Wadsworth.

Merton, Robert K. 1957. *Social Theory and Social Structure*. New York: Free Press.

_____. 1976. "Discrimination and the American Creed." Pages 189–216 in *Sociological Ambivalence and Other Essays*. New York: Free Press.

Migration Policy Institute. 2011. "Shaping Brazil: The Role of International Migration," by Ernesto Friedrich Amaral and Wilson Fusco. www.migrationinformation.org/Profiles/display.cfm?id=311.

National Center for Education Statistics. 2008. "Table 234: Number of Degree-Granting Institutions and Enrollment in these Institutions." Digest of Education Statistics. http://nces.ed.gov/programs/digest/d08/tables/dt08_234.asp.

National Center for Education Statistics. 2009. Digest of Educational Statistics. nces.ed.gov/programs/digest.

_____. 2010. Public School Graduates and Dropouts From the Common Core of Data: School Year 2007–08 (June). http://nces.ed.gov/pubs2010/2010341.pdf.

National Center for Public Policy and Higher Education. 2008 (August 2008). "Policy Alert: Is College Opportunity Slipping Away?" www.highereducation.org/reports/reports_center_2008/shtml.

National Park Service. 2009. "We Shall Overcome: Historic Places of the Civil Rights Movement." www.nps.gov/nr/travel/civil rights/.

Norris, Michelle. 2002. "Comedy and Race in America: Three Comedians Who Get Serious Laughs from Thorny Issues." *All Things Considered*, December 9–11. wwwnpr.org/programs/atc/features/2002/dec/comedians.

Obama, Barack. 2004. *Dreams from My Father*. New York: Crown

Ogbu, John U. 1990. "Minority Status and Literacy in Comparative Perspective." *Daedalus* 119(2):141–168.

Office of Management and Budget. 2011. "Revisions for the Classification of Federal Data on Race and Ethnicity." U.S. Bureau of the Census. www.census.gov/population/www/socdemo/race/Ombdir15.html.

Omi, Michael, and Howard Winant. 1986. *Racial Formation in the United States: From the 1960s to the 1980s*. New York: Routledge & Kegan Paul.

Osava, Maria. 2010. "'Colonisation Made Us Poor,' Say Indigenous Peoples." InterPress Service (January 14). http://ipsnews.net/news.asp?idnews=49999.

Page, Clarence. 1996. *Showing My Colors: Impolite Essays on Race and Identity*. New York: Harper Collins.

Parent, Anthony S., Jr., and Susan Brown Wallace. 2002. Pp. 451–458 in *The Social Construction of Race and Ethnicity in the United States*, ed. by J. Ferrante and P. Brown Jr. Upper Saddle River, NJ: Prentice-Hall.

Rawley, James A. 1981. *The Transatlantic Slave Trade: A History*. New York: Norton.

Saboia, Ana Lucia, and João Saboia. 2007. "White, Black and Browns in Labour Market in Brazil: A Study About Inequalities." (November). www.cigss.umontreal.ca/Docs/SSDE/pdf/Saboia.pdf.

Schermerhorn, R. A. 1978. *Comparative Ethnic Relations: A Framework for Theory and Research*. Chicago: University of Chicago Press.

Smith, Amy Erica. 2011. "Affirmative Action in Brazil." *Americas Quarterly* (January 13). www.americasquarterly.org/node/1939.

Steele, Claude. 1995. "Black Students Live Down to Expectations." *New York Times* (August 31):A25.

Telles, Edward E. 2004. *Race in Another America: The Significance of Skin Color in Brazil*. Princeton, NJ: Princeton University Press.

Toro, Luis Angel. 1995. "'A People Distinct from Others: Race and Identity in Federal Indian Law and the Hispanic Classification in OMB Directive No. 15." *Texas Tech Law Review* 26:1219–1274.

U.S. Bureau of the Census. 1910. "Color or Race, Nativity, and Parentage." www2.census.gov/prod2/decennial/documents/36894832v1ch03.pdf.

_____. 2003. *The Arab Population: 2000*. www.census.gov/prod/2003pubs/c2kbr-23.pdf.

_____. 2009. American Community Survey. http://www.census.gov/acs/www/

_____. 2010. *Income, Poverty, and Health Insurance Coverage in the United States: 2009* (September 2010). www.census.gov/prod/2010pubs/p60-238.pdf.

U.S. Bureau of Labor Statistics. 2010. "Labor Force Characteristics of Foreign-Born Workers." (March 19). www.bls.gov/news.release/forbrn.nr0.htm.

_____. 2011. Statistical Abstract of the United States. www.census.gov/compendia/statab/cats/population.html.

U.S. Department of State. 2010. *Brazil—International Religious Freedom Report*. Bureau of Democracy, Human Rights, and Labor. www.state.gov/g/drl/rls/irf/2010/148738.htm.

Verkuyten, Maykel. 2005. "Ethnic Group Identification and Group Evaluation Among Minority and Majority Groups." *Journal of Personality and Social Psychology* 88(1): 121–138.

Waters, Mary C. 1994. "Ethnic and Racial Identities of Second Generation Black Immigrants in New York City." *International Migration Review* 28(4):795–820.

Weber, Max. 1922. Economy and Society, vol. 2, ed. Guenther Roth and Claus Wittich, trans. Ephraim Fischof, 1978. Berkeley: University of California Press.

Wirth, Louis. 1945. "The Problem of Minority Groups." Pages 347–372 in *The Science of Man*, ed. by R. Linton. New York: Columbia University Press.

Wooddell, George, and Jacques Henry. 2005. "The Advantage of a Focus on Advantage: A Note on Teaching Minority Groups." *Teaching Sociology* 33(3): 301–309.

Woods, Tiger. 2006. Quoted in excerpts from *Who's Afraid of a Large Black Man?* by C. Barkley. www.wnyc.org/books/45779.

World Lingo 2011. www.worldlingo.com/en/company/case_studies.html.

Chapter 10

American Psychiatric Association. 2009. "What is Sexual Orientation?" www.apa.org/topics/sorientation.html.

Ault, Amber, and Stephanie Brzuzy. 2009. "Removing Gender Identity Disorder from the Diagnostic and Statistical Manual of Mental Disorders: A Call for Action." *Social Work* (April). http://findarticles.com/p/articles/mi_hb6467/is_2_54/ai_n31528170/.

Barton, Gina. 2005. "Prisoner Sues for the Right to Sex Change." *Milwaukee Journal Sentinel Online* (January 22). www3.jsonline.com/story/index.aspx?id=295581.

Baumgartner-Papageorgiou, Alice. 1982. *My Daddy Might Have Loved Me: Student Perceptions of Differences between Being Male and Being Female*. Denver: Institute for Equality in Education.

Bearak, Barry. 2009. "Inquiry about Sprinter's Sex Angers South Africans." *New York Times* (August 26): A6.

Bem, Sandra Lipsitz. 1993. *The Lenses of Gender: Transforming the Debate on Sexual Inequality*. Binghamton, NY: Vail-Ballou.

Boxer, Barbara (U.S. Senator). 2007. "Historical Timeline for Women's History." http://boxer.senate.gov/whm/time_1.cfm.

Bumiller, Elizabeth. 2010 (March 3). "Repeal of 'Don't Ask, Don't Tell' Policy Filed in Senate." www.nytimes.com/2010/03/04/us/politics/04military.html.

Busch, Bill. 2003. Quoted in B. Syken's "Football in Paradise." *Sports Illustrated* (November 3).

Cahill, Betsy, and Eve Adams. 1997. "An Exploratory Study of Early Childhood Teachers' Attitudes toward Gender Roles." *Sex Roles: A Journal of Research* 36(7–8): 517–530.

CBS 60 Minutes. 2010 (January 17). "American Samoa: Football Island." www.cbsnews.com/stories/2010/01/14/60minutes/main6097706.shtml.

F

G

H

I

J

T

U

V

W